Chemical Principles in the Laboratory

Thirteenth Edition

Emil J. Slowinski
Macalester College
St. Paul, Minnesota

Wayne C. Wolsey
Macalester College
St. Paul, Minnesota

Robert C. Rossi
Macalester College
St. Paul, Minnesota

Australia • Brazil • Canada • Mexico • Singapore • United Kingdom • United States

Chemical Principles in the Laboratory,
Thirteenth Edition
Emil J. Slowinski, Wayne C. Wolsey,
and Robert C. Rossi

SVP, Product: Cheryl Costantini

VP, Product: Thais Alencar

Senior Product Director, Portfolio Product
Management: Mark Santee

Portfolio Product Director: Maureen McLaughlin

Portfolio Product Manager: Helene Alfaro

Portfolio Product Assistant: Ellie Purgavie

Learning Designer: Michael Jacobs

Senior Content Manager: Aileen Mason

Subject Matter Expert: Theresa Dearborn

Digital Project Manager: Nikkita Kendrick

Director, Product Marketing: Danaë April

Product Marketing Manager: Andrew Stock

Content Acquisition Analyst: Ann Hoffman

Production Service: Lumina Datamatics Ltd.

Compositor: Lumina Datamatics Ltd.

Designer: Chris Doughman

Cover Image: ULTRA.F/Photodisc/Getty Images

About the cover: The different colors in fireworks
are brought about by adding specific elements
to the pyrotechnic mixtures from which they are
made. This is because when elements become
extremely hot, they emit characteristic colors
of light. The color specificity results from the
unique electronic structure of each element
and the quantum mechanical rules governing
how electrons can change the energy level they
occupy, the underlying quantum mechanics of
which are the subject of Experiments 11 and 44.

For product information and technology assistance, contact us at
**Cengage Customer & Sales Support, 1-800-354-9706 or
support.cengage.com.**

For permission to use material from this text or product, submit all
requests online at **www.copyright.com**.

Library of Congress Control Number: 2022921945

ISBN: 978-0-357-85127-2

Cengage
200 Pier 4 Boulevard
Boston, MA 02210
USA

Cengage is a leading provider of customized learning solutions. Our
employees reside in nearly 40 different countries and serve digital learners
in 165 countries around the world. Find your local representative at
www.cengage.com.

To learn more about Cengage platforms and services, register or access
your online learning solution, or purchase materials for your course,
visit **www.cengage.com**.

Printed in the United States of America
Print Number: 01   Print Year: 2023

Preface

About the Authors

Emil J. Slowinski was a DeWitt Wallace Professor of Chemistry at Macalester College. He earned a B.S. degree from Massachusetts State College in 1946 and a Ph.D. in physical chemistry from the Massachusetts Institute of Technology in 1949. He taught at Swarthmore College, 1949–1952; the University of Connecticut, 1952–1964; and Macalester College, 1964–1988. His sabbatical leaves were at Oxford University in 1960 and the University of Warsaw in 1968. He was a coauthor, with Bill Masterton and/or Wayne Wolsey, of more than 25 books in various areas of general chemistry. He was actively involved in all editions of *Chemical Principles in the Laboratory* up through the Ninth Edition, and even after retiring from active writing offered insights, advice, and support to his coauthors. Alas, he passed away in 2015, but his years of effort remain pivotal to this manual.

Wayne C. Wolsey, an inorganic chemist, received his B.S. from Michigan State University in 1958 and his Ph.D. from the University of Kansas in 1962. He joined the Macalester College faculty in 1965 and is now retired, but maintains his active involvement with *Chemical Principles in the Laboratory*. His last three sabbaticals were spent at the Oak Ridge National Laboratory. In 2001–2002, he investigated various complexing agents for their effectiveness in dissolving calcium oxalate kidney stones, in collaboration with a former student, now a urologist. He has received various awards, including Minnesota College Science Teacher of the Year in 1989, Macalester's Thomas Jefferson Award in 1993, designation as a MegaMole contributor to Minnesota Chemical Education in 1997, and an award from the Minnesota State AAUP Conference in 2001 for his support of academic freedom and shared governance. He remains professionally active in a number of scientific organizations.

Robert C. Rossi is an entrepreneur in technology, education, and applied science. He obtained a B.S. degree in chemical engineering from the University of Wisconsin—Madison in 1993 and upon graduating joined the Peace Corps, serving as a teacher in the Fiji Islands. He then taught and carried out applied photoelectrochemistry and semiconductor physics research at the California Institute of Technology, earning a Ph.D. in 2001. After several years teaching as a visiting professor at Carleton College, he moved to Macalester College, where he served as the Laboratory Supervisor in the Chemistry Department from 2003 to 2014. In 2011 he became a coauthor of *Chemical Principles in the Laboratory*, first writing for the Tenth Edition.

While a great deal of this manual remains influenced by his handiwork, dating back as far as to his first days teaching labs at Macalester College, Professor Emeritus Emil Slowinski is, unfortunately, no longer with us save in spirit and in our hearts. We miss both his experimental perspicacity and his knack for clear, direct prose aimed at students, but we have done our humble best to emulate both in putting forth this new edition.

To the Instructor

If this is the first time you are using this manual, we have done what we can to make the transition to a new manual as easy as possible. The *Instructor's Manual*, available from this product's Resources section in the Cengage Instructor Center, contains a list of required equipment and chemicals, directions for preparing solutions, and suggestions for dealing with the disposal of chemical waste from each experiment. It details the time required to do the experiments and the approximate cost per student. It also offers comments and suggestions for each experiment that may be helpful, sample data and calculations, and answers to the Advance Study Assignment questions (see next paragraph).

Each experiment in this manual includes an "Advance Study Assignment," or "ASA," designed to assist students in preparing for the experiment and particularly (where applicable) in making the calculations required for it. In such cases, the ASAs offer sample data and (in some detail) step the students through how that data can be used to obtain the desired results. If students work through the ASA for an experiment before coming to lab, they will be well prepared to make the necessary calculations on the basis of the data they obtain in the laboratory; we encourage you to employ the ASAs as pre-lab exercises.

New in This Edition

In preparing this edition we extensively edited and revised each experiment to improve the clarity of the instructions and explanations and to simplify the procedures and math. We endeavored to—borrowing a turn of phrase from fellow technical author Byron Bird—"eschew obfuscation" (reject the use of words that hide meaning, such as these), and instead use the simplest, clearest language possible. Mindful that few, if indeed any, of the students in most general chemistry courses these days go on to become chemists, the reliance on terms specific to the field has been reduced. Such terms are introduced, but not used throughout the text when they are not central to a given experiment and a simple replacement is available in layperson's terms. As an example, "bubbling (technical term: effervescence)" appears with its parenthetical once in the introduction to an experiment, but is thereafter referred to as simply "bubbling." The experiments whose steps are the most complicated, unforgiving, or dangerous if performed out of sequence have had their instructions changed to a bulleted (Experiment 10) or numbered (Experiment 29) step-by-step format and their analyses streamlined and simplified.

We know these changes will be appreciated by some and cause frustration to others. We are "old school" ourselves, and it was with some reluctance that we set forth on "modernizing" the writing and expectations in this manual. We are, however, very happy with the results, and it is our hope that our fellow traditionalists will be as well—we have tried to strike the right balance between adapting to new realities and sticking to fundamental imperatives that have long driven good chemistry lab pedagogy. We have seen an increasing need for these changes in our own labs and students, and are making these changes in the expectation that longtime users of this manual are experiencing the same. Language and math proficiency as well as time expectations at the college level are declining.

Throughout the manual, we made it clearer when amounts do not require measurement: quantities expressed in words rather than numbers ("about one milliliter") need not be measured out and can be (and are best) approximated. At the same time, we underlined exact numbers throughout the text. On the report pages we now indicate the blanks that need to be filled in while still in the laboratory with a <u>double underline</u>, provide more space to record answers and show work, and have clarified many of the questions.

We modified seven experimental procedures to make them safer, and simplified the math in ten. The vast majority of the experimental protocols have been improved: simplified, clarified, shortened, or made to consume less resources, as well as modified to make them more flexible with respect to the lab equipment used. Six experiments have been significantly renamed to make it clearer to what key chemistry topic(s) they relate.

We thoroughly rewrote and modified the experiments on electrochemistry with the express goal of making this topic more understandable for beginning students: emphasizing clarity and comprehension over dense coverage of every nuance of this complicated subject. We also completely rewrote the introductions to both of the "Resolving Matter into Pure Substances" experiments to make them more accessible to beginning students.

Appendix D now offers improved and updated information on equipment and techniques; we expanded it to incorporate autopipets and heat sources, and made it less equipment-specific. It now offers more tips and tricks, along with the reasons for them, and we hope that both students and instructors might learn from these. We added more detailed information on complexing agents to Appendix B1, and updated Appendices G1 and G2 to reflect changes in the spreadsheet packages they introduce.

General Outline of the Text

This manual consists of 44 experiments and 10 appendices. No laboratory sequence is likely to carry out every experiment; rather, instructors will select specific experiments that fit with their course objectives and pair them with appropriate appendices (which contain reference data and instructions for carrying out common operations). The experiments are designed to be independent of one another. In each, we introduce the theory involved, state in detail the procedures to be followed, and describe how to draw conclusions from observations and collected data. Each experiment includes report pages that offer an organized place to record data and observations as well as work through the analysis. Finally, each experiment has an Advance Study Assignment (ASA) associated with it, intended to prepare students for the experiment as well as guide them through the process of analyzing the experimental results. These provide sample data and show (in some detail) how that data can be used to obtain the desired insights from the experiment. Throughout the text, **boldface** is used to highlight important safety information, while <u>double underlined</u> entries on the report pages are those which should be filled in *before leaving the lab*.

Note: This text incorporates some unique (for now...we hope it proves innovative!) symbology that deserves to be summarized here:

- Text is *italicized* for emphasis
- Text in **boldface** contains important safety information
- <u>Underlined</u> *text* indicates key terms; underlined *numerals* are exact (have no uncertainty); underlined chemicals inside of square brackets indicate concentrations at *equilibrium*, measured in moles per liter but used *without units*
- 100 °C and 373 K are temperatures; +10 C° and +10 K° are temperature *differences*
- := means "equal to by definition", ≈ means "almost equal to", and ~ means "roughly equal to"
- Concentrations of a chemical (such as NaCl) are indicated by placing the chemical inside square brackets: [NaCl]
- {Units} are specified inside braces (curly brackets), especially in table headings and in axis labels on graphs
- Subscripted digits in numerical values are insignificant (see Appendix E)

⚛ Cengage WebAssign **Ancillary Package**

WebAssign for *Chemical Principles in the Laboratory*, Thirteenth Edition, puts powerful tools in the hands of instructors with a flexible and fully customizable online instructional solution, enabling them to deploy assignments (such as the Advance Study Assignments), instantly assess individual student and class performance, and help students master the course concepts. With WebAssign's powerful digital platform and *Chemical Principles in the Laboratory*'s specific content, instructors can tailor their course with a wide range of assignment settings, add their own questions and content, and connect with students effectively using communication tools. To learn more about the content in WebAssign for *Chemical Principles in the Laboratory*, refer to the Educator Guide document found in the Cengage Instructor Center.

Acknowledgments

As with any endeavor, there are many people who contribute to a project such as this. We value input and queries from users of the manual, and these have guided many changes in this edition. We would like to thank in particular our former student Ben Tokheim of Southwest Minnesota State University (then at Luther College) for helpful feedback on the prototype of Experiment 44, and Richard Molinelli of Western Connecticut State University for bringing to our attention issues with the report pages that have been more thoroughly addressed in this edition. We appreciate the assistance of many members of the Cengage staff, in the form of editorial and technical support in preparing this edition, and would like to thank in particular Aileen Mason for jumping into the breach as our Content Manager and doing an amazing job. We owe thanks for the smashing cover design to Chris Doughman and Michael Jacobs, the latter also providing great advice on how to clearly convey concepts as well as a much-needed dose of levity from time to time. The diligent copyediting work of Ann V. Paterson and thorough proofreading and thoughtful suggestions provided by Brendan Z. Foster are most appreciated. We gratefully acknowledge the sustained efforts of Neha Bhargava and her many colleagues at Lumina Datamatics, who painstakingly turned our revisions into the finished product you see before you. The cooperation of our chemistry colleagues at Macalester College over the years has also been essential.

We appreciate detailed reviews of the 12th Edition by Mitch Menzmer, Ph.D. of Southern Adventist University, Isaac Hon of Albertus Magnus College, Dr. Arnulfo Mar of the University of Texas—Rio Grande Valley, Dr. Jason Brinkley of Rock Valley College, and Bhanu Priya Viraka Nellore of Mississippi Valley State University, whose thoughtful feedback informed changes to several experiments, as well as the broader insights provided by the reviews of Amy Lumley of Coffeyville Community College and Pamela Auburn of Lone Star College.

It has been a great experience being involved with this laboratory manual over its many editions. We appreciate the support of our users. We encourage any comments, questions, or suggestions you may have! Please send them to wolsey@macalester.edu and cpl@rrts.us.

Finally, we acknowledge the patient support of our families as we prepared this edition.

Wayne C. Wolsey
Robert C. Rossi

Contents

Preface to the Student

The field of chemistry grew out of efforts to understand how one material can be changed into another, as well as what is and is not possible in this regard. Chemistry became a science when the results of experiments began to be carefully documented, quantified, shared, and studied for patterns. Attempts to explain and systematize these results led to concepts such as elements and chemical reactions, and eventually to theories that could actually explain, and—critically—even predict, experimental results.

In spite of its many successful theories, chemistry remains an experimental science. As our world has grown increasingly digital, many initiatives have emerged based on the hope that students might learn chemistry by working with software and observing reactions on a computer screen rather than actually carrying them out in a laboratory. While we are proponents of innovation and follow these efforts closely, even dabbling in them ourselves, we remain convinced that chemistry is, at its core, a hands-on laboratory science, in the vast majority of cases best learned by carrying out experiments first-hand, with real chemicals. Your instructors must agree—because teaching "wet" laboratory sections is a difficult and expensive undertaking, not the path of least resistance. Please keep in mind that a lot of resources are going into giving you the opportunity to learn chemistry in this way; we urge you to make the most of it!

It is not easy to do good experimental work. It requires experience, thought, and care. The effort you put into your laboratory sessions can pay off in many ways. You can gain a better understanding of how the chemical world works, manual dexterity in manipulating apparatus, experience applying mathematics to complex systems, and, perhaps most importantly, a way of thinking that allows you to better analyze many problems in and out of science. Perhaps you will find you enjoy doing chemistry and go on to a career as a chemist. We should make clear, however, that while professional chemists rely on the skills that you will learn from the experiments in this manual, the experiments they conduct differ in several important respects. First, the experiments in this manual have all been carried out many times, and their design tested and refined, such that they offer reasonably reproducible results—at least to the extent possible with an experimental science. In research chemistry, a great deal of thought and effort (and some measure of luck!) must go into the design of experiments if they are to have any hope of producing a reproducible result; we have done that work for you. All of the chemicals that go into an experiment must also be correctly prepared, and that work will largely be done for you by your instructors or their colleagues. Second, the "correct" outcome of these experiments is known: if something else is obtained (because we can be confident in the experimental design and the chemicals used), it is almost always the result of an error made in the course of carrying out the experiment. When doing research-level chemistry, we must always consider whether the experimental design or some other unknown factor is responsible for an interesting result, and the experiment must be repeated multiple times to ensure that it is reproducible. So even if you have great success in carrying out the experiments in this manual, do not become overconfident: you will emerge well-prepared to do what professional chemists do . . . but you will not have done it yet!

In writing this manual, we have attempted to illustrate many established principles of chemistry with experiments that are as engaging as possible. These principles are basic to the science but are usually not intuitively obvious. With each experiment we introduce the theory involved, state in detail the procedures used, describe how to draw conclusions from your observations, and, in an Advance Study Assignment (ASA), ask you to answer questions similar to those you will encounter in the experiment. Each ASA is designed to help prepare you for and assist you in the analysis of the experiment it is associated with. Where applicable, we furnish you with sample data and show (in some detail) how that data can be used to obtain the desired results. The Advance Study Assignments generally include the guiding principles as well as the specific relationships to be employed. If you work through the steps in each ASA, you should understand how to proceed when you are called upon to analyze the data you collect in the laboratory. *Before coming to lab, you should read over the experiment and do the Advance Study Assignment.* If you prepare for lab, you will get more out of it!

Each experiment also includes report pages on which your instructor may direct you to record and analyze your data. These help ensure you record essential data, as well as step you through its analysis. On the report pages, <u>double underlining</u> indicates things you must determine while still in the lab; do not leave lab without having *all* of those blanks filled in!

To give some experiments a bit of a challenge, we ask you to work with chemical "unknowns," whose identity is unknown to you (but known to your instructors). Your instructors will generally assess such experiments, at least in part, based on your success in determining the nature of the unknown.

This edition includes Appendix H, on the "Statistical Treatment of Laboratory Data." When professional scientists collect experimental data, especially when performing chemical analyses to determine chemical compositions, they usually measure the same quantity multiple times and summarize their results by reporting an average (the arithmetic mean) and a measure of how consistent the data are (through a statistical parameter called the standard deviation). In several experiments of this type, you will be asked to calculate these quantities from your results. Your instructor may decide to make these calculations an optional part of the experiments, but these values can be obtained on many standard calculators or with any spreadsheet (such as Excel or Google Sheets, to which introductions are offered in Appendices G1 and G2).

Note: The following conventions are used throughout the text:

- Text is *italicized* for emphasis

- Text in **boldface** contains important safety information

- <u>Underlined</u> *text* indicates key terms; underlined *numerals* are exact (have no uncertainty); underlined chemicals inside of square brackets indicate concentrations at *equilibrium*, measured in moles per liter but used *without units*

- 100 °C and 373 K are temperatures; $+10$ C° and $+10$ K° are temperature *differences*

- := means "equal to by definition", ≈ means "almost equal to", and ~ means "roughly equal to"

- Concentrations of a chemical (such as NaCl) are indicated by placing the chemical inside square brackets: [NaCl]

- {Units} are specified inside braces (curly brackets), especially in table headings and in axis labels on graphs

- Subscripted digits in numbers are insignificant (see Appendix E)

Safety in the Laboratory

Read This Section Before Performing Any of the Experiments in This Manual

A chemistry laboratory can be, and should be, a safe place to work. Yet each year in academic and industrial laboratories accidents occur that in some cases seriously injure—or even kill—chemists. Most of these accidents could have been foreseen and prevented had those involved used proper judgment and taken proper precautions.

The experiments in this manual have been selected at least in part because they can be done safely. Instructions in the procedures should be followed carefully and in the order given. Sometimes even a change in the concentration or amount of one chemical is enough to make a chemical reaction occur in a very different way, such as at a much faster rate, which can be dangerous. Do not deviate from the procedure specified when performing experiments unless specifically told to do so by your instructor. **Boldface** is used to highlight information with important safety consequences.

Eye protection: One of the simplest, and most important, things you can do in the laboratory to avoid injury is to protect your eyes by wearing safety glasses or their equivalent. Your instructor will tell you what eye protection to use, and you should use it. Glasses worn up in your hair may be fashionable, but they will not protect your eyes. If you use contact lenses, it is even more important that you wear safety glasses as well.

Chemical safety: Many chemicals are toxic, which means that they can act as poisons or carcinogens (causes of cancer) if they get into your digestive or respiratory system. *Never* taste lab chemicals, and avoid getting them on your skin. If contact should occur, wash off the affected area promptly, with plenty of water. Also, wash your hands *every time* you finish working in the laboratory. Never use your mouth as a pump in a chemistry laboratory: use a rubber bulb or other device designed for the purpose. Avoid breathing gases given off by chemicals or reactions. If directed to smell anything, do so cautiously. Use a fume hood (a laboratory workspace that pulls gases away as you work) when the directions call for it.

Some chemicals, such as concentrated acids or bases, or gases such as chlorine and bromine, are corrosive, which means that they can cause chemical burns and eat through your clothing. Where such chemicals are being used, we note the potential danger with a **CAUTION:** box at that point in the procedure. Be particularly careful when carrying out that step. Always read the label on a bottle before using it; there is a great difference between the properties of 1 molar sulfuric acid, 1 M H_2SO_4(aq), and those of concentrated (18 M) H_2SO_4(aq).

A number of chemicals used in this manual are flammable, which means they are easily set on fire. These include heptane, ethanol, and acetone. Keep ignition sources, such as hotplates and laboratory burners, well away from any open containers of such chemicals, and be careful not to spill them on a laboratory bench where they might be easily ignited.

When disposing of chemical waste from an experiment, use good judgment. Some dilute, nontoxic solutions can be poured down the sink. Insoluble or toxic materials should be put in waste containers provided for that purpose. Your lab instructor may give you instructions for treatment and disposal of the products from specific experiments: it is critical that you follow them.

Safety equipment: In the laboratory there are various pieces of safety equipment, which may include safety showers, emergency eyewashes, fire extinguishers, and fire blankets. Learn where these items are located, so that you will not have to waste precious time looking for them if you ever need them in a hurry.

Laboratory attire: Be thoughtful when dressing for lab. Long, flowing fabrics should be avoided, as should bare feet. Sandals and open-toed shoes offer less protection than regular shoes. Keep long hair tied back, out of the way of flames and chemicals. Tight-fitting jewelry (especially rings) can worsen the harm caused by chemical exposure and should not be worn in lab.

If an accident occurs: During a laboratory course accidents sometimes occur. For the most part these will not be serious and might involve a spilled chemical, a beaker of hot water that gets tipped over, a dropped test tube, or a small fire.

A common response in such a situation is panic. Try not to respond to an otherwise minor accident by doing something irrational, such as running from the laboratory, when a safe way to address the accident is available. If a fellow student has an accident, watch for signs of panic and calmly tell them what to do; if it seems necessary, help them do it. Call your instructor for assistance.

Most chemical spills are best handled by quickly absorbing wet material with a paper towel and then washing the area with water from the nearest sink. Use an eyewash if you get something in your eye. In the case of a severe hazardous chemical spill on your clothing or shoes, use an emergency shower and *immediately* take off the affected clothing. In the case of a fire in a beaker, on a bench, or on your clothing or that of another student, do not panic and run, which could make the fire worse and wastes valuable time. Smother the fire with an extinguisher, a blanket, or water, as seems most appropriate at the time. If the fire is in a piece of equipment or on the lab bench and does not appear to require instant action, have your instructor put the fire out. Glassware sometimes breaks in chemical laboratories. It is essential that broken glass be cleaned up properly and completely—ask your instructor for guidance. If you cut yourself on anything in lab, or are otherwise injured, tell your instructor, who will assist you in properly treating the injury. Sometimes, in chemical environments, the gravity and proper course of action for a given injury differs from that outside the laboratory, so please do not keep an injury to yourself.

Understand what you are doing: If the directions in an experiment are not clear to you, do not try to do that part of the experiment by imitating what another student seems to be doing. Ask your instructor, a teaching assistant, or a fellow student for guidance. Do not hesitate to ask others for help in understanding unfamiliar phrases, terms, or expressions.

Although we have spent considerable time here describing some of the safety issues you should be concerned within a chemistry laboratory, this does not mean you should work in fear. Chemistry is not a dangerous activity when practiced properly. Chemists as a group live longer than other professionals, in spite of their exposure to potentially dangerous chemicals. In this manual we have attempted to describe safe procedures and to use chemicals that are safe when used properly. Many thousands of students have safely performed these experiments, and you can too. However, we authors cannot be in the laboratory when you carry out the experiments to be sure that you observe the necessary precautions. You and your laboratory instructor must therefore see to it that the experiments are done properly, and assume responsibility for any accidents or injuries that may occur.

Disclaimer: Chemistry experiments using chemicals and laboratory equipment can be dangerous, and misuse may cause serious bodily harm. Cengage Learning encourages you to speak with your instructors and become acquainted with your school's laboratory safety regulations before attempting any experiments. Cengage Learning and the authors have provided, for your convenience only, safety information intended to serve as a starting point for good practices. Cengage Learning and the authors make no guarantee or representations as to the accuracy or sufficiency of such information and/or instructions.

Experiment 1

The Densities of Liquids and Solids

Given a sample of a pure substance, many of its characteristics can be determined—including its temperature, mass, color, and volume. The mass and volume of different samples of a given pure substance are related: if the mass is divided by the volume, the result is the same for each sample, and is called the <u>density</u>. That is, for different-sized samples A, B, and C of the same pure substance at constant temperature and pressure,

$$\frac{\text{mass}_A}{\text{volume}_A} = \frac{\text{mass}_B}{\text{volume}_B} = \frac{\text{mass}_C}{\text{volume}_C} = \text{a constant} = \text{the density of the pure substance}$$

The density of a pure substance is independent of the size of the sample and is one of the fundamental properties of that substance. The density of pure liquid water is 1.00000 g/cm^3 at $4\,°C$ and one atmosphere pressure, and slightly less than that at room temperature (0.9970 g/cm^3 at $25\,°C$). Densities of pure liquids and solids range from values a bit less than 1 g/cm^3 to values that are much larger. Osmium metal has a density of 22.5 g/cm^3 and is the densest material known at ordinary pressures.

Density is useful in determining what will float on what. For example, the density of ice is 0.92 g/cm^3: this is less than the density of liquid water, so ice floats in water. Although most types of wood are less dense than water and will float, there are a few that are denser than water and will sink.

In any density determination, two quantities must be determined—the mass and the volume of the same amount of a substance. With liquids and solids, mass can be determined by weighing a sample of the substance on a balance. (In the process of "weighing" you actually determine the mass that experiences the same gravitational force as that experienced by what is being weighed.) The mass of a sample of liquid in a container can be found by subtracting the mass of the empty container from the mass of the container plus the liquid. This is because (except in some very extreme situations) mass is an <u>additive property</u>, meaning that the mass of two things weighed together is equal to the sum of their two masses when weighed separately.

The volume of a liquid can be determined by means of a calibrated container. In the laboratory a graduated cylinder is often used for routine measurements of volume. However, liquid volume measurements precise enough for meaningful density determinations are made using a <u>pycnometer</u>, which is a container having a precisely defined volume. The volume of a solid can be determined by direct measurement if the solid has a regular geometric shape. This is not usually the case with ordinary solid samples, however. A convenient way to determine the volume of an irregularly shaped solid is to accurately measure the volume of liquid <u>displaced</u> (pushed out of the way) when the solid is immersed in a liquid. The volume of the solid will equal the volume of the liquid that it displaces. Volume is not reliably an additive property, but in this case (that of two immiscible, incompressible substances) it is.

In this experiment you will determine the density of an unknown liquid and the density of an unknown solid. First you will weigh an empty flask and its stopper (which together serve as a pycnometer). You will then fill the flask completely with water, measuring the mass of the filled and stoppered flask. From the difference in these two masses you will find the mass of water, and then, from the known density of water, determine the volume of the flask. You will empty and dry the flask, fill it with the unknown liquid, and weigh again. From the mass of the unknown liquid and the volume of the flask you will determine the density of the liquid. To determine the density of an unknown solid metal, you will add the metal to the dry, empty pycnometer flask and weigh them together. This will allow you to find the mass of the metal. You will then fill the flask with water, leaving the metal in the flask, and weigh again. The increase in mass is that of the added water; from that increase, and the density of water, you can calculate the volume of water added. The volume of the metal must equal the volume of the flask minus the volume of water. From the mass and volume of the metal you will calculate its density. The calculations involved are outlined in detail in the Advance Study Assignment (ASA) at the end of this experiment.

Experimental Procedure

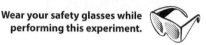

Ideally, before you begin this experiment, you will be shown how to properly operate the balances in your laboratory. If that is not possible, consult the section on mass in Appendix D for an overview.

A. Determining the mass of a coin

Use a balance to determine the mass of a coin to the nearest milligram, that is, to ± 0.001 g. Record this mass on the report page. Remove the coin and ensure the balance reads zero once again; if it does not, press the TARE button to make it do so. Weigh a different coin on the same balance, recording its mass. Add these two individual masses together and record the sum on the report page. Again remove the coin and ensure the balance reads zero. Now weigh both coins together, recording the total mass. Remove one of the coins, press the TARE button, and put the coin back on the balance. (You do not need to record this result, but make sure you understand it!) When you are satisfied that your results are those you would expect and are ready to move on, obtain a glass-stoppered flask, which will serve as a pycnometer, and samples of an unknown liquid and an unknown metal.

B. Determining the density of a liquid

If your pycnometer flask is not clean and dry, clean it with detergent solution and water, rinse it with a few milliliters of acetone, and dry it by letting it stand for a few minutes in the air or by *gently* blowing compressed air into it for a few moments (until it no longer smells of acetone).

Weigh the dry flask with its stopper, to the nearest milligram. Fill the flask with deionized water until the liquid level is nearly to the *top* of the ground glass surface in the neck. Put the stopper in the flask in order to drive out *all* the air and any excess water. Gradually work the stopper into the flask until it is firmly seated in position. Wipe any water from the outside of the flask with a towel and soak up all excess water from around the top of the stopper. Your goal is to end up with the flask *completely* filled with water, with no air bubbles in it at all, and completely dry on the outside.

Weigh the flask again. Given the density of water at the temperature of the laboratory (see Appendix A) and the mass of water in the flask, you should be able to determine the volume of the flask very precisely.

Empty the flask, dry the inside (follow the instructions at the start of Part B), and fill it with your unknown liquid. Stopper and dry the outside of the flask as you did when working with the water, and then weigh the stoppered flask full of the unknown liquid, making sure its outer surface is completely dry. This measurement, used with those you made previously, will allow you to accurately determine the density of your unknown liquid.

C. Determining the density of a metal

Pour your sample of liquid from the pycnometer flask back into its container. Rinse the flask with a small amount of acetone and dry it thoroughly inside (follow the instructions at the start of Part B) and out. Add as much of your metal sample to the flask as you can without making it difficult to insert the stopper all the way. (The more metal you use, the more accurate your results are likely to be.) Weigh the flask, with its stopper and the dry metal in it, to the nearest milligram.

Leaving the metal in the flask, fill the flask with water as well and then replace the stopper. Roll the metal around in the flask to make sure that no air remains trapped between the metal pieces. Refill the flask if necessary, and then weigh the dry, stoppered flask filled with water and the metal sample. (It must be dry on the outside and not have any air bubbles trapped inside it!) Properly done, the measurements you have made in this experiment will allow you to calculate the density of your metal sample to within 0.1%.

> **Disposal of reaction products:** Drain the water from the flask, then pour the wet metal out onto a paper towel. Dry the metal as best you can, then put the metal back into its container. Dry the flask and return it along with its stopper, your metal sample, and your unknown liquid sample.

Name _____ Section _____

Experiment 1

Data and Calculations: The Densities of Liquids and Solids

A. Determining the mass of a coin

Mass of first coin _____ g   Mass of second coin _____ g

Sum of these masses: _____ g

Mass of first and second coins weighed together _____ g
Are these results consistent with mass being an additive property?

B. Determining the density of a liquid

Unknown liquid ID code _____

Mass of empty flask and stopper _____ g

Mass of stoppered flask filled with water _____ g

Mass of stoppered flask filled with unknown liquid _____ g

Temperature in the laboratory _____ °C

Mass of water required to fill flask _____ g

Density of water at laboratory temperature (See Appendix A;
note that a density in g/mL is identical to a density in g/cm^3,
because 1 mL := 1 cm^3.)

_____ g/cm^3

Volume of flask = volume of water _____ cm^3

Mass of unknown liquid _____ g

Density of unknown liquid _____ g/cm^3

To how many significant figures can the unknown liquid's
density be properly reported? (See Appendix E.) _____

(continued on following page)

C. Determining the density of a metal

Metal ID code _____

Mass of stoppered flask filled with dry metal _____ g

Mass of stoppered flask filled with water and metal _____ g

Mass of metal _____ g

Mass of water _____ g

Volume of water _____ cm^3

Volume of metal _____ cm^3

Density of metal _____ g/cm^3

To how many significant figures can the density of the metal be
properly reported? _____

Explain why the value obtained for the density of the metal is likely to have a larger percent error
(uncertainty divided by the density itself) than the density determined for the unknown liquid.

Experiment 1

Advance Study Assignment: The Densities of Liquids and Solids

1. Finding the volume of a flask

A student obtained a clean and dry glass-stoppered pycnometer flask. She weighed the flask and stopper on an analytical balance and found the total mass to be 34.166 g. She then filled the flask with water and obtained a mass for the full stoppered flask of 68.090 g. From these data, and the fact that at the temperature of the laboratory the density of water was 0.997 g/cm³, find the volume of the stoppered flask.

a. First you need to obtain the mass of the water in the flask. This is found by recognizing that mass is an additive property: that is, that the mass of something is equal to the sum of the masses of its parts. For the filled and stoppered flask,

mass of filled stoppered flask = (mass of empty stoppered flask) + (mass of water)

so   mass of water = (mass of filled stoppered flask) − (mass of empty stoppered flask)

mass of water = _____ g − _____ g = _____ g

b. The density of a pure substance is equal to its mass divided by its volume:

$$\text{density} := \frac{\text{mass}}{\text{volume}} \quad \text{or} \quad \text{volume} = \frac{\text{mass}}{\text{density}}$$

The volume of the flask is equal to the volume of the water it contains. Because you know the mass and density of the water, you can find its volume: this is also the volume of the flask.

$$\frac{\text{volume}}{\text{of flask}} = \frac{\text{volume}}{\text{of water}} = \frac{\text{mass of water}}{\text{density of water}} = \frac{\text{_____ g}}{\text{_____ g/cm}^3} = \text{_____ cm}^3$$

2. Determining the density of a liquid

Having obtained the volume of the flask, the student emptied the flask, dried it, and filled it with an unknown liquid whose density she wished to determine. The mass of the stoppered flask when completely filled with the unknown liquid was 63.682 g. Find the density of the unknown liquid.

a. First you need to find the mass of the unknown liquid by calculating the difference between the mass of the full flask and that of the empty flask (from Question 1):

mass of unknown liquid = _____ g − _____ g = _____ g

mass of stoppered   mass of empty
flask filled with   stoppered flask
unknown liquid

[We have labelled the two blanks in the middle of the line above with what goes in each of them. We ask *you* to do this for Question 3(a).]

(continued on following page)

b. Because the volume of the unknown liquid equals that of the flask, you know both the mass and volume of the unknown liquid and can find its density using the equation in Question 1(b). Calculate the density of the unknown liquid. Show your work in the space below, indicating the general formula used and then filling in the values, as done for you in the problems above.

density of unknown liquid = _____ g/cm^3

3. Determining the density of a metal

The student then emptied the flask and dried it once again. To the empty flask she added pieces of a metal until the flask was almost full. She weighed the stoppered flask and its metal contents and found the mass to be 177.611 g. She then added water to the metal-filled flask, replacing all the air between the metal pieces, stoppered it, and obtained a total mass of 179.837 g for the flask, stopper, metal, and water. Find the density of the metal.

a. To find the density of the metal you need to know its mass and volume. You can obtain its mass by taking advantage of the fact that mass is an additive property. As demonstrated in Question 2(a), label the two middle blanks with what goes in them:

mass of metal = _____ g − _____ g = _____ g

b. To determine the volume of the metal, note that the volume of the flask must equal the volume of the metal plus the volume of water in the filled flask containing both metal and water. If you can find the volume of the water, you can obtain the volume of metal because the two volumes add up to a known total volume: the volume of the flask. To obtain the volume of the water, first calculate its mass:

mass of water = mass of (flask + stopper + metal + water) − mass of (flask + stopper + metal)

mass of water = _____ g − _____ g = _____ g

The volume of water is found from its density, as in Question 1(b):

$$volume\ of\ water = \frac{mass\ of\ water}{density\ of\ water} = \frac{_____\ g}{_____\ g/cm^3} = _____\ cm^3$$

c. From the volume of the water, calculate the volume of metal:

volume of metal = (volume of flask) − (volume of water)

volume of metal = _____ cm^3 − _____ cm^3 = _____ cm^3

From the mass and volume of the metal, find its density, using the equation in Question 1(b):

density of metal = _____ g/cm^3

Now go back to Question 1 and check that you have reported the proper number of significant figures in each of the results you calculated in this assignment. Use the rules on significant figures as given in your course, chemistry text, or Appendix E.

Experiment 2

Resolving Matter into Pure Substances, 1. Paper Chromatography

Chemicals differ in the extent to which they associate with different substances, and this provides a common basis for separating mixtures. In <u>chromatography</u>, a mixture is exposed to one material (the <u>mobile phase</u>) that flows through another (the <u>stationary phase</u>, which does not move) and each component in the mixture associates with these two phases to differing extents. The more a chemical prefers the mobile phase, the more quickly it will flow with it. If a mixture component does not associate with the mobile phase at all, it will remain where it started.

In <u>paper chromatography</u>, the stationary phase is a sheet of filter paper, while the mobile phase is a liquid that is gradually absorbed up through the paper by <u>capillary action</u>: the tendency of liquids to flow spontaneously into tiny passages, such as the spaces between the fibers in paper. (Capillary action is what causes a dry sponge or paper towel to soak up water.) A small amount of the mixture to be separated is placed in a localized spot near an edge of the filter paper, and that edge is then exposed to the mobile phase such that the paper starts absorbing it. The liquid gradually soaks its way through the filter paper, carrying with it the components in the mixture: to an extent reflecting how much each component prefers to be in the liquid as opposed to on the filter paper.

There are many other kinds of chromatography in common use, including gas, column, and thin-layer chromatography. Gas chromatography, as its name implies, uses a gas as the mobile phase: it separates components based largely on how easily they enter the gas phase, which temperature has a powerful effect on. The many components in gasoline (a very complex mixture) can be isolated by gas chromatography, as shown in Figure 2.1.

In this experiment you will use paper chromatography to separate a mixture of water-soluble metallic ions. You will apply samples containing a few micrograms of ions as small spots on a sheet of filter paper and use a blend of acid, water, and alcohol as the mobile phase (technical term: eluent). Most of the ions

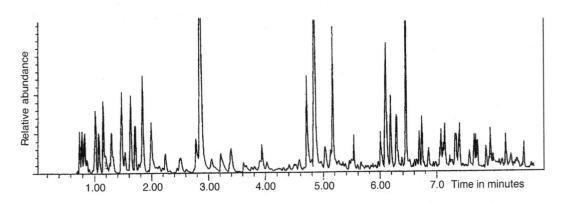

Figure 2.1   This is a gas chromatogram of a sample of unleaded gasoline. Each peak corresponds to a different molecule, indicating gasoline has many components: at least fifty, each of which can be identified. The molar masses vary from about fifty to about one hundred and fifty, with the largest peak, at about three minutes, being due to toluene, $C_6H_5CH_3$. The sample size for this chromatogram was less than 10^{-6} g (less than one microgram, 1 µg)! Gas chromatography offers the best method for separating complex liquid mixtures whose components enter the gas phase when heated.

Chromatogram courtesy of Prof. Becky Hoye at Macalester College.

have visible colors in this mobile phase, so their positions can be observed directly as the chromatography proceeds. At the end of the separation process you will apply a staining agent to the filter paper, which reacts with all of the ions to produce colored products (changing the color of some already-colored ions). By observing the color and position of the chromatographed spots associated with known ions, and comparing those against the spots obtained from an unknown mixture of ions, you will be able to figure out which ions are present in the unknown.

It is possible to describe the position of spots such as those you will observe in terms of a ratio called the "retention factor", R_f. In this experiment the mobile phase (the solvent) will flow past the starting point of the spots a certain distance, denoted L, while each ion will move a certain distance from its starting point, denoted D. These can be measured in any length units desired, so long as they are both measured in the *same* units. Their units then cancel and the ratio D/L is dimensionless. The ratio is called the "R_f value" (or the "retention factor") for that ion:

$$R_f := \frac{D}{L} = \frac{\text{distance ion moves}}{\text{distance solvent moves}} \tag{1}$$

R_f for a specific ion will depend on the mobile and stationary phases used, as well as the temperature, but it is reasonably independent of concentration and what other ions are present. Retention factors can have values between zero and one. $R_f = 0$ indicates no association with the mobile phase, and thus that the component did not move at all, while $R_f = 1$ indicates the component remained completely in the mobile phase, moving through the stationary phase as quickly as the mobile phase did.

Experimental Procedure

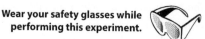

Wear your safety glasses while performing this experiment.

A. Preparing the stationary phase

Obtain an unknown mixture and a piece of filter paper 19 cm long and 11 cm wide. Along the 19-cm edge, draw a pencil line about one centimeter from that edge. Starting 1.5 cm from the end of the line, mark the line at 2-cm intervals. Label the segments of the line as shown in Figure 2.2, with the formulas of the ions to be studied and the known and unknown mixtures. (*This must all be done in pencil, not pen.*)

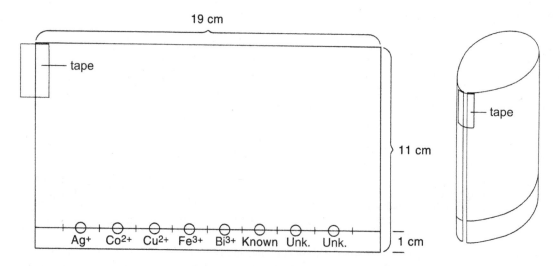

Figure 2.2   This figure shows how the filter paper should be set up for this experiment.

B. Preparing the samples

Put two or three drops of 0.10 M solutions of the following compounds in separate labeled or otherwise identifiable small (13 × 100 mm) test tubes, one solution to a tube:

$$AgNO_3 \quad Co(NO_3)_2 \quad Cu(NO_3)_2 \quad Fe(NO_3)_3 \quad Bi(NO_3)_3$$

In solution these substances exist as ions. The metallic ions are Ag^+, Co^{2+}, Cu^{2+}, Fe^{3+}, and Bi^{3+}, respectively, and each drop of solution contains very roughly six hundred micrograms (micrograms is abbreviated as μg, and 1 μg $= 10^{-6}$ g) of metal ion. Into a sixth small test tube put 2 drops of each of the five solutions; swirl until the solutions are well mixed. This mixture will be your known, because you know it contains all of the metal ions.

C. Putting the samples on the stationary phase

Your instructor will furnish you with a fine capillary tube, which will serve as an applicator. Test the application procedure by dipping the applicator into one of the colored solutions and touching it for just a very short time to a round piece of filter paper. The liquid from the applicator should form a spot no larger than 8 mm in diameter. Practice making spots until you can reproduce the spot size each time.

Clean the applicator by dipping the tip about one centimeter into a beaker of deionized water and then touching the round filter paper to remove the liquid. In this case, continue to hold the applicator against the paper until all the liquid in the capillary tube has been absorbed into the filter paper. Repeat this cleaning procedure one more time. Dip the applicator into one of the metal ion solutions and put a spot on the line on the rectangular filter paper in the region labeled for that ion. Clean the applicator twice, then repeat the procedure with another solution. Continue this approach until you have put a spot for each of the five metal ions and the known and unknown mixtures on the paper, cleaning the applicator between solutions. Dry the paper completely by moving it in the air or holding it briefly in front of a hair dryer or heat lamp (low setting). Apply the known and unknown mixtures three more times to the same spots; the known and unknown mixtures are less concentrated than the metal ion solutions, so this procedure will increase the amount of each ion present in the spots. Be sure to dry the spots between applications, because otherwise they will get larger. Do not heat the paper more than necessary, just enough to dry the spots.

D. Performing the chromatography

Pour 15 mL of the mobile phase into a 600-mL beaker and cover it with a watch glass. The mobile phase is a mixture of hydrochloric acid, HCl(aq), in water with ethanol and butanol, which are organic solvents.

Check to make sure that the spots on the filter paper are all dry. Place a 4- to 5-cm length of tape along the upper end of the left edge of the paper, as shown in Figure 2.2, so that about half of the tape is on the paper. Form the paper into a cylinder by attaching the tape to the other edge in such a way that the edges are parallel but *do not overlap*. When you are finished, the pencil line at the bottom of the cylinder should form a circle (approximately) and the two edges of the paper should not quite touch. Stand the cylinder up on the lab bench to check that such is the case; readjust the tape if necessary. *Do not* tape the lower edges of the paper together.

Place the cylinder in the mobile phase in the 600-mL beaker, with the sample spots down near (*but not below!*) the liquid surface. The paper should not touch the walls of the beaker. Cover the beaker with the watch glass. The solvent will gradually rise up the filter paper by capillary action, carrying along the metal ions at different rates. After the process has gone on for a few minutes, you should be able to see colored spots rising on the paper, showing the positions of some of the metal ions.

E. Determining ion colors

While the chromatography is proceeding, test the effect of the mobile phase and the staining agent on the different metal ions. Put an 8-mm spot of each metal ion solution on the same clean piece of round filter paper, labeling each spot in pencil and cleaning the applicator between solutions. Dry the spots as before. Some of them will have a little color; record those colors on the report page. Next, place your applicator into the mobile phase and touch it to each spot. Note the colors of the wet spots on the report page. Dry the spots again, noting again their colors. Put the filter paper on a paper towel, and, using the spray bottle filled with the staining agent, spray the paper evenly with the staining agent, getting the paper moist but not really wet. The staining agent is a solution containing potassium ferrocyanide and potassium iodide. It forms colored precipitates or reaction products with many metal ions, including all of those used in this experiment. Note the colors obtained with each of the metal ions. Considering that each spot contains such a small amount of metal ion, the tests are surprisingly definitive.

F. Analyzing the results

When the mobile phase has risen as high as time permits, but not higher than within about two centimeters of the top of the filter paper (which typically requires between seventy-five minutes and two hours), remove the filter paper cylinder from the beaker and take off the tape. Draw a pencil line along the <u>solvent front</u>: the furthest up the paper the mobile phase reached, which may not be a straight line. Dry the paper with gentle heat until it is dry. Note any metal ions that must be in your unknown by virtue of your being able to see their colors. Then, with the filter paper set out flat on a paper towel, spray it with the staining agent. Any metal ions you identified in your unknown before staining should be observed, as well as any that require staining for detection.

Measure the distance from the straight line on which you applied the spots to the solvent front, which is distance L in Equation 1. Then measure the distance from the pencil line to the center of the spot made by each of the metal ions, when pure and in the known; this is distance D. Calculate the R_f value for each metal ion. Then calculate R_f values for the metal ions in the unknown. How do the R_f values compare?

Disposal of reaction products: When you are finished with this experiment, pour the remaining mobile phase into a waste container, not down the sink. Wash your hands before leaving the laboratory.

Take it further (optional): These days many foods and beverages, and almost all candies, are colored with food dyes and lakes (dyes bound to solid binders): often a mixture of several, which should be listed in the ingredients list on the label. Use paper chromatography to separate such mixtures, realizing that no staining agent will be needed, and a lake's color will only move if the dye is released from the binder. The mobile phase for this experiment works well, but so can rubbing alcohol, or in some cases even salt water.

Experiment 2

Data and Calculations: Resolving Matter into Pure Substances,
1. Paper Chromatography

	Ag^+	Co^{2+}	Cu^{2+}	Fe^{3+}	Bi^{3+}
Colors observed (or "colorless")					
Dry	____	____	____	____	____
After exposure to the mobile phase?	____	____	____	____	____
After drying again?	____	____	____	____	____
After staining?	____	____	____	____	____

Pure metal ions

Remember to specify units when reporting distances!

	Ag^+	Co^{2+}	Cu^{2+}	Fe^{3+}	Bi^{3+}
Distance solvent moved (L)	____	____	____	____	____
Distance ion moved (D)	____	____	____	____	____
R_f	____	____	____	____	____

Known mixture

	Ag^+	Co^{2+}	Cu^{2+}	Fe^{3+}	Bi^{3+}
Distance solvent moved (L)	____	____	____	____	____
Distance metal ion moved (D)	____	____	____	____	____
R_f	____	____	____	____	____

(continued on following page)

Unknown mixture

Colors observed

    Dry   _____   _____   _____   _____   _____

    After staining   _____   _____   _____   _____   _____

Distance solvent moved (L)   _____   _____   _____   _____   _____

Distance metal ion moved (D)   _____   _____   _____   _____   _____

R_f _____   _____   _____   _____   _____

Metal ion this spot is caused by   _____   _____   _____   _____   _____

Unknown ID code _____

Experiment 2

Advance Study Assignment: Resolving Matter into Pure Substances,
1. Paper Chromatography

1. A student chromatographs a mixture containing five components, and after developing the spots with a staining agent observes the following:

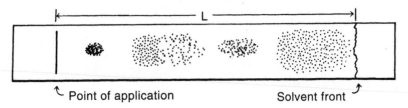

What are the five R_f values?

2. Explain, in your own words, why samples can often be separated into their components by chromatography.

(continued on following page)

3. The solvent front in this experiment moves 3 cm in about ten minutes. Why should the experiment not be stopped at that time? (Why wait much longer, for the solvent front to move 10. cm?)

4. In this experiment it takes 10. microliters (microliters is abbreviated as μL, and $\underline{1}$ μL $= \underline{10^{-6}}$ L) of solution to produce a spot 8 mm in diameter. (We write 10. to indicate that the zero in this number is significant; that is, it could take between 9 and 11 microliters to form an 8 mm spot, rather than anything between 0 and 20 microliters, which is what "10 μL" would indicate.) If the $Fe(NO_3)_3$ solution contains 5.59 g of Fe^{3+} per liter, how many micrograms of Fe^{3+} ion are there in one such spot?

_____ micrograms of Fe^{3+}

Experiment 3

Resolving Matter into Pure Substances, 2. Fractional Crystallization

Understanding how specific chemicals behave is fundamental to the study of chemistry. In gaining such understanding, and applying it, it helps to have access to pure chemicals, rather than mixtures. This requires that chemists develop and use means of separating mixtures and purifying chemicals. Most such methods take advantage of differences in the <u>physical properties</u> of distinct chemicals, such as their boiling or melting points, solubilities, and densities, though differences in chemical reactivity can also be used.

A common technique for separating mixtures of solids is <u>fractional crystallization</u>. It takes advantage of solubility differences to separate chemicals into solid and liquid phases. As a simple example, consider some sand mixed with salt. Because the salt is soluble in water but the sand is insoluble, the two components of this mixture can be separated by adding enough water to the mixture to dissolve the salt, then physically separating the <u>solid phase</u> (the sand) from the <u>liquid phase</u> (the salty water). Solid salt can then be recovered by <u>evaporating</u> the water (moving it into the gas phase from the liquid phase). Fractional crystallization can also be used to separate soluble solids from one another, particularly if their solubilities differ appreciably at some attainable temperature.

The goal of this experiment is to separate a mixture of silicon carbide, potassium nitrate, and copper sulfate, using water as the solvent. Silicon carbide, SiC, is a black solid. It is insoluble in water. Potassium nitrate, KNO_3, and copper sulfate, $CuSO_4 \cdot 5H_2O$, are water-soluble solids with solubilities that change with temperature, as indicated in Figure 3.1. The copper sulfate used in this experiment is blue as a solid, as well as when dissolved in solution. Its solubility increases fairly rapidly with temperature. Potassium nitrate is a white solid, clear and colorless in solution. Its solubility in water increases even more rapidly, by a factor of 20, between $0\,°C$ and $100\,°C$.

You will be given a sample: a mixture containing SiC and KNO_3, as well as a smaller amount of $CuSO_4 \cdot 5H_2O$. Dissolving both the KNO_3 and $CuSO_4 \cdot 5H_2O$ in hot water will leave only the SiC as a solid. You will use a material full of microscopic holes (technical term: <u>porous</u>), specifically filter paper, to separate out the SiC by <u>vacuum filtration</u>: actively sucking the liquid through the filter, leaving the solid on top of it. Then you will cool the hot liquid containing the two salts to near ice temperature. Because the KNO_3 is much less soluble at this temperature, and because the amount of $CuSO_4 \cdot 5H_2O$ is small, most of the KNO_3 will crystallize out of the solution as a solid, leaving most of the $CuSO_4 \cdot 5H_2O$ in the liquid. Filtering this cold mixture isolates most of the KNO_3 as a solid.

The solid KNO_3 recovered from this process will be impure, contaminated with a small amount of $CuSO_4 \cdot 5H_2O$. You will use a process called <u>recrystallization</u> to purify it, adding a bit of water to the solid: enough to dissolve all of the remaining $CuSO_4 \cdot 5H_2O$. Recrystallization takes advantage of the fact that pure crystals are more <u>stable</u> (more energetically favorable) than impure crystals are under similar conditions. Given time and opportunity, impure crystals will dissolve in favor of pure ones—almost all of the remaining $CuSO_4 \cdot 5H_2O$ will end up in the liquid, and the solid KNO_3 left behind will be purified.

To measure the purity of the KNO_3 you isolate, you will take advantage of the deep blue color $CuSO_4$ takes on in the presence of aqueous ammonia, $NH_3(aq)$. The more intense the blue color, the more $CuSO_4$ is present.

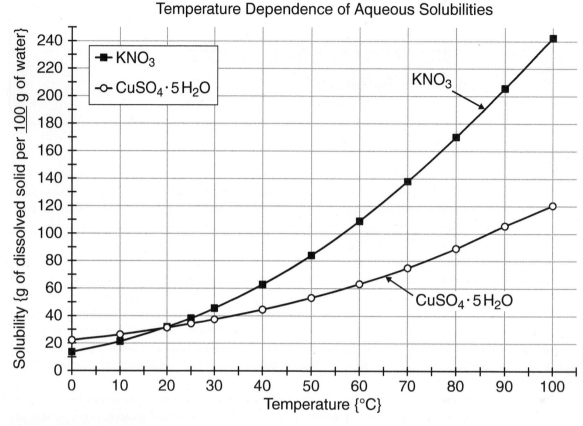

Figure 3.1 The solubility of potassium nitrate, KNO_3, and copper(II) sulfate pentahydrate, $CuSO_4 \cdot 5H_2O$, in water both change a lot with temperature. While both solubilities increase with increasing temperature, the solubility of potassium nitrate is lower than that of copper(II) sulfate pentahydrate at low temperatures but increases more quickly, such that in hot water the solubility of KNO_3 is quite a bit higher than that of the copper salt. *Note*: Consult Appendix E for help reading this graph.

Wear your safety glasses while performing this experiment.

Experimental Procedure

Note: In this experiment, determine masses to the nearest tenth of a gram: that is, to ±0.1 g. Use top-loading balances if they are available.

Obtain a Büchner funnel, a suction flask, and a sample (about twenty grams) of your unknown solid mixture. Weigh a clean and dry (approximately 150-mL) beaker and record its mass on the report page. Add your sample to the beaker and weigh again. Record the result on the report page, then add about forty milliliters of deionized water, which will be enough to dissolve the soluble solids.

A. Isolating silicon carbide, SiC

Using a hotplate or a laboratory burner, gently heat the contents of your beaker to 50 °C (between 40 °C and 60 °C), while stirring the mixture. (See Appendix D for more details on how to use your heat source.) When the blue and white solids are all dissolved, pour the contents of the beaker into a Büchner funnel while gentle suction is being applied (see Figure 3.2 and Appendix D). Transfer as much of the black solid carbide as you can to the funnel, using your rubber policeman. Turn off the vacuum line, break the suction by twisting the vacuum hose as you pull it off the suction flask, and remove the funnel from the top of the suction flask by

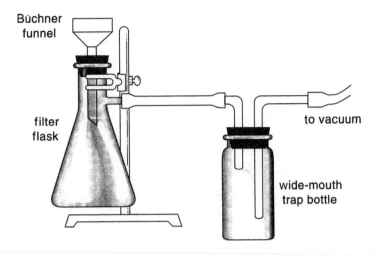

Figure 3.2 To separate a solid from a liquid using a Büchner funnel, put a piece of circular filter paper in the funnel. Thoroughly moisten the filter paper with deionized water from a wash bottle, but do not add more water than the filter itself will absorb. Just before pouring the mixture of solid and liquid into the filter, turn on the suction. Keep the suction on while filtering the mixture. Consult Appendix D for more details.

rotating and lifting it at the same time. Then transfer the blue liquid (technical term: <u>filtrate</u>) to the (cleaned) beaker and add 15 drops of 6 M nitric acid, HNO_3, which will help ensure the copper sulfate remains in solution in later steps. **CAUTION:** **HNO_3 is a strong acid; avoid contact with it**. Reassemble the funnel, turn the vacuum line back on, and wash the SiC on the filter paper with deionized water. Continue to apply suction for a few minutes, to dry the SiC. While you are waiting, weigh a clean, dry watch glass and record its mass. Turn off the vacuum line, break the suction again, and rotate the (cup of) the Büchner funnel to free it from the rest of the apparatus. Use your spatula to help you transfer the SiC crystals from the funnel onto the watch glass so that they may further dry in the air. Remove and throw away the used filter paper. When you are finished with the rest of the experiment, you will weigh the dry SiC.

B. Isolating potassium nitrate, KNO_3

Heat the blue liquid in the beaker to the boiling point, and then boil it gently until white crystals of KNO_3 are visible in the liquid. **CAUTION:** **The hot liquid may have a tendency to boil unevenly (technical term: bump), causing a splash risk, so do not heat it too strongly**. While the liquid is being heated, prepare some ice-cold deionized water by putting your wash bottle in an ice-water bath; also record the mass of a single circle of filter paper that fits your Büchner funnel.

When white crystals are clearly present in the boiling liquid (the solution may appear cloudy at that point), stop heating and add 12 mL of room-temperature deionized water to the solution. Stir the mixture with a glass stirring rod to dissolve the solids, including any on the walls; if necessary, warm the solution—but do not boil it again.

Cool the solution to room temperature in a water bath, and then to nearly 0 °C in an ice bath. White crystals of KNO_3 will come out of solution. Stir the cold mixture of crystals and solution for several minutes. Check the temperature of the mixture with your thermometer. Continue stirring until the temperature of your mixture gets down to 3 °C, or even a bit lower (measured *while the mixture is being stirred*).

Assemble the Büchner funnel and put the filter paper into it, but do not turn on the vacuum yet. Chill the funnel by filling it with ice-cold deionized water from your wash bottle, and, after about a minute, pulling the water through it by turning on the suction. Separate the solid KNO_3 from the mixture in the beaker using the Büchner funnel, while everything is still cold. Your rubber policeman will be helpful when you transfer the last of the crystals. Press the crystals dry with a clean piece of filter paper, and continue to apply suction while you clean and dry your beaker. Turn off the vacuum line, break the suction by twisting the vacuum hose as you pull it off the suction flask, and rotate the (cup of) the Büchner funnel to free it from the rest of the apparatus. Then transfer the filter paper and the KNO_3 crystals from the funnel into the beaker. Weigh the beaker and its contents, then remove the filter paper and discard it.

> **Disposal of reaction products:** By this procedure you have separated most of the KNO_3 in your sample from the $CuSO_4$, which is present in the liquid in the suction flask. This liquid may now be discarded, so dispose of it as directed by your instructor.

C. Analyzing the purity of KNO₃

The KNO_3 crystals you have prepared contain a small amount of $CuSO_4$ as an impurity. To find the amount of $CuSO_4$ present, weigh out 0.5 g (between 0.4 and 0.6 g) of the crystals into a tared 50-mL beaker. Dissolve the crystals in 3.0 mL of deionized water, and then add 3.0 mL of 6 M ammonia, NH_3. The copper impurity will turn deep blue in the NH_3. Pour the solution into a small (13 × 100 mm) test tube. Compare the intensity of the blue color with that in a series of standard solutions prepared by your instructor. Estimate the relative concentration of $CuSO_4 \cdot 5H_2O$ in your product and record it on the report page.

D. Recrystallizing KNO₃

Add just enough room-temperature deionized water to the larger beaker containing impure KNO_3 that its contents no longer act like a paste when stirred, but instead flow freely: a <u>slurry</u>. This will not be enough water to completely dissolve the KNO_3, but recrystallization takes advantage of the fact that impure crystals dissolve more easily than pure crystals do. By thoroughly stirring the slurry for a few minutes, you will give all the KNO_3 crystals contaminated with $CuSO_4$ a chance to dissolve, and reform as pure crystals, leaving the $CuSO_4$ impurity in solution. (It may look like nothing is happening as you stir, but it is!)

Continuing to stir, cool the beaker in an ice bath until your thermometer indicates the temperature is below 3 °C. Then filter the slurry through an ice-cold Büchner funnel as you did before, transferring as much of the solid as possible to the funnel. Again press the crystals dry with a piece of filter paper and continue to apply suction while you clean and dry the beaker. Transfer the filter paper and the (now purified) KNO_3 crystals back into the beaker. Weigh the beaker and its contents.

Determine the amount of $CuSO_4$ impurity in your recrystallized sample as you did with the first batch, recording the concentration indicated by the analysis. The recrystallization should have significantly improved the purity of your KNO_3.

Weigh your now-dry SiC on its watch glass. Show your samples of SiC and KNO_3 to your instructor for evaluation. Then return the samples.

Experiment 3

Data and Calculations: Resolving Matter into Pure Substances, 2. Fractional Crystallization

Mass of empty beaker   _____ g

Mass of sample plus beaker   _____ g

Mass of sample   _____ g

A. Isolating silicon carbide, SiC

Mass of watch glass   _____ g

Mass of watch glass and dry SiC   _____ g

Mass of SiC recovered
= (mass of watch glass and dry SiC) − (mass of watch glass)   _____ g

Percent mass of sample recovered as SiC $= \dfrac{\text{mass of SiC}}{\text{mass of sample}}$ _____ $\%_{\text{mass}}$

B. Isolating potassium nitrate, KNO$_3$

Mass of dry filter paper   _____ g

Mass of beaker plus filter paper and impure KNO$_3$ _____ g

Mass of impure KNO$_3$ recovered   _____ g

Percent (by mass) of sample recovered as (impure) KNO$_3 = \dfrac{\text{mass of KNO}_3}{\text{mass of sample}} \times 100\%$ _____ $\%_{\text{mass}}$

(continued on following page)

C. Analyzing the purity of KNO₃

Percent (by mass) of $CuSO_4 \cdot 5\,H_2O$ present in impure KNO_3 _____ %$_{mass}$

D. Recrystallizing KNO₃

Mass of beaker plus filter paper and purified KNO_3 _____ g

Mass of purified KNO_3 recovered
    (Assume each sheet of filter paper has the same mass; you weighed one in Part B.)   _____ g

Percent (by mass) of sample recovered as purified $KNO_3 = \dfrac{\text{mass of } KNO_3}{\text{mass of sample}} \times 100\%$ _____ %$_{mass}$

Percent (by mass) of $CuSO_4 \cdot 5\,H_2O$ in purified KNO_3 _____ %$_{mass}$

Did the recrystallization step further purify the KNO_3? What were the costs of taking this extra step (what did you have to give up, or do, to get this result)?

Experiment 3

Advance Study Assignment: Resolving Matter into Pure Substances,
2. Fractional Crystallization

1. Using Figure 3.1, determine the following:
   (See "Interpreting a Graph" in Appendix E for help.)

   a. How many grams of $CuSO_4 \cdot 5H_2O$ will dissolve in 100 g of H_2O at $100\,°C$?

   _____ g $CuSO_4 \cdot 5$ H_2O

   b. How many grams of water are required to dissolve 3.9 g of $CuSO_4 \cdot 5H_2O$ at $100\,°C$? [*Hint*: Your
       answer to Problem 1(a) gives you the needed conversion factor for (g $CuSO_4 \cdot 5H_2O$) to (g H_2O).]

   _____ g H_2O

   c. How many grams of water are required to dissolve 37 g of KNO_3 at $100\,°C$?

   _____ g H_2O

   d. How many grams of water are required to completely dissolve a mixture containing 37 g of KNO_3
       and 3.9 g of $CuSO_4 \cdot 5H_2O$, assuming that the solubility of one substance is not affected by the presence
       of another? (This is generally a surprisingly good assumption when the substances have no ions in
       common! It means you need only the larger of the two amounts of water needed to completely dissolve
       the individual components, *not their sum*.)

   _____ g H_2O

(continued on following page)

2. A solution at 100 °C containing 3.9 g of $CuSO_4 \cdot 5H_2O$ and 37 g of KNO_3 in 27.2 g of water is cooled to 0 °C.

   a.   How much KNO_3 remains in solution?

   _____ g KNO_3

   b.   How much KNO_3 crystallizes out?

   _____ g KNO_3

   c.   How much $CuSO_4 \cdot 5H_2O$ crystallizes out?

   _____ g $CuSO_4 \cdot 5H_2O$

   d.   What percent (by mass) of the KNO_3 in the sample is recovered?

   $$\%\text{recovery, by mass} = \frac{\text{mass of } KNO_3 \text{ recovered}}{\text{mass of } KNO_3 \text{ present in sample}} \times 100\%$$

   _____ $\%_{\text{mass}}$

Experiment 4

Determining a Chemical Formula

When atoms of different elements combine to form a chemical <u>compound</u>, they typically do so in a ratio that is a counting number (1, 2, 3, and so on). Examples of <u>chemical formulas</u> with such ratios include $CaCl_2$, KBr, C_2H_6, and Fe_2O_3. When more than two elements are present in a compound, its chemical formula still indicates the atom ratios. Thus the formula C_2H_6O indicates that the carbon, hydrogen, and oxygen atoms occur in that compound in the ratio 2:6:1. Many compounds have more complex formulas, but the same principles apply.

Determining the formula of a compound involves finding the mass of each of the elements in a weighed sample of that compound. For example, if a 40 g sample of the compound NaOH is separated into its elements, 23 g of sodium, 16 g of oxygen, and 1 g of hydrogen are obtained. Because (see Appendix C) the molar mass of sodium atoms is 23 g/mol, that of oxygen atoms is 16 g/mol, and that of hydrogen atoms is 1 g/mol, this indicates that the sample of NaOH contains equal numbers of Na, O, and H atoms. Therefore the atom ratio Na:O:H is 1:1:1, and the formula of the compound is NaOH. In terms of moles, one mole of NaOH, 40 g, contains one mole of Na, 23 g; one mole of O, 16 g; and one mole of H, 1 g. The atom ratio in a compound is equal to the mole ratio, which can be obtained by chemical analysis.

In this experiment you will use these principles to find the formula of the compound with the general formula $Cu_xCl_y \cdot z\,H_2O$, where x, y, and z are counting numbers that, when known, establish the formula of the compound. (In expressing the formula of a <u>hydrated</u> compound such as this one, in which water molecules remain intact within the compound, the formula of the water molecules, H_2O, is retained in the formula of the compound instead of the elements being written separately as H_{2z} and O_z.)

The compound you will study is called "copper chloride hydrate" (or "hydrated copper chloride"). It is stable, can be obtained in pure form, has a characteristic blue-green color that changes as the compound is changed chemically, and is relatively easy to decompose into the elements and water. In this experiment you will first drive out the water, which is called the <u>water of hydration</u>, from an accurately weighed sample of the compound. This occurs when you gently heat the sample to a little over 100 °C. As the water is driven out, the color of the sample changes from blue-green to a tan-brown color similar to that of dry dirt. The compound that results is <u>anhydrous</u> (not hydrated: that is, free of water) copper chloride. If you subtract its mass from that of the hydrate, you can determine the mass of the water that was driven off, and, using the molar mass of water, find the moles of H_2O that were in the sample.

In the next step you need to find either the mass of copper, Cu, or the mass of chlorine, Cl, in the now-anhydrous copper chloride, Cu_xCl_y. It turns out to be much easier to determine the mass of the copper, and to then calculate the mass of chlorine by difference. You will do this by dissolving the anhydrous sample in water, which results in a green solution containing copper and chloride ions. To that solution you will add some aluminum metal wire. In contact with a solution containing copper ions, aluminum metal will react chemically with those ions, converting them to copper metal. The copper metal appears on the wire as the reaction proceeds, and has the red-orange color typical of metallic copper. When the reaction is complete, you will remove the excess Al, separate the copper from the liquid, and weigh the dried metal. From its mass you will calculate the moles of copper in the sample. You will calculate the mass of chlorine by subtracting the mass of copper from that of the anhydrous copper chloride, and from that value determine the moles of chlorine. The mole ratio of $Cu : Cl : H_2O$ reveals the formula of the compound.

Experimental Procedure

Wear your safety glasses while performing this experiment.

Accurately determine the mass of a clean, dry crucible, without a cover, on an analytical balance. Place about one gram of the unknown hydrated copper chloride in the crucible. With your spatula, break up any chunks larger than a pea by pressing them against the wall of the crucible. Then weigh the crucible and its contents accurately. Record your results on the report page.

If you are using a laboratory burner as your heat source: Place the uncovered crucible on a clay triangle supported by an iron ring. Light your laboratory burner away from the crucible and adjust the burner so you have a small, quiet blue flame. Holding the bottom of the burner in your hand, gently heat the crucible as you move the burner back and forth with the flame about five centimeters underneath it. Do not heat the crucible too strongly, or you will cause dehydrating copper chloride to jump out of the crucible; be patient. As the sample warms, you will see that the blue-green crystals begin to change to brown around the edges. Continue gentle heating, slowly converting all of the hydrated crystals to the anhydrous brown form. After all of the crystals appear to be brown, continue heating gently (moving the burner back and forth under the crucible) for an additional two minutes. Remove the burner, cover the crucible to minimize rehydration, and let it cool for at least 10 minutes.

If you are using a hotplate as your heat source: Place the crucible directly on the hotplate and turn it on at 50% power, or set it to a surface temperature half of its maximum. Leave it there until the blue-green crystals in the crucible all become brown, and then for about two minutes more. Turn off the hotplate and cover the crucible. Use tongs or a thickly folded paper towel to move the covered crucible onto a casserole dish or watch glass to cool for at least 10 minutes.

Remove the cover and slowly roll the brown crystals around in the crucible. If any blue-green color remains, repeat the heating process. Finally, weigh the cool, uncovered crucible and its contents accurately.

Transfer the brown crystals in the crucible to an empty 50-mL beaker. Rinse out the crucible with two 6-mL (taking significant figures into account, that is 6 ± 1 mL) portions of deionized water, and add the rinsings to the beaker. Swirl the beaker gently to dissolve the brown solid. The solid will change back to blue-green as the copper ions are rehydrated, then dissolve to yield a blue solution. Cut off about twenty centimeters of thin aluminum metal, Al(s), wire (≈ 0.25 g) and form it into a loose spiral coil around a pen or pencil; then remove it and shape it so it will fit entirely inside the 50-mL beaker. Put the coil into the solution so that it is completely immersed. Within a few moments you will observe some bubbles of hydrogen gas, H_2, and the formation of copper metal, Cu(s), on the Al wire. As the copper ions are converted to copper metal, the blue color of the solution will fade. The Al metal wire provides the electrons needed to turn the copper ions into atoms, and in the process some of it is converted into aluminum ions. (The hydrogen gas is formed as the aluminum also shoves electrons onto H^+ ions in the copper solution, forming bubbles of H_2 gas.)

When the reaction is complete, the solution will be colorless, and most of the copper metal that was produced will be on the Al wire. (This will take roughly thirty minutes, but watch the color of the solution, not the time, to know when it is complete! Swirling the beaker or stirring its contents periodically will help the reaction reach completion, but be gentle enough that you do not break the wire.) Add 5 drops of 6 M HCl to dissolve any insoluble aluminum salts and clear up the solution. Use your glass stirring rod to move the aluminum wire around and remove the copper from the wire as completely as you can. Slide the unreacted aluminum wire up the wall of the beaker with your stirring rod, and, while the wire is hanging from the rod, rinse off any remaining Cu particles with water from your wash bottle. If necessary, complete the removal of the Cu with a drop or two of 6 M HCl added directly to the wire. Put the wire aside; it has done its duty.

In the beaker you now have the metallic copper produced in the reaction, in a solution containing an aluminum salt. You will now use a Büchner funnel to separate the copper from the solution. Weigh a dry piece of filter paper that will fit in the Büchner funnel and record its mass. Put the paper in the funnel, moisten it with a little water to ensure a good seal, and then apply light suction. Before the filter paper dries out, pour just the liquid in your beaker (technical term: decant) into the funnel, leaving behind as much of the solid copper as you can. (See Appendix D for more details.) Wash the copper metal thoroughly with deionized water, breaking up any large clumps of copper with your stirring rod. Transfer the copper and rinse liquid to the Büchner funnel. Wash any remaining copper into the funnel with water from your wash bottle. *All* of the copper must be transferred to the funnel. Rinse the copper on the paper once again with water. Turn off the vacuum line and break the suction by twisting off the vacuum hose. Add about ten milliliters of 95% ethanol to the funnel, enough to completely cover the copper, and after about one minute apply suction. Pull air through the funnel for at least 5 minutes, which should dry the copper and the filter paper. While you are waiting, determine and record the mass of a clean watch glass. Use your spatula to help you transfer the paper and the copper from the filter onto the watch glass. Give the paper a few more minutes in the air to dry, then weigh the watch glass, filter paper, and copper metal together and record the result.

Disposal of reaction products: Dispose of the liquid waste and copper produced in this experiment as directed by your instructor. The aluminum wire can be thrown in the trash, or recycled by placing it *inside* an aluminum can or a ball of aluminum foil at least the size of your fist.

Experiment 4

Data and Calculations: Determining a Chemical Formula

Molar masses: copper, Cu _____ g/mol   chlorine, Cl _____ g/mol

hydrogen, H _____ g/mol   oxygen, O _____ g/mol

Mass of crucible   _____ g

Mass of crucible and hydrated sample   _____ g

Mass of hydrated sample   _____ g

Mass of crucible and dehydrated sample   _____ g

Mass of dehydrated sample   _____ g

Mass of filter paper   _____ g

Mass of watch glass   _____ g

Mass of watch glass, filter paper, and copper metal   _____ g

Mass of copper metal   _____ g

(continued on following page)

Moles of copper metal   _____ moles

Mass of water lost upon dehydrating sample   _____ g

Moles of water present in hydrated sample   _____ moles

Mass of chlorine in sample (by difference)   _____ g

Moles of chlorine in sample   _____ moles

Mole ratio of chlorine to copper in sample $= \dfrac{\text{moles of chlorine}}{\text{moles of copper}}$ _____ :1

Mole ratio of water to copper in hydrated sample $= \dfrac{\text{moles of water}}{\text{moles of copper}}$ _____ :1

Formula of dehydrated sample (round each ratio to the nearest counting number)   _____

Formula of hydrated sample (use coefficients rounded to
the nearest counting number)   _____

Name _____ **Section** _____

Experiment 4

Advance Study Assignment: Determining a Chemical Formula

1. To find the mass of one mole of an element, look up its molar mass in a table (consult Appendix C or a Periodic Table). For a compound, the molar mass is equal to the sum of the molar masses of the atoms in it. Find the molar mass of

   Cu _____ g/mol   Cl _____ g/mol   H _____ g/mol

   O _____ g/mol H_2O _____ g/mol

2. If you can find the ratio of the moles of the elements in a compound to one another, you can find the formula of the compound. In a certain compound of copper and oxygen, Cu_xO_y, a sample weighing 0.9573 g is found to contain 0.7616 g of copper, Cu.

   a. How many moles of copper are in the sample? $\left(\text{moles of Cu} = \dfrac{\text{mass of Cu in sample}}{\text{molar mass of Cu}} \right)$

   _____ moles

   b. How many grams of oxygen are in the sample? (The mass of the sample equals the mass of Cu plus the mass of O.)

   _____ g

   c. How many moles of oxygen atoms (O) are in the sample?

   _____ moles

(continued on following page)

d. What is the mole ratio (moles of Cu/moles of O) in the sample?

_____ :1

e. What is the formula of the oxide? (The atom ratio equals the mole ratio, expressed using the smallest counting numbers possible.)

_____

f. What is the molar mass of the copper oxide?

_____ g/mol

Experiment 5

Identifying a Compound by Mass Relationships

When chemical reactions occur, there is a relationship between the masses of the reactants and products that follows directly from the balanced equation for the reaction and the molar masses of the substances involved. In this experiment you will use this relationship to identify an unknown substance.

Your unknown will be one of the following four compounds, all of which are salts:

$$NaHCO_3 \qquad Na_2CO_3 \qquad KHCO_3 \qquad K_2CO_3$$

sodium   sodium carbonate   potassium   potassium
hydrogencarbonate   hydrogencarbonate   carbonate

In the first part of this experiment you will heat a weighed sample of your compound in a crucible. If your sample is a <u>carbonate</u> (it contains the CO_3^{2-} ion), no chemical reaction will occur, but any small amount of water sticking to the surface of it (technical term: <u>adsorbed on</u> it) will be driven off. If your sample is a <u>hydrogencarbonate</u> (it contains the HCO_3^- ion), it will decompose to the corresponding carbonate by the following reaction, using $NaHCO_3$ as the example:

$$2\,NaHCO_3(s) \quad \rightarrow \quad Na_2CO_3(s) \quad + \quad H_2O(g) \quad + \quad CO_2(g) \qquad \textbf{(1)}$$

sodium   sodium carbonate   water vapor   carbon
hydrogencarbonate   (product)   (product)   dioxide gas
(reactant)   (product)

In this case there would be a significant decrease in mass, because some of the products of the reaction, which started out as part of your hydrogencarbonate reactant, will be driven off as gases. If such a mass decrease occurs, you can be sure that your sample is a hydrogencarbonate.

In the second part of this experiment, you will treat the solid carbonate in the crucible with aqueous hydrochloric acid, $HCl(aq)$. There will be considerable bubbling (technical term: <u>effervescence</u>) as CO_2 gas is given off (technical term: evolved); the reaction that occurs is, using Na_2CO_3 as the example,

$$Na_2CO_3(s) \quad + \quad 2\,H^+(aq) \quad + \quad 2\,Cl^-(aq) \quad \rightarrow \quad 2\,NaCl(s) \quad + \quad H_2O(\ell) \quad + \quad CO_2(g) \qquad \textbf{(2)}$$

sodium carbonate   dissolved hydrochloric acid   sodium   liquid water   carbon dioxide
(reactant)   (reactant)   chloride   (product)   gas (product)
                                                     (product)

(HCl exists in solution as ions, H^+ and Cl^-, so that is how it appears in Reaction 2.) You will then heat the crucible strongly to drive off any excess HCl and any water that is present, obtaining pure, dry, solid sodium chloride, $NaCl(s)$, or potassium chloride, $KCl(s)$, as your final product.

To identify your unknown, you will need to calculate the molar masses of the possible reactants and final products. For each of the possible unknowns there will be a different relationship between the mass of the original sample and the mass of the chloride salt produced in Reaction 2. If you know your sample is a carbonate, you need only be concerned with the mass relationships in Reaction 2 and should use the mass of the carbonate after it has been heated (dried) as the mass of your original compound. If you have a hydrogencarbonate, the overall reaction your sample undergoes will be the sum of Reactions 1 and 2, and you should use the mass of your sample before heating as the mass of your original compound.

From your experimental data you will calculate the change in mass that occurred when you formed the chloride from the compound you started with. That difference divided by the mass of the original salt differs for each of the possible starting compounds and will not change with the mass of the sample. Call that ratio Q:

$$Q := \frac{(\text{mass of chloride} - \text{original mass})}{\text{original mass}} \tag{3}$$

Calculating the theoretical values of Q for each of the possible compounds and comparing them to the Q value obtained for your unknown should allow you to determine its identity. Because the mass change after the first heating will tell you whether your compound is a carbonate or a hydrogencarbonate, the Q value is only really needed to figure out which of the two possible compounds of that type yours may be.

Experimental Procedure

Wear your safety glasses while performing this experiment.

Clean your crucible and its cover by rinsing them with deionized water and then drying them with a towel. Support the crucible on a clay triangle in an iron ring if you are heating with a laboratory burner, or place it directly on a hotplate. Place the cover on top, but in a slightly off-center position (slightly <u>ajar</u>) such that gases can escape. Heat the crucible and cover strongly for 5 minutes (use the maximum heat setting on a hotplate), then allow them to cool to room temperature. If you are using a hotplate, you will need to move the cover and crucible off the hotplate to cool; use tongs or a thickly folded paper towel to move them into a casserole dish or onto a watch glass. It will take about ten minutes for the crucible and cover to cool completely, but you *must* wait until they are really at room temperature or your mass measurement will be artificially lowered by <u>convection currents</u>: by heated, rising air. While you are waiting, obtain an unknown sample to use in this experiment and record your observations of it on the lower half of the back of the report page.

Once they are cool, weigh the crucible and cover together on an analytical balance. Record the combined mass on the report page. With a spatula, transfer about half a gram of your unknown to the crucible. Weigh the filled crucible with the lid on it, recording the total mass on the report page.

Put the crucible on the clay triangle or hotplate, with the cover slightly ajar again, such that the lid does not perfectly seal the crucible: gases need an escape route, but you want to keep liquids and solids that might spatter inside the crucible. Gradually heat the crucible until you are applying the maximum possible heat to it. (If you are using a hotplate, just turning it on to maximum after you place the crucible on top works fine.) Maintain this level of heating for at least 5 minutes. Then allow the crucible to cool for at least 10 minutes and weigh it, with its cover and contents, on an analytical balance, recording the mass. (While waiting, calculate the mass of unknown that was in your crucible before heating.)

At this point the sample in the crucible is a dry carbonate, because the heating process converts any hydrogencarbonate to carbonate.

Put the crucible on the lab bench, leaving the cover off. *One drop at a time*, add about twenty-five drops of 6 M hydrochloric acid, HCl, to the sample. As you add each drop, you will probably observe bubbling as carbon dioxide gas, $CO_2(g)$, is produced. Let the bubbling slow before adding the next drop, to keep the sample in the lower part of the crucible and avoid losing any of it. You do not want the reaction to foam up over the edge! By the time you have added all of the HCl, the bubbling should have stopped and the solid should be completely dissolved. If any solid does remain, heat the crucible gently and/or add 6 more drops of 6 M HCl to get it to dissolve.

Again place the cover on the crucible slightly ajar, to allow gases to escape during the next heating operation. Heat the crucible, more gently this time, for about ten minutes, to slowly evaporate the water and excess HCl. (If using a hotplate, set it to about half of its maximum power.) If you heat too strongly, spattering will occur and you may lose some sample. Once the sample is no longer a liquid, apply maximum heat to it for at least 10 minutes.

Allow the crucible to cool for at least 10 minutes. Weigh it, with its cover and contents, on the analytical balance, recording the total mass on the report page.

Disposal of reaction products: Dispose of what remains of your sample by dissolving it in water and pouring it down the drain, unless your instructor directs otherwise.

Experiment 5

Data and Calculations: Identifying a Compound by Mass Relationships

Molar masses: Na _____ g/mol   K _____ g/mol   H _____ g/mol

C _____ g/mol   O _____ g/mol   Cl _____ g/mol

$NaHCO_3$ _____ g/mol Na_2CO_3 _____ g/mol   NaCl _____ g/mol

$KHCO_3$ _____ g/mol K_2CO_3 _____ g/mol   KCl _____ g/mol

Mass of crucible and cover   _____ g

Mass of crucible and cover plus unknown before heating   _____ g

Mass of unknown before heating   _____ g

Mass of crucible, cover, and unknown after heating   _____ g

Loss of mass of sample upon heating   _____ g

Unknown is a: carbonate   hydrogencarbonate   (Circle your choice)

How do you know?

Mass of crucible, cover, and solid chloride   _____ g

Mass of solid chloride   _____ g

Change in mass when original compound was converted to a chloride   _____ g

Experimentally observed $Q = \dfrac{\text{change in mass}}{\text{original mass}} = $ _____ $=$ _____

(continued on following page)

Theoretical values of Q, as obtained by Equation 3

Possible sodium compound _____ | Possible potassium compound _____

1 mole of compound → _____ moles NaCl | 1 mole of compound → _____ moles KCl

_____ g of compound → _____ g NaCl | _____ g of compound → _____ g KCl

Change in mass _____ g | Change in mass _____ g

$$Q = \frac{\text{change in mass}}{\text{original mass}} = $$ | $$Q = \frac{\text{change in mass}}{\text{original mass}} = $$

Identity of unknown _____   Unknown ID code _____

Observations of the unknown (describe it, in detail):

Experiment 5

Advance Study Assignment: Identifying a Compound
by Mass Relationships

1. Using the method of this experiment, a student attempted to identify which of the four possible salts they were given. They found that when they heated a sample of their unknown weighing 0.5032 g, the mass barely changed, dropping to 0.5023 g. When this was then converted to a chloride, the mass went up, to 0.5549 g.

   a. Did the student have a carbonate as their unknown?   yes / no (circle one)

   Please provide your reasoning here:

   b. What are the two salts the student might have had as their unknown?

   _____ or _____

   c. Write the balanced chemical equation for the overall reaction that occurs when each of these two original compounds is converted to a chloride. If the compound is a hydrogencarbonate, use the sum of Reactions 1 and 2. If the sample is a carbonate, use Reaction 2. Write the equation for the Na (sodium) salt and then the one for the K (potassium) salt.

   d. How many moles of the chloride salt would be produced from one mole of the original compound?

   _____ moles of chloride salt

(continued on following page)

e. How many grams of the chloride salt would be produced from one molar mass of the original compound? (Filling in the table at the top of the report page now will help!)

Molar masses: NaHCO₃ _____ g/mol   Na₂CO₃ _____ g/mol   NaCl _____ g/mol

KHCO₃ _____ g/mol   K₂CO₃ _____ g/mol   KCl _____ g/mol

If a sodium salt, _____ g of the original compound → _____ g of the chloride

If a potassium salt, _____ g of the original compound → _____ g of the chloride

f. What is the theoretical value of Q, as found by Equation 3,

if the student has the Na (sodium) salt? _____

if the student has the K (potassium) salt? _____

g. What was the student's observed value of Q for their unknown?

$$Q_{observed} = \text{_____}$$

h. Which compound did they have as their unknown? _____

Experiment 6

Properties of Hydrates

Most solid chemical compounds will contain some water if they have been exposed to the atmosphere for any length of time. In most cases the water is present in very small amounts and is merely sticking to (technical term: adsorbed on) the surface of the solid. Other solid compounds contain larger amounts of water, chemically bound into the structure of the crystals themselves. These compounds are called hydrates and are usually ionic salts (though most proteins also do this). The water that is bound in these salts is called water of hydration, and it is usually attached to the negatively charged ions (technical term: anions) in the salt.

The water molecules in a hydrate are relatively easy to remove. In many cases, just heating a hydrate to a temperature somewhat above the boiling point of water will drive off the water of hydration. Hydrated barium chloride is typical; it is converted to anhydrous (non-hydrated) $BaCl_2$ if heated above 115 °C:

$$BaCl_2 \cdot 2\,H_2O(s) \rightarrow BaCl_2(s) + 2\,H_2O(g) \text{ at } t \geq 115\,°C \tag{1}$$

In a dehydration reaction such as Reaction 1, the crystal structure of the solid changes, and the color of the salt may also change.

Some hydrates lose water of hydration to the air, even at room temperature, in a process called efflorescence. The amount of water lost depends on the amount of water in the air, as measured by its humidity, and the temperature. In moist air, cobalt(II) chloride ($CoCl_2$) is fully hydrated and exists as $CoCl_2 \cdot 6\,H_2O$, which is red. In dry air the salt loses its water of hydration and is found as anhydrous $CoCl_2$, which is blue. At intermediate humidities the stable form is the violet dihydrate, $CoCl_2 \cdot 2\,H_2O$. Cobalt chloride can therefore provide a visual indication of the humidity in the air, based on its color.

Some anhydrous ionic compounds absorb water from the air or other sources so strongly that they can be used to dry liquids or gases. These substances are called dessicants, and are said to be hygroscopic. A few ionic compounds can take up so much water from the air that they dissolve in the water they absorb (technical term: deliquesce); for example, sodium hydroxide, NaOH, will do this, and so it is said to exhibit deliquescence.

Some compounds give off (technical term: evolve) water on being heated but are not true hydrates, because the water is produced by decomposition of the compound rather than by loss of water of hydration. Organic compounds, particularly carbohydrates, often behave this way. Decompositions of this sort are not reversible; adding water to the product will not regenerate the original compound. True hydrates typically undergo reversible dehydration. Adding water to anhydrous $BaCl_2$ will cause formation of $BaCl_2 \cdot 2\,H_2O$, or if enough water is added you will get a solution containing Ba^{2+} and Cl^- ions. Many ionic hydrates are soluble in water and are usually prepared by crystallization from solutions having water as the solvent. If you order barium chloride from a chemical supplier, you will probably get crystals of $BaCl_2 \cdot 2\,H_2O$, which is a stable, stoichiometrically pure compound. The amount of bound water may depend on the way the hydrate is prepared, but in general the moles of water per mole of compound is either a counting number (1, 2, 3, and so on) or an odd multiple of one-half (such as ½, ³⁄₂, or ⁵⁄₂).

In this experiment you will study some of the properties of hydrates. You will identify the hydrates in a group of compounds, observe the reversibility of their hydration reactions, and test some substances for efflorescence or deliquescence. Finally, you will be asked to determine the amount of water lost by a sample of unknown hydrate upon heating. From this amount, given the formula or the molar mass of the anhydrous sample, you will be able to calculate the formula of the hydrate itself.

Wear your safety glasses while performing this experiment.

Experimental Procedure

A. Identifying hydrates

Place about half a gram of each of the compounds in the following list in dry test tubes, one compound to a tube. (If using 13 × 100 mm test tubes, this is a depth of about one centimeter in the tube; if using 18 × 150 mm test tubes, it is enough to fill the rounded bottom of the tube.) Carefully observe the behavior of each compound when you heat the test tube it is in gently with a burner flame or by holding it over a hotplate surface set to maximum. (See Appendix D for more details.) If droplets of water condense on the cool upper walls of the test tube, this is evidence the compound may be a hydrate. Note the nature and color of the solid that remains after heating. Let the tube cool and try to dissolve the remaining solid in a few milliliters of water, warming very gently if necessary. A true hydrate will tend to dissolve in water, producing a solution with a color very similar to that of the original hydrate. If the compound is a carbohydrate, it will give off water on heating and will tend to char (darken due to heat "cooking" it, as a marshmallow does over a fire); when put in water, the water may become caramel colored but the solid will not dissolve completely.

nickel chloride	sucrose
potassium chloride	calcium carbonate
sodium tetraborate (borax)	barium chloride

B. Reversibility of hydration

Throughout this process, note any color changes, as well as any other visible changes you observe. Gently heat a few crystals, ~0.3 g, of cobalt(II) chloride hexahydrate, $CoCl_2 \cdot 6H_2O$, in an evaporating dish until the color change appears to be complete. After cooling, dissolve the dried solid in the evaporating dish in a few milliliters of water, then heat the resulting solution and boil it to dryness once again. **CAUTION:** **Heat only gently once the liquid is gone!** Put the evaporating dish on the lab bench and allow it to cool.

C. Deliquescence and efflorescence

Place a few crystals of each of the compounds in the following list on separate watch glasses and put them next to the dish of $CoCl_2$ prepared in Part B. Depending on their composition and the humidity (amount of moisture in the air), the samples may gradually lose water of hydration to, or pick up water from, the air. They may also remain unaffected. To establish whether the samples gain or lose mass, weigh each of them on a balance to ± 0.01 g. Record their masses. Weigh them again after about an hour to detect any change in mass. Observe the samples occasionally during the laboratory period, noting any changes in color, crystal structure, or degree of wetness.

$Na_2CO_3 \cdot 10H_2O$ ("washing soda")	$KAl(SO_4)_2 \cdot 12H_2O$ ("alum")
$CaCl_2$ (anhydrous)	$CuSO_4$ (anhydrous)

D. Percent by mass of water in a hydrate

Clean a porcelain crucible and its cover with 6 M HNO_3. **CAUTION:** **This is a strong acid; avoid contact with it.** Any stains not removed by this treatment will not interfere with this experiment. Rinse the crucible and its cover with deionized water. Put the crucible with its cover on top, but in a slightly off-center position (slightly ajar) such that gases can escape, on a clay triangle and heat with a burner flame or directly on a hotplate surface. Heat the crucible and cover, gently at first, but then at maximum heat for about two minutes. Allow the crucible and cover to cool completely (if using a hotplate, carefully transfer them onto a watch glass to cool, using tongs or a folded paper towel), and then weigh them to ± 0.001 g.

Obtain a sample of unknown hydrate and place about a gram of the hydrate in the crucible. Weigh the crucible, cover, and sample together, again to ± 0.001 g. Put the crucible on the clay triangle or hotplate surface, again with the cover in an off-center position to allow the escape of water vapor. Heat again, gently at first and then strongly, keeping the crucible and its contents at maximum heat for at least 5 minutes. Then center the cover on the crucible, such that it seals well, and let it cool to room temperature. (Again, if using a hotplate, transfer the covered crucible onto a watch glass to cool.) Weigh the cooled crucible along with its cover and contents.

Examine the dehydrated solid in the crucible. Add water until the crucible is two-thirds full and stir. Warm gently if the solid does not dissolve. Does the solid appear to be soluble in water?

Disposal of reaction products: Dispose of the materials from this experiment as directed by your instructor.

Experiment 6

Data and Calculations: Properties of Hydrates

A. Identifying hydrates

	Is water released upon heating?	Appearance of heated solid?	Is the heated solid water-soluble?	Do you think this is a hydrate?
Nickel chloride	═══════	═══════	═══════	───────
Potassium chloride	═══════	═══════	═══════	───────
Sodium tetraborate	═══════	═══════	═══════	───────
Sucrose	═══════	═══════	═══════	───────
Calcium carbonate	═══════	═══════	═══════	───────
Barium chloride	═══════	═══════	═══════	───────

B. Reversibility of hydration

Summarize your observations of $CoCl_2 \cdot 6H_2O$:

Do the dehydration and hydration of $CoCl_2$ appear to be reversible?

C. Deliquescence and efflorescence

	Mass (sample + glass) {g}		Observations	Conclusion(s)
	Initial	Final		
$Na_2CO_3 \cdot 10H_2O$	═══════	═══════	═══════	───────
$KAl(SO_4)_2 \cdot 12H_2O$	═══════	═══════	═══════	───────
$CaCl_2$	═══════	═══════	═══════	───────
$CuSO_4$	═══════	═══════	═══════	───────
$CoCl_2$	═══════	═══════	═══════	───────

(continued on following page)

D. Percent by mass of water in a hydrate

Mass of crucible and cover
(*Note*: Warm objects will *not* weigh accurately on a balance!)

_____ g

Mass of crucible, cover, and solid hydrate

_____ g

Mass of crucible, cover, and heated solid

_____ g

Calculations and Results

Mass of solid hydrate

_____ g

Mass of heated solid

_____ g

Mass of H_2O lost

_____ g

Percent by mass of H_2O in the unknown hydrate

_____ $\%_{mass}$ H_2O

Molar mass of anhydrous salt (if given to you)

_____ g/mol

Moles of water per mole of unknown hydrate

_____

Unknown ID code

_____

Experiment 6

Advance Study Assignment: Properties of Hydrates

1. A student weighs a clean and dry crucible with its cover, obtaining a mass of 23.599 g. She adds a sample of a pink manganese(II) sulfate hydrate. She then obtains a total mass of 26.742 g for the crucible, cover, and sample. She heats the crucible to drive off the water of hydration, keeping the crucible at maximum heat for about ten minutes with the cover slightly ajar (open). She then lets the crucible cool completely and finds the crucible, cover, and contents now weigh 25.310 g. In the process the sample was converted to off-white anhydrous $MnSO_4$.

   a. What was the mass of the hydrate sample used?

   _____ g hydrate

   b. What is the mass of the anhydrous $MnSO_4$?

   _____ g $MnSO_4$

   c. What mass of water was driven off?

   _____ g H_2O

   d. What is the percent by mass of water in the hydrate?

   $$\%_{mass} \text{ water} = \frac{\text{mass of water in sample}}{\text{mass of hydrate sample}} \times 100\%$$

   _____ $\%_{mass}$ H_2O

(continued on following page)

e. How many grams of water would there be in 100.0 g of this hydrate? How many moles?

_____ g H_2O; _____ moles H_2O

f. How many grams of $MnSO_4$ are there in 100.0 g of this hydrate? How many moles? (What percentage of the mass of the hydrate is $MnSO_4$? Convert the mass of $MnSO_4$ to moles. The molar mass of $MnSO_4$ is 151.00 g/mol.)

_____ g $MnSO_4$; _____ moles $MnSO_4$

g. How many moles of water are present per mole of $MnSO_4$?

_____ mol H_2O/mol hydrate

h. What is the formula of the hydrate?

_____

Experiment 7

Analyzing an Unknown Chloride

One of the important applications of precipitation reactions (reactions that produce a solid product) is in quantitative chemical analysis, whose goal is to determine amounts in a chemical sample. Many substances that precipitate from solution have such low solubilities that the precipitation reaction by which they are formed can be considered to proceed to completion: silver chloride is an example. If a solution containing Ag^+ ion is added to a sample solution containing Cl^- ion, the ions will react to form AgCl:

$$Ag^+(aq) + Cl^-(aq) \rightarrow AgCl(s) \tag{1}$$

Not all precipitation reactions are fast; but most are, including this one. Because Ag^+ and Cl^- react in a 1:1 ratio, Reaction 1 will continue to occur until the total number of Ag^+ ions added becomes equal to the total number of Cl^- ions initially present in the sample. This serves as the basis for the chloride analysis you will carry out in this experiment.

A convenient method for determining the amount required for complete reaction with a sample is called titration. A solution of known concentration (the titrant) is added in measured amounts to the unknown sample until the reaction is complete.

In the titration you will do in this experiment, the titrant is a solution of silver nitrate, $AgNO_3$, of known molarity; it is added from a calibrated volume dispensing device called a buret (see Appendix D) to a solution containing a measured amount of unknown. The titration is stopped when a color change occurs in the solution, indicating that the moles of Ag^+ added have just become equal to the moles of Cl^- initially present in the sample. This is called the endpoint of the titration. The color change is caused by a chemical, called an indicator, that is added to the sample at the beginning of the titration. The volume of $AgNO_3$ titrant needed to reach the endpoint can be measured accurately with the buret, and the moles of Ag^+ added can then be calculated from the known molarity of the $AgNO_3$ titrant solution.

In the Fajans method for the volumetric analysis of chloride, which you will employ in this experiment, the indicator used is called dichlorofluorescein. After enough silver nitrate has been added to react with all of the chloride ions, and the concentration of Ag^+ in solution begins to rise, positively charged silver ions begin to stick to the surface of (technical term: adsorb on) the very tiny (technical term: colloidal) silver chloride particles. The silver chloride particles then have a positive charge, which attracts the negatively charged dichlorofluorescein ions onto their surfaces. When this occurs, the silver chloride takes on a pink color in the presence of the yellowish, fluorescent dichlorofluorescein, indicating that the endpoint of the titration has been reached. Because AgCl gradually darkens in the presence of light, your titration solutions will eventually become dark purple in color—more quickly in brighter light. You will therefore not be able to compare against the endpoint color of your titrated solutions for very long after each titration is complete.

In this experiment, you will start with a 100.-mL volumetric flask containing an unknown chloride solution. You will dilute this solution to the 100.-mL mark, and after mixing, pipet out 10.0-mL samples of the diluted solution into three separate flasks for titration with a standardized titrant solution of $AgNO_3$, meaning one whose molarity is both accurately and precisely (see Appendix H) known. Using the volume of titrant required to reach the endpoint in each titration, you will obtain three replicate measurements (independent measurements of the same quantity) of the molarity of the diluted chloride solution in the volumetric flask. Given the molarity of $AgNO_3$ in the titrant, $[AgNO_3]_{\text{titrant}}$,

$$\text{moles } Ag^+ = \text{moles } AgNO_3 = \frac{\text{concentration}}{\text{of titrant}} \times \frac{\text{volume of}}{\text{titrant used}} = [AgNO_3]_{\text{titrant}} \times V_{\text{titrant}} \tag{2}$$

where the volume of $AgNO_3$ titrant used is expressed in liters, and the square brackets around $AgNO_3$ indicate the concentration of the titrant in moles per liter of solution. At the endpoint of the titration,

$$\text{moles of } Ag^+ \text{ added} = \text{moles of } Cl^- \text{ present in sample of diluted unknown} \tag{3}$$

$$\text{moles of Cl}^- \text{ present in sample of diluted unknown} = \frac{\text{chloride concentration}}{\text{in titrated sample}} \times \frac{\text{volume of}}{\text{titrated sample}}$$

$$= [\text{Cl}^-]_{\text{sample}} \times V_{\text{sample}} \qquad \text{(4)}$$

Experimental Procedure

Wear your safety glasses while performing this experiment.

Obtain a buret, a 100.-mL volumetric flask containing your concentrated unknown chloride solution, and a 10.0-mL volumetric pipet. *Being careful to not overshoot,* add deionized water to the volumetric flask, up to the mark on the narrow neck: squeeze your wash bottle more gently or switch to using a dropper as the bottom of the meniscus (the curved top surface of the liquid) approaches the mark, and view the approach with the mark on the flask at the same level as your eyes. (See Appendix D.) Place the stopper in the flask, and, holding the stopper in place with the thumb, index finger, or palm of the same hand holding the flask, invert the flask 13 times to thoroughly mix the solution, allowing the trapped air bubble to travel the entire height of the flask each time. Use the pipet (see Appendix D) to transfer 10.0 mL of your now-diluted chloride solution to a 250-mL Erlenmeyer flask, and add about fifty milliliters of deionized water. Add 1 drop of dichlorofluorescein indicator.

Without diluting it in any way, transfer about one hundred milliliters of the standardized $AgNO_3$ titrant solution into a clean, *dry* 125-mL Erlenmeyer flask. This will be your total supply for the entire experiment, so do not waste it. Silver nitrate is expensive! Silver ion solutions leave long-lasting black stains on skin and clothing, so take care not to spill your titrant. If called for, clean your buret thoroughly with detergent solution, with the aid of a buret brush, and rinse it with deionized water. If your buret starts out wet with anything other than the titrant, pour 2 or 3 successive portions of a few milliliters of the $AgNO_3$ titrant into the buret and tip it back and forth to rinse the inside walls. Allow the $AgNO_3$ to drain out of the buret tip each time after doing so. Fill the buret with $AgNO_3$ titrant. Open the stopcock momentarily to flush any air bubbles out of the buret tip. Be sure that your stopcock fits snugly and that the buret does not leak. (See Appendix D for additional details.)

Read the initial buret level to ± 0.01 mL. You may find it useful when making readings to put a white card, with a thick black stripe or black tape on it, behind the meniscus. If the black line is held just below the level to be read, its reflection in the surface of the meniscus will help you obtain an accurate reading. (Note that if you use such a tool, it is critical that you use it the same way each time you take a measurement for a given titration: it will change what you read!) Begin to add the $AgNO_3$ titrant solution to the chloride solution while swirling the flask. Tiny colloidal particles of white AgCl will begin to form, giving a slight cloudiness to the yellowish solution. At the beginning of the titration, you can add the titrant (the $AgNO_3$ solution) fairly rapidly, a few milliliters at a time, swirling the flask to mix the solution. You will find that a pinkish color forms in the solution and disappears as you mix by swirling.

Gradually decrease the rate at which you add the $AgNO_3$ solution as the pink color persists longer. At some stage you may find it convenient to set your buret stopcock to deliver titrant slowly—drop by drop—while you swirl the flask. The endpoint of the titration is indicated by the pink color no longer going away with swirling. Right at the endpoint you may note that the solution takes on a peach color as the lasting pink and some yellow color are both present. The key to this titration is to focus on the hue (color) of the solution rather than the color's intensity. If you are careful, you can hit the endpoint within one drop of titrant. When you have reached the endpoint, stop the titration and record the buret level, again to ± 0.01 mL.

Analyze two more 10.0-mL diluted chloride samples in the same way. You can use the same 250-mL Erlenmeyer flask, provided you rinse it well with deionized water. It is fine if it is wet, as long as what it is wet with contains no chloride ion! Note the volume of $AgNO_3$ used in the previous titration and refill the buret as necessary, to ensure you will not go below the 50.00-mL mark at the bottom of the buret.

Calculate the concentration of chloride ion in the diluted solution in your 100.-mL volumetric flask, as independently indicated by each of your three titrations. Also calculate the average (mean) and the standard deviation of these three results. (Consult Appendix H for details.)

Take it further (optional): Find the mass percent of Cl^- or NaCl in a commercial food product, such as a bouillon cube or canned soup. Compare your results with those on the nutrition label.

Disposal of reaction products: Silver is environmentally toxic. Pour all titrated solutions and remaining titrant into the waste container(s) provided, unless your instructor directs otherwise.

Experiment 7

Data and Calculations: Analyzing an Unknown Chloride

Molarity of standardized $AgNO_3$ titrant solution: _____ M

Titration #	1	2	3
Initial buret reading	_____ mL	_____ mL	_____ mL
Final buret reading	_____ mL	_____ mL	_____ mL
Volume of $AgNO_3$ titrant used to titrate sample	_____ mL	_____ mL	_____ mL
Moles of $AgNO_3$ used to titrate sample	_____ mol	_____ mol	_____ mol
Moles of Cl^- present in sample	_____ mol	_____ mol	_____ mol
Volume of diluted chloride solution pipetted	_____ mL	_____ mL	_____ mL
Total moles of Cl^- in 100.-mL volumetric flask	_____ mol	_____ mol	_____ mol
Cl^- concentration in diluted solution in volumetric flask	_____ M	_____ M	_____ M

Average (mean) chloride concentration: _____ M

Standard deviation in chloride concentration: _____ M

Unknown ID code _____

Experiment 7

Advance Study Assignment: Analyzing an Unknown Chloride

1. A student performed this experiment and obtained the following chloride concentration values: 0.02813 M, 0.02828 M, and 0.02708 M.

   a. What is the average (mean) concentration?

   _____ M

   b. What is the standard deviation of these three concentration results?

   _____ M

2. A 10.0-mL sample of a diluted chloride solution required 13.89 mL of 0.02014 M $AgNO_3$ to reach the Fajans endpoint.

   a. How many moles of Cl^- ion were present in the sample? (Use Equations 2 and 3.)

   _____ moles Cl^-

   b. What was the concentration of chloride in the diluted solution? (Use Equation 4)

   _____ M

3. How would the following errors affect the concentration of Cl⁻ obtained in Question 2(b)? Give your reasoning in each case.

   a. The student read the molarity of $AgNO_3$ as 0.02104 M instead of 0.02014 M.

   b. The student was past the endpoint of the titration when he took the final buret reading.

   c. The buret was wet with deionized water before the titrant was added, and it was not rinsed with the titrant before being filled with it.

Experiment 8

Using the Ideal Gas Law—Atmospheric Pressure and the Absolute Zero of Temperature

Gases differ from liquids and solids in that much of the ordinary physical behavior of gases can be described by a single mathematical relationship common to all gases. Such a relationship does not exist for liquids or solids, so information about their physical properties must be obtained by direct experimentation on the specific liquid or solid of interest.

The <u>ideal gas law</u> describes the physical behavior of most gases, under most conditions:

$$PV = nRT \tag{1}$$

In this relationship P is the pressure the gas experiences, V is the volume it occupies, n is the number of gas particles present (usually in moles), T is the temperature of the gas on an absolute temperature scale, and R is a fixed value (for a given set of units) called the *gas constant*. The units chosen for the variables P, V, n, and T determine the numerical value of R; or, if one starts with a chosen value of R, that choice determines what units must be used for the variables.

In many cases, one or more of the variables in the ideal gas law will be known to be unchanging (or "constant", or "fixed") in a given situation. For example, if the gas is held in a sealed container, and is not undergoing a chemical reaction, the number of gas particles, n, will be unchanging (constant). If the gas is also in a fixed-temperature environment, such as a water bath, T will be constant. In such situations the constant terms can be grouped on one side of the ideal gas law to reveal the relationship between the remaining variables: in this example, with n, R, and T all constant, the product of P and V should remain equal to the constant product nRT as P and V change. That means that for most gases under such conditions, doubling the pressure, P, will cut the volume of the gas, V, in half.

The Celsius temperature scale uses the freezing point of water as its zero point: that is, water freezes at $0\,^\circ\text{C}$. In contrast, an <u>absolute temperature scale</u> uses as its zero point the lowest possible temperature (technical term: <u>absolute zero</u>), such that on that scale negative temperatures are not possible. Any non-absolute temperature scale can be converted to an absolute temperature scale by adding a (constant) correction factor to it. So, for a temperature on the Celsius scale, t, an absolute temperature T can be calculated as $t + k$, where k is called the <u>zero point</u> of the <u>Kelvin temperature scale</u>, which is the absolute temperature scale based on the Celsius degree.

The gas we encounter most often is the air around us. Its physical behavior is like that of other gases, so under ordinary conditions it obeys the ideal gas law, Equation 1. The pressure of the air is called the atmospheric pressure. It varies a little from day to day, and at sea level it averages out to equal the pressure exerted by a column of mercury 76 cm high, or by a column of water 10 m high; this is 1×10^5 Pascals, or in common American units, 15 psi (pounds per square inch, lbf/in^2). Ordinarily we are not aware of this substantial pressure because it is nicely balanced by our bodies, but if it decreases suddenly, as it does in a tornado, it will definitely get our attention: it can, among other things, cause a house to fly apart! Atmospheric pressure is measured with a <u>barometer,</u> of which there are several types.

In the first part of this experiment you will try to verify the zero point of the Kelvin temperature scale by heating a sample of air, keeping its volume constant while measuring the increase in pressure. Assuming the validity of the ideal gas law, you will calculate the temperature at which the pressure would become zero. That is the absolute zero of temperature, and the zero point of the Kelvin temperature scale.

In the rest of this experiment you will determine the atmospheric pressure using a <u>U-tube manometer:</u> a vertical U-shaped tube filled with liquid that will allow you to monitor the pressure and volume of an air sample as it is slightly compressed at constant temperature. This will allow you to experimentally observe how changes in volume and pressure are related to the atmospheric pressure.

Experimental Procedure

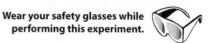

Obtain a 1000-mL beaker, a 500-mL Erlenmeyer flask fitted with a stopper with a glass tubing insert, a sensitive thermometer, a glass U-tube manometer, a straight section of glass manometer tubing, a short segment of flexible tubing, and a dropper or pipet.

A. Verifying the absolute zero of temperature

In this part of the experiment you will examine how the pressure of a sample of air of fixed volume increases when you raise the temperature. Given that $PV = nRT$ for the air, if the volume and amount of air remain constant then the ideal gas law (Equation 1) simplifies to $P = kT$, where k is a constant equal to nR/V and the pressure is proportional to the absolute (Kelvin) temperature, T. Your goal in this part of the experiment is to establish the conversion between the Celsius temperature, t, and the Kelvin temperature, T.

Assemble the apparatus shown in Figure 8.1. If a magnetic stirrer is available, put a magnetic stirring bar in the 1000-mL beaker and center the beaker on the stirrer. Clamp a *dry* 500-mL Erlenmeyer flask in the beaker as shown. (If the flask is not dry, add a few milliliters of acetone and shake well. Pour out the acetone and gently blow compressed air into the flask until it is thoroughly dry.) Fill the beaker with room-temperature water: ideally, water that has been stored overnight in the lab. Start the stirrer, if you have one. If you do not have a magnetic stirrer available, stirring with a stirring rod works, and in fact you should do that even in addition to using a magnetic stirrer. Connect the short piece of flexible tubing to the glass tubing in the stopper, and then moisten the rubber stopper and insert it firmly into the flask. Clamp the U-tube manometer into position. You can rest the bottom of the U-tube on the lab bench surface, if that is convenient. Using your dropper or pipet, add a few milliliters of water to the U-tube manometer, so that the level in each arm is about four centimeters above the lowest part of the "U". Connect the other end of the flexible tubing to the left arm of the manometer. Mark the water level in the left arm with a piece of tape.

Stir the water around the flask with a stirring rod (as well as with the magnetic stirrer, if you have one). If the system has no leaks, and the water in the beaker is at room temperature, with stirring the levels will remain steady for several minutes. If necessary, adjust the position of the tape once the levels hold. Once the level remains constant after 2 minutes of stirring, record the levels in the left and right arms of the manometer to ± 0.1 cm. Record the temperature of the water to $\pm 0.1\,°C$, and record the atmospheric pressure as indicated by a barometer, in mm Hg.

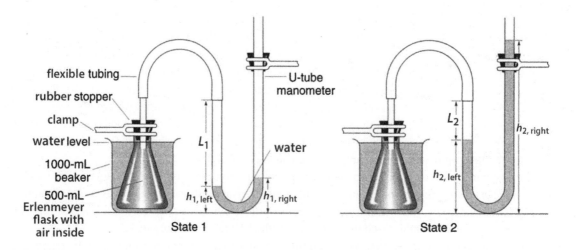

Figure 8.1   The apparatus shown in this figure (the same apparatus, in different states) is used for both determining the atmospheric pressure and verifying the absolute zero of temperature.

Using your dropper or pipet, add a few milliliters of room-temperature water to the right arm of the manometer, enough to raise the level in the *left* arm by about three centimeters. Add some warm tap water (at about 40 °C) to a beaker, and very slowly add some of this warm water to the 1000-mL beaker. The level in the left arm will go back down as the air in the flask warms and expands. Add warm water as necessary until the level in the left arm is back at the location you marked with tape. If the level goes too low, add a little room-temperature water to the right arm to bring it to the proper position. Wait a minute or so, with stirring, to make sure the levels hold steady and the temperature is not changing. Read the two levels and the temperature.

Repeat the steps in the preceding paragraph at two or three higher temperatures, being sure to wait long enough for the levels and temperature to become steady. At the higher temperatures, the water in the 1000-mL beaker will tend to cool, so you may need to add warm water to make it hold its temperature. Your final level in the right arm should be near the top of the manometer. You should have at least four runs for which you have recorded water levels and temperature.

You can process the data you have obtained by making a graph of total pressure, P_{total}, versus the Celsius temperature, t. The total pressure of the gas will equal the measured atmospheric pressure plus the difference, Δh, between the level in the right arm of the manometer and the level in the left arm. You have measured pressures in cm of water, but atmospheric pressure is usually reported in mm Hg. It is convenient to use cm H_2O as your pressure unit, converting cm Hg to cm H_2O using a conversion factor:

$$P_{atm} \text{ in cm } H_2O = P_{atm} \text{ in cm Hg} \times \left(\frac{\text{density of Hg}}{\text{density of } H_2O}\right) = P_{atm} \text{ in cm Hg} \times \left(\frac{13.57 \, g/mL}{1.000 \, g/mL}\right)$$

$$= 13.57 \times P_{atm} \text{ in mm Hg} \times \left(\frac{1 \text{ cm}}{10 \text{ mm}}\right)$$

$$= 1.357 \times P_{atm} \text{ in mm Hg} \qquad \qquad \textbf{(2)}$$

$$\text{In units of cm } H_2O, \quad P_{total} = P_{atm} + \Delta h \qquad \qquad \textbf{(3)}$$

Using Equations 2 and 3, make a table of temperatures t in °C and corresponding values of P_{total} in cm H_2O. Then use graph paper or a computer to make a graph of total pressure versus Celsius temperature (see Appendix E). The resulting plot of P vs t should be a straight line, with the general equation $y = mx + b$, or, with our variables, $P = mt + b$, where m is the slope of the line and b is a constant. Using your graph, find the equation of the line. Then find the temperature t_0 in °C at which P equals zero. The zero point of the Kelvin temperature scale (on the Celsius scale) is the negative of t_0. The Advance Study Assignment steps you through this calculation.

B. Determining the atmospheric pressure

This part of the experiment uses the same apparatus as the first part. Disconnect the flexible tubing from the manometer and pour out the water. Take the flask out of the beaker and pour the (warm) water out of the beaker. Clamp the flask back in the now-empty beaker with the stopper firmly inserted. Fill the beaker with room temperature water. Check with your thermometer to see that the water is indeed at room temperature. In this part of the experiment it is critical that the temperature of the flask remain constant.

Using your dropper or pipet, add a few milliliters of room-temperature water to the manometer, so that the levels are about four centimeters above the bottom of the "U". Then connect the flexible tubing to the left arm of the manometer. The apparatus at that point should look like the left-hand image in Figure 8.1.

You have now isolated the air in the flask in a fixed initial state: State 1. That state is defined by its volume, pressure, and temperature. The volume, V_1, of the air is essentially that of the flask, but you will measure it in Part C. The pressure, P_1, is equal to the atmospheric pressure, P_{atm}, plus the difference between the heights of the liquid levels in the manometer, $h_{1, \, right} - h_{1, \, left}$ (see the section on manometers in Appendix D):

$$P_1 = P_{atm} + h_{1, \, right} - h_{1, \, left} \qquad \qquad \textbf{(4)}$$

As in the first part of this experiment, cm H_2O is the most convenient pressure unit to use.

As before, use a piece of tape to mark the water level in the left arm. Stir the water in the beaker for 3 minutes and confirm that the level remains unchanged. If it does change, add a little warm or cold water to the 1000-mL beaker to bring it to room temperature and stir for at least 2 more minutes. Keep doing this until the level is steady. Then measure the heights $h_{1, \text{right}}$ and $h_{1, \text{left}}$ with a ruler or meter stick and record them to the nearest 0.1 cm. Also record the distance from the left water level to the top of the left arm, L_1, which you will need for Part C.

Now compress the air in the flask just a little, by adding room-temperature water to the right arm of the manometer. Add the water slowly, using your dropper or pipet. You will see that both levels go up, but the right-hand level goes up faster than the left, because the pressure of the air in the flask increases as its volume is reduced. Continue adding water until the level is near the top of the right arm. Move the tape to the new water level in the left arm and confirm it remains unchanged after you stir for at least 2 minutes. Once it is stable, measure the heights $h_{2, \text{right}}$ and $h_{2, \text{left}}$ in the right and left arms of the manometer and record them, along with the length of the empty section of the left arm, L_2. The air sample is now in State 2 (see Figure 8.1).

In State 2, the pressure P_2 of the air is equal to P_{atm} plus $h_{2, \text{right}} - h_{2, \text{left}}$. The volume V_2 is slightly less than V_1, by an amount equal to the volume of the air in the manometer that was displaced by water as the air was compressed. Call that change in volume ΔV; you will measure it in Part C, below.

Summarizing, for States 1 and 2:

$$P_1 = P_{\text{atm}} + h_{1, \text{right}} - h_{1, \text{left}} \tag{4}$$

$$P_2 = P_{\text{atm}} + h_{2, \text{right}} - h_{2, \text{left}} \tag{5}$$

$$\Delta P = P_2 - P_1 \qquad \Rightarrow \qquad P_2 = P_1 + \Delta P \qquad V_2 = V_1 - \Delta V$$

Because the temperature and number of gas particles are constant, the ideal gas law indicates that

$$P_1 V_1 = P_2 V_2 \text{ (for a fixed amount of gas at constant temperature)}$$

or, substituting in for P_2 and V_2,

$$P_1 V_1 = (P_1 + \Delta P)(V_1 - \Delta V) \tag{6}$$

Solving this equation for P_1 indicates that

$$P_1 = (V_1 \Delta P - \Delta V \Delta P)/\Delta V \tag{7}$$

Equations 7 and 4 can be used to calculate P_{atm} once you determine V_1 and ΔV.

C. Calibrating the system

To determine the volumes V_1 and ΔV, first remove the Erlenmeyer flask from the beaker and take out the rubber stopper. Disconnect the flexible tubing from the manometer, but leave it attached to the stopper. In a sink, fill the Erlenmeyer flask completely with water, all the way to the very top, and then push the stopper back into it. (This approach should prevent any air from remaining trapped in the Erlenmeyer flask.) Next, attach the straight piece of glass manometer tubing to the end of the flexible tubing and hold it vertically, above the stopper and Erlenmeyer flask. Add water to the glass tube until it is filled to a depth L_2 above where it meets the flexible tubing. Then measure how much *more* water is needed to fill it to a depth L_1: this is ΔV, which you should record. An easy way to do this is to measure and record the mass of a beaker full of water, then transfer the required amount of water from that beaker into the tube. Weigh the beaker again, and the mass difference will indicate the mass of water that corresponds to ΔV. It is precise enough to use 1.00 g/mL for the density of water, so the volume ΔV in milliliters is numerically equal to the mass of the water in grams. To determine V_1, measure the total amount of water in the Erlenmeyer flask, flexible tubing, and glass tube when it is filled to L_1. You can do this based on the mass of the water, or its volume; it should be a bit more than 500 mL.

Using the data you have obtained and the relevant equations, calculate P_1 and P_{atm}. The units of P_{atm} will be cm H_2O, which you can convert to cm Hg by dividing by (13.57 cm H_2O/cm Hg). Report the value of P_{atm} in both units. Compare the value you obtain with that indicated by a barometer.

Complete all of your calculations before leaving the lab. When you are finished, turn off and return all borrowed equipment.

Experiment 8

Data and Calculations: Using the Ideal Gas Law—Atmospheric Pressure and the Absolute Zero of Temperature

A. Verifying the absolute zero of temperature

From the laboratory barometer:

P_{atm} in mm Hg _____ P_{atm} in cm Hg _____ P_{atm} in cm H_2O _____

To convert P_{atm} from cm Hg to cm H_2O, multiply by (13.57 cm H_2O/cm Hg).

$$P_{total} = P_{atm} + h_{right} - h_{left} \text{ (for } P \text{ in cm } H_2O)$$

Manometer heights {cm H_2O}		Temp, t {°C}	P_{total} {cm H_2O}
$h_{1, right}$ _____	$h_{1, left}$ _____	t_1 _____	P_1 _____
$h_{2, right}$ _____	$h_{2, left}$ _____	t_2 _____	P_2 _____
$h_{3, right}$ _____	$h_{3, left}$ _____	t_3 _____	P_3 _____
$h_{4, right}$ _____	$h_{4, left}$ _____	t_4 _____	P_4 _____
$h_{5, right}$ _____	$h_{5, left}$ _____	t_5 _____	P_5 _____

Using graph paper or a computer, make a graph of P_{total} vs t. (Appendices E, G1, and/or G2 may help with this.) The points should fall on a straight line, with the equation $P = mt + b$, where m is the slope and b is a constant. If you need to find the equation by hand, see Appendix E for the method. If you find the equation using a computer, print out the graph with the line and the equation shown on it and include it with this report. In any case, report your result here:

Equation for P_{total} vs t: _____

Find the temperature t_0, in °C, at which P_{total} becomes zero:

$t_0 = $ _____ °C   Let $k = -t_0$,   and $T = t + k$

On the Kelvin scale, $T = 0$ at absolute zero, which is t_0 °C on the Celsius scale.

In your equation for P_{total} vs t, substitute $(T - k)$ for t and show that, with your value of k, the equation reduces to $P_{total} = mT$.

(continued on following page)

B. Determining the atmospheric pressure

Manometer Heights {cm H_2O}

$h_{1, \text{right}}$ ———— $h_{1, \text{left}}$ ———— L_1 ————

$h_{2, \text{right}}$ ———— $h_{2, \text{left}}$ ———— L_2 ————

Using Equations 4 and 5, evaluate P_1, P_2, and ΔP:

$P_1 = P_{\text{atm}} +$ ———— cm H_2O $P_2 = P_{\text{atm}} +$ ———— cm H_2O $\Delta P =$ ———— cm H_2O

C. Calibrating the system

Mass of beaker filled with water   ———— g

Mass of beaker with water remaining after water level
in tube was raised from L_2 to L_1 ———— g

Mass of water required to raise level in glass tube from L_2 to L_1 ———— g = ΔV in mL = ΔV in cm^3

Mass (or volume) of water in Erlenmeyer flask, flexible tubing, and glass tube filled to L_1

———— g (or mL) = V_1 in mL = V_1 in cm^3

Using Equation 7, evaluate P_1 and P_{atm}:

$P_1 =$ ———— cm H_2O $P_{\text{atm}} =$ ———— cm H_2O

Express P_{atm} in cm Hg and compare your value to that obtained from the laboratory barometer.
(*Hint*: 13.57 cm H_2O = 1 cm Hg; show your work!)

$P_{\text{atm}} =$ ———— cm Hg $P_{\text{barometer}} =$ ———— cm Hg

Experiment 8

Data and Calculations: Using the Ideal Gas Law—Atmospheric Pressure and the Absolute Zero of Temperature

For use as needed.

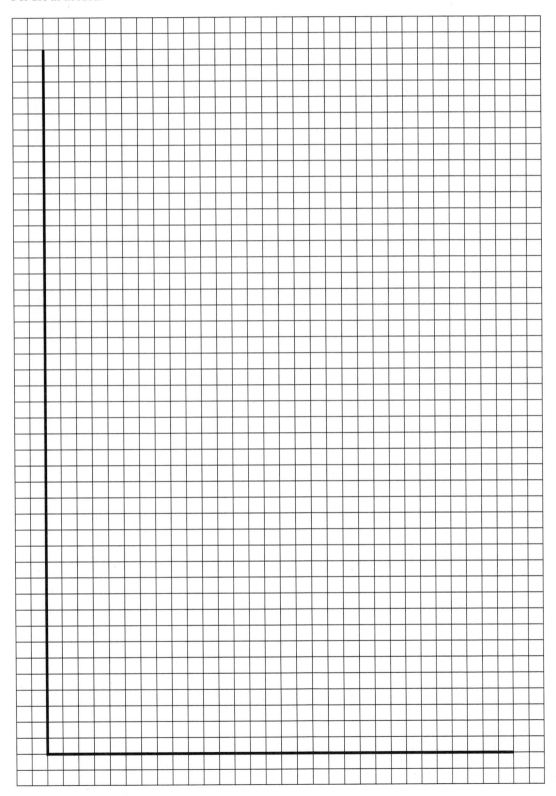

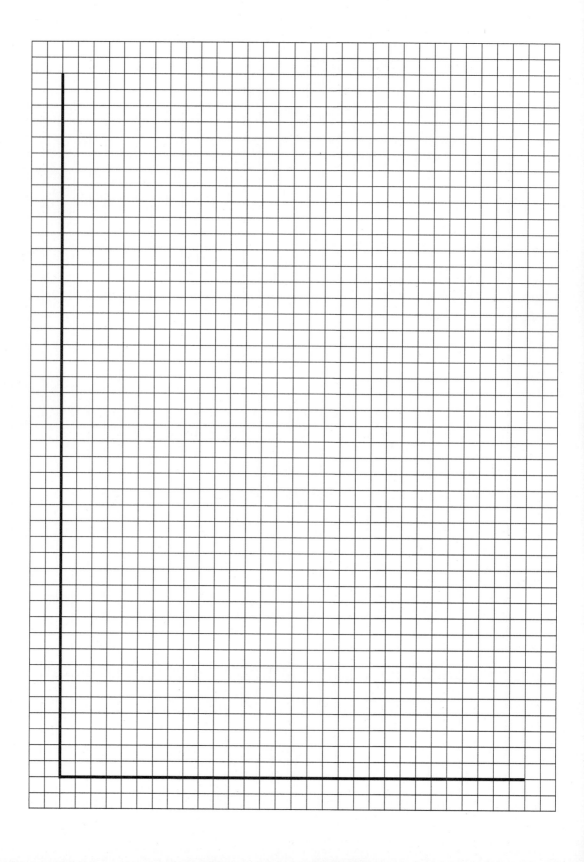

Experiment 8

Advance Study Assignment: Using the Ideal Gas Law—Atmospheric
Pressure and the Absolute Zero of Temperature

1. The atmospheric pressure on a spring day in Minnesota was found to be 726 mm Hg.

   a. What would that pressure be in cm H_2O?
      (Use Equation 2, or 13.57 cm H_2O = $\underline{1}$ cm Hg = $\underline{10}$ mm Hg)

      _____ cm H_2O

   b. About how many meters of water would it take to exert that pressure?

      _____ m H_2O

2. In this experiment a student found that when they increased the temperature of a 454-mL sample of air from 22.8 °C to 30.9 °C, the pressure of the air went from 1021 cm H_2O up to 1049 cm H_2O. Because the air expands linearly with temperature, the equation relating P to t is of the form

$$P = mt + b \qquad\qquad (8)$$

   where m is the slope of the line and b is a constant.

   a. What is the slope of the line? (*Hint*: Find the change in P divided by the change in t.)

      $m =$ _____ cm $H_2O/C°$

   b. Find the value of b. (*Hint*: Substitute known values of P and t into Equation 8 and solve for b.)

      $b =$ _____ cm H_2O

(*continued on following page*)

c. Express Equation 8 in terms of the values of m and b.

$$P =$$

d. At what temperature t will P become zero?

$$P = 0 \text{ at } t = \underline{\hspace{4cm}} °C = t_0 = -k$$

(Please note that you are unlikely to get results anywhere near this good when actually carrying out this experiment, as it involves a very large extrapolation.)

e. The temperature t_0 in Part (d) is the absolute zero of temperature. The Kelvin temperature scale is one on which that temperature is 0 K. On that scale, $T = t + k$. Show that, on the Kelvin scale, the equation you obtained in Part (c) reduces to $P = mT$.

Experiment 9

Determining the Molar Mass of an Easily Vaporized Liquid Using the Ideal Gas Law

One of the important applications of the ideal gas law is in experimentally determining the molar masses of gases. The molar mass of a gas can be determined based on the mass of a sample of the gas under known conditions of temperature and pressure at which the gas obeys the <u>ideal gas law</u>,

$$PV = nRT \tag{1}$$

If the pressure P is in atmospheres, the volume V is in liters, the temperature T is in Kelvins (K), and the amount n is in moles, then the appropriate value of the gas constant R to use is 0.08206 L·atm/(mole·K).

From measured values of P, V, and T for a sample of gas, Equation 1 can be used to find the moles of gas in the sample. The molar mass, MM, is equal to the mass m of the gas sample divided by the moles of gas it is composed of, n:

$$n = \frac{PV}{RT} \qquad MM = \frac{m}{n} \tag{2}$$

This experiment involves determining the molar mass of an easily vaporized liquid using Equation 2. A small amount of the liquid is introduced into a weighed flask. The flask is then placed in boiling water, where the liquid vaporizes completely, driving out the air and filling the flask with pure <u>vapor</u> (evaporated liquid) at atmospheric pressure and the temperature of the boiling water. By cooling the flask so that this vapor condenses, the mass of the vapor can be measured and used to calculate a value for the liquid's molar mass, MM.

Experimental Procedure*

Wear your safety glasses while performing this experiment.

Obtain a special round-bottom flask, a tiny stopper, a large stopper with a glass cap in it, and an unknown liquid sample. Support the flask on a cork ring or an evaporating dish, or in a beaker, at all times, so it does not roll away on you! With the tiny stopper loosely inserted in the neck of the flask, weigh the empty, dry flask on an analytical balance. Use a tared cork ring or evaporating dish to support the flask on the balance pan.

Pour about five milliliters of your unknown liquid sample into the flask. **Do not reinsert the tiny stopper!** Add about three hundred milliliters of water to the 600-mL beaker and assemble the apparatus shown in Figure 9.1, with the beaker centered on the hotplate and the floating flask kept centered in the beaker and a half centimeter above its bottom by the flask's neck being under the clamped-in-place cap and large stopper. Add a few boiling chips to the water in the 600-mL beaker and heat the water to the boiling point. (See Appendix D for heating advice.) Watch the liquid level in your flask: it should gradually drop as vapor escapes through the cap, even before the water boils. After all the liquid has disappeared, the water starts to boil, and no more vapor comes out of the cap, lower the heat but continue to boil the water gently for 5 to 8 minutes. Measure and record the temperature of the boiling water. Shut off the hotplate and wait until the water has just stopped boiling, then remove the cap and immediately insert the tiny stopper used previously: tightly, this time.

Loosen the clamp holding the large stopper in place. Remove the flask from the beaker of water, holding it by the neck, which will be cooler. Immerse the flask in a beaker of cool water to a depth of about five centimeters. After holding the flask in the water for about two minutes to allow it to cool, carefully remove the tiny stopper *for not more than a second or two* to allow air to enter, and then tightly reinsert the tiny stopper.

*See W. L. Masterton and T. R. Williams, J. Chem. Educ. *36*, 528 (1959). For an alternate apparatus, see Instructor's Manual.

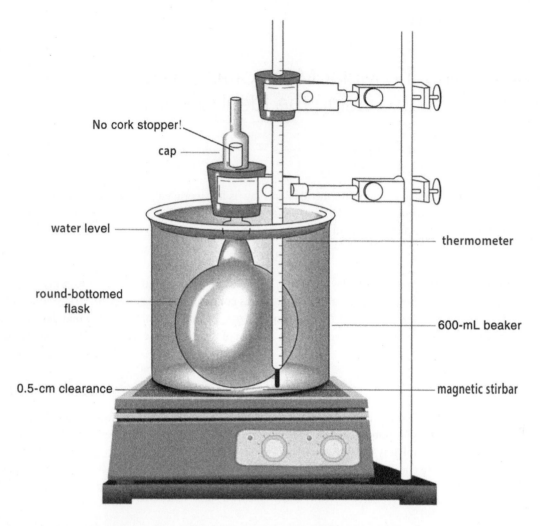

Figure 9.1 In the apparatus used in this experiment, the round-bottom flask floats in place, held down by the cap, rather than being rigidly connected to the large stopper. The thermometer should not touch the sides or bottom of the 600-mL beaker, and the magnetic stirbar must turn freely.

(As the flask has cooled, the vapor inside has condensed and the pressure has dropped, which is why air rushes in when the tiny stopper is removed.)

Dry the flask with a towel to remove the surface water and allow it to come to room temperature. Loosen the tiny stopper again momentarily to equalize any remaining pressure differences, then reweigh the flask. Its mass will have changed, because some of the air that was in it when you weighed it before has been replaced with the amount of your unknown liquid required to fill the flask completely with vapor at the temperature of the boiling-water bath. Read the atmospheric (ambient) pressure and record it.

Pour the remaining 5 mL of your liquid sample into the flask and repeat the heating process, again using the apparatus in Figure 9.1. Repeat the pressure equalization and weighing steps for this replicate measurement.

You may be given the volume of the flask by your instructor, or you may be directed to measure its volume by weighing the flask, stoppered and full of water, on a balance (top-loading, if available). *Do not fill the flask with water unless you are specifically told to do so.*

When you have completed this experiment, return the flask: do not attempt to wash or clean it in any way. Also return the stopper and cap, along with any remaining unknown sample (in its original container).

Experiment 9

Data and Calculations: Determining the Molar Mass of an Easily Vaporized Liquid Using the Ideal Gas Law

	Trial 1		Trial 2
Unknown ID code	=====	→	=====
Mass of flask and stopper	===== g	→	===== g
Mass of flask, stopper, and condensed vapor	===== g		===== g
Mass of flask, stopper, and water (*see directions; only if told to do so!*)	===== g	←	===== g
Temperature of boiling-water bath	===== °C		===== °C
Atmospheric (ambient) pressure	===== mm Hg		===== mm Hg

Calculations and Results

	Trial 1		Trial 2
Pressure of vapor, P	_____ atm		_____ atm
Volume of flask (volume of vapor), V	_____ L	→	_____ L
Temperature of vapor, T	_____ K		_____ K

(continued on following page)

	Trial 1	**Trial 2**

Mass of vapor, m _____ g   _____ g

Moles of vapor, n _____ mol   _____ mol

Molar mass of unknown, as found
by substitution into Equation 2   _____ g/mol   _____ g/mol

Experiment 9

Advance Study Assignment: Determining the Molar Mass of an Easily
Vaporized Liquid Using the Ideal Gas Law

1. A student found the mass of an empty, stoppered flask to be 53.256 g. She then added about five milliliters
    of an unknown liquid and heated the flask (in a boiling-water bath) to 98.8 °C. After all the liquid had
    vaporized, she removed the flask from the bath, stoppered it, and let it cool. Once it reached room
    temperature, she momentarily removed the stopper, then replaced it and weighed the flask containing air
    and condensed vapor, obtaining a mass of 53.870 g. The volume of the flask was known to be 326.1 mL.
    The absolute atmospheric (ambient) pressure in the laboratory that day was 728 mm Hg.

    a. What was the pressure of the vapor in the flask, in atm? (The vapor was pushed on by the atmosphere,
        so it was at the pressure of the atmosphere.)

    $P =$ _____ atm

    b. What was the temperature of the vapor, in Kelvins, when it was in the boiling water bath? What was
        the volume of the flask, in liters?

    $T =$ _____ K $V =$ _____ L

    c. What was the mass of condensed vapor that was present in the flask?

    $m =$ _____ grams

    d. How many moles of condensed vapor were present? (Use Equation 2)

    $n =$ _____ moles

(continued on following page)

    e. What is the mass of one mole of vapor? (Use Equation 2; this is the molar mass of the vapor)

$MM = $ _____ g/mole

2. How would each of the following procedural errors affect the results of this experiment? (Would the experimental molar mass come out higher, lower, or unchanged as a result of the error?) Give your reasoning in each case.

    a. Not all of the liquid was vaporized when the flask was removed from the boiling-water bath.

    b. The flask was not dried before the final weighing with the condensed vapor inside.

    c. The flask had not cooled completely to room temperature in the cool water bath, but the tiny stopper was not removed again before the final weighing.

    d. The flask was left open to the atmosphere while it was being cooled, and the tiny stopper was inserted just before the final weighing.

    e. The flask was stoppered and removed from the boiling-water bath before the vapor had reached the temperature of the boiling water. All the liquid had vaporized.

Experiment 10

Analyzing an Aluminum–Zinc Alloy*

Some metals react with solutions of strong acids to produce hydrogen gas and a solution of a salt of the metal. Small amounts of hydrogen are commonly prepared by the action of hydrochloric acid on metallic zinc, according to the net ionic equation

$$Zn(s) + 2\,H^+(aq) \rightarrow Zn^{2+}(aq) + H_2(g) \tag{1}$$

Reaction 1 indicates that one mole of zinc produces one mole of hydrogen gas. If the hydrogen gas is collected under known conditions, the mass of zinc in a pure sample can be determined by measuring the amount of hydrogen it produces in its reaction with acid.

Because aluminum reacts with strong acids in a similar way,

$$2\,Al(s) + 6\,H^+(aq) \rightarrow 2\,Al^{3+}(aq) + 3\,H_2(g) \tag{2}$$

it is also possible to determine the amount of aluminum in a pure sample by measuring the amount of hydrogen it produces when reacting with an acid solution. In this case two moles of aluminum produce three moles of hydrogen gas.

Because the amount of hydrogen produced by one gram of zinc is not the same as the amount produced by one gram of aluminum,

$$1 \text{ mole Zn} \rightarrow 1 \text{ mole } H_2 \;\Rightarrow\; 65.4 \text{ g Zn} \rightarrow 1 \text{ mole } H_2 \;\Rightarrow\; 1.00 \text{ g Zn} \rightarrow 0.0153 \text{ mole } H_2 \tag{3}$$

$$2 \text{ moles Al} \rightarrow 3 \text{ moles } H_2 \;\Rightarrow\; 54.0 \text{ g Al} \rightarrow 3 \text{ moles } H_2 \;\Rightarrow\; 1.00 \text{ g Al} \rightarrow 0.0556 \text{ mole } H_2 \tag{4}$$

it is possible to react a known mass of an <u>alloy</u> (a mixture of metals) containing only zinc and aluminum with acid, determine the amount of hydrogen gas produced, and calculate the percentages of zinc and aluminum in the alloy using Relations 3 and 4. The object of this experiment is to carry out such an analysis.

In this experiment you will react a weighed sample of an aluminum–zinc alloy with an excess of acid and collect the hydrogen gas produced in the reaction over water (as shown in Figure 10.1), determining its volume from the mass of water the gas displaces. If the volume, temperature, and contribution to the total pressure (the <u>partial pressure</u>) of the hydrogen gas, P_{H_2}, are known, the ideal gas law can be used to calculate how many moles of hydrogen were produced by the sample, n_{H_2}:

$$P_{H_2}V = n_{H_2}RT, \quad n_{H_2} = \frac{P_{H_2}V}{RT} \tag{5}$$

The total pressure P of gas in the flask is equal to the partial pressure of the hydrogen, P_{H_2}, plus the partial pressure of the water vapor, P_{H_2O}, in the collected gas:

$$P = P_{H_2} + P_{H_2O} \tag{6}$$

The gas collected in this experiment is saturated with water vapor, so under these conditions P_{H_2O} is equal to the vapor pressure of water, VP_{H_2O}, at the temperature of the experiment. This value is constant at a given temperature, and can be found in Appendix A. The total gas pressure P in the flask is very nearly equal to the atmospheric (ambient) pressure, P_{atm}.

Substituting these values into Equation 6 and solving for P_{H_2} yields

$$P_{H_2} = P_{atm} - VP_{H_2O} \tag{7}$$

*W. L. Masterton, J. Chem. Educ. *38*, 558 (1961). For a source of alloy samples, see Instructor's Manual.

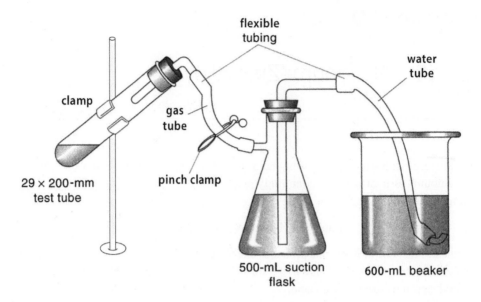

Figure 10.1 This experiment uses the water displacement method to determine the amount of hydrogen gas generated in the chemical reaction.

Using Equations 5 and 7, you will be able to calculate n_{H_2}, the moles of hydrogen produced by your weighed alloy sample. You can then calculate the mass percentages of Al and Zn in the alloy. For a sample containing m_{Al} grams of Al and m_{Zn} grams of Zn, combining Equations 3 and 4 indicates that

$$n_{H_2} = (m_{Al} \times 0.0556) + (m_{Zn} \times 0.0153) \tag{8}$$

For a one-gram sample of alloy, m_{Al} and m_{Zn} represent the mass fractions of Al and Zn, that is, $\%_{Al}/100$ and $\%_{Zn}/100$. Therefore

$$N_{H_2} = \left(\frac{\%_{Al}}{100} \times 0.0556\right) + \left(\frac{\%_{Zn}}{100} \times 0.0153\right) \tag{9}$$

where N_{H_2} is the moles of H_2 produced *per gram* of alloy. Because we know the alloy contains *only* aluminum and zinc,

$$\%_{Al} + \%_{Zn} = 100\% \quad \Rightarrow \quad \%_{Zn} = 100\% - \%_{Al} \tag{10}$$

so that Equation 9 can be written in the form

$$N_{H_2} = \left(\frac{\%_{Al}}{100} \times 0.0556\right) + \left(\frac{100 - \%_{Al}}{100} \times 0.0153\right) \tag{11}$$

Using algebra, this can be solved directly for $\%_{Al}$:

$$N_{H_2} = \left(\frac{\%_{Al}}{100} \times 0.0556\right) + \left(\left[1 - \frac{\%_{Al}}{100}\right] \times 0.0153\right)$$

$$N_{H_2} = \left(\frac{\%_{Al}}{100} \times 0.0556\right) + \left(0.0153 - \frac{\%_{Al}}{100} \times 0.0153\right)$$

$$N_{H_2} = 0.0153 + \left(\frac{\%_{Al}}{100} \times 0.0556 - \frac{\%_{Al}}{100} \times 0.0153\right)$$

$$N_{H_2} - 0.0153 = \left(\frac{\%_{Al}}{100} \times 0.0403\right)$$

$$\frac{\%_{Al}}{100} = \frac{(N_{H_2} - 0.0153)}{0.0403} \tag{12}$$

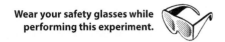

Experimental Procedure

- Obtain a 500-mL suction flask, a large (29 × 200 mm) test tube, a gas tube assembly, and a water tube assembly. Assemble this equipment as shown in Figure 10.1.

- Obtain a sample of Al-Zn alloy and a gelatin capsule. Record the Unknown ID code of your alloy on the bottom of the back of the report page.

Note: Use care in both of the following weighing steps, because the tiny sample size makes it such that even a small measurement error can lead to a significant experimental error.

- Weigh the gelatin capsule on an analytical balance, to ± 0.0001 g, recording the result on the report page.

- Add 0.2 ± 0.05 g of your alloy to the capsule, tearing the alloy with your fingers if the pieces are too big to fit. Seal the capsule with *all* of the alloy *completely inside the capsule*, and then record the mass of the capsule with the alloy in it, to ± 0.0001 g, on the report page.

- Fill both the suction flask and the beaker about two-thirds full of room-temperature water.

- Moisten the stopper on the water tube and insert it firmly into the suction flask.

- Open the pinch clamp and apply suction to the gas tube, pulling water into the flask from the beaker until the water level in the flask is a few centimeters below the side-arm.

- Close the pinch clamp to prevent siphoning. The water tube should now be full of water, with no air bubbles in it.

- Gently remove the water tube from the beaker and set it down on the lab bench. After you do this, water should not leak out the end. (If it does, you must find and fix the leak!)

- Pour out the water remaining in the beaker, letting it drain for a second or two. Without drying it, weigh the empty beaker on a balance (top-loading, if available), to ± 0.1 g.

- Return the beaker to your lab station and gently put the water tube back into it.

- Pour 10. mL (that is, 10 ± 1 mL) of 6 M HCl, hydrochloric acid, into the large test tube.

CAUTION: **6 M HCl is a solution of a gaseous strong acid. Avoid skin contact! If your skin starts to itch, you have probably made contact with HCl. Wash the itchy area with soap immediately. HCl gas will be present above 6 M HCl, particularly when the solution is hot; avoid exposure to these fumes.**

- Drop the gelatin capsule into the HCl solution; if it sticks to the walls of the test tube, poke it down into the acid with your stirring rod.

- Insert the gas line stopper firmly into the test tube and open the pinch clamp. If a little water goes into the beaker at that point, pour that water out, letting the beaker drain for a second or two.

CAUTION: **Make sure the test tube is not aimed toward anyone, and that the pinch clamp has been removed.**

- Within a few minutes, the acid solution will dissolve the wall of the capsule and begin to react with the alloy. Removing the test tube from the clamp and agitating it gently can speed things up, **but grip it only above the liquid level,** as the part of the test tube in contact with the acid will become quite hot as the reaction proceeds. **Stop agitating the tube once the reaction picks up speed, or it may get going too fast and bubble over!** The hydrogen gas that is formed will go into the suction flask and displace water from the flask into the beaker. The volume of water displaced will equal the volume of gas produced.

- As the reaction proceeds you will probably observe a dark foam, which contains particles of unreacted alloy. The foam may carry some of the alloy up the sides of the test tube. Remove the test tube from the clamp and tilt and/or rotate it gently to make sure that all of the alloy gets into the acid solution. The reaction should be over in five to ten minutes. At that time the liquid solution will again be clear, the foam will be essentially gone, the capsule will be completely dissolved, and no unreacted alloy should remain.

- When the reaction is over, close the pinch clamp, and then gently remove the water tube from the beaker and set it on the lab bench again.

- Weigh the beaker with the displaced water in it to $\pm$ 0.1 g, recording the result on the report page.

- Measure and record the temperature of the displaced water.

- Record the atmospheric (ambient) pressure in the lab.*

- Pour the used acid solution into a waste container.

- Reassemble the apparatus and conduct a second (replicate) trial, using the same alloy.

Take it further (optional): Using the procedure and apparatus in this experiment, find the percent by mass of aluminum in a soft-drink or beer can. These cans are made of aluminum-magnesium alloys (different ones in the bottom, sides, and top of the can), so a rigorous analysis requires replacing the molar mass of zinc, Zn, with that of magnesium, Mg, in Equation 3 and the formulas derived from it as well as using alloy samples from a specific part of the can. Like Zn, Mg reacts with acid in a 1:1 ratio.

> **Disposal of reaction products:** HCl needs to be neutralized before it can be disposed of, and large amounts of Zn and Al harm the aquatic environment; so, pour any liquids containing acid and/or metal into a waste container unless directed to do otherwise by your instructor. The displacement water used in this experiment can be dumped down the drain, provided nothing went wrong and it is not contaminated with metal or acid. If you had your test tube overflow into your suction flask, please consult with your lab instructor.

---

*The pressure of the gas in the flask will differ slightly from the atmospheric pressure, because the water levels inside and outside the flask are not quite equal. The error arising from this is so small that it is not worth attempting to correct for, though.

Experiment 10

Data and Calculations: Analyzing an Aluminum–Zinc Alloy

	Trial 1	Trial 2
Mass of gelatin capsule	_____ g	_____ g
Mass of alloy sample plus capsule	_____ g	_____ g
Mass of drained beaker	_____ g	_____ g
Mass of beaker plus displaced water	_____ g	_____ g
Atmospheric (ambient) pressure, P_{atm}	_____ mm Hg	_____ mm Hg
Water (and H_2) temperature, t	_____ °C	_____ °C
Mass of alloy sample	_____ g	_____ g
Mass of displaced water	_____ g	_____ g
Volume of displaced water (density of water, $\rho = 1.00$ g/mL)	_____ mL	_____ mL
Volume of collected gas, V	_____ L	_____ L
Temperature of collected gas, T	_____ K	_____ K
Vapor pressure of water at t, VP_{H_2O}, from Appendix A	_____ mm Hg	_____ mm Hg

(continued on following page)

	Trial 1	**Trial 2**

Partial pressure of H_2, P_{H_2} _____ mm Hg   _____ mm Hg

_____ atm   _____ atm

Moles of H_2 from entire sample, n_{H_2} _____ moles   _____ moles

Moles of H_2 per gram of sample, N_{H_2} _____ moles/g   _____ moles/g

$\%_{Al}$ (from Equation 12)   _____ $\%_{mass}$ Al   _____ $\%_{mass}$ Al

Unknown ID code   _____

Experiment 10

Advance Study Assignment: Analyzing an Aluminum–Zinc Alloy

A student obtained the following data in one trial of this experiment:

Mass of gelatin capsule	0.1134 g
Mass of alloy sample plus capsule	0.3218 g
Mass of drained beaker	145.3 g
Mass of beaker plus displaced water	406.1 g
Atmospheric (ambient) pressure, P_{atm}	723 mm Hg
Water (and H_2) temperature, t	25 °C

1. What was the mass of the alloy sample?

_____ g

2. What was the mass of the displaced water?

_____ g

3. What was the volume of the displaced water? (Use $\rho = 1.00$ g/mL as the density of water.)

_____ mL

4. Calculate the volume of the gas, V, generated by the reaction, in liters:

_____ L

5. What was the temperature of the gas, T, in Kelvins, assuming it was at the same temperature as the displaced water?

_____ K

6. Look up the vapor pressure of water, VP_{H_2O}, at temperature t, in Appendix A:

_____ mm Hg

7. Subtract the vapor pressure of water, VP_{H_2O}, from the atmospheric (ambient) pressure, P_{atm}, to get the partial pressure of H_2, P_{H_2}; then convert this value to atmospheres:

_____ mm Hg

_____ atm

8. How many moles of H_2, n_{H_2}, were generated by the sample? (Use Equation 5.)

_____ moles

9. Divide this by the mass of the alloy sample to obtain N_{H_2}, the moles of H_2 per gram of sample:

_____ moles/g

10. Finally, use Equation 12 to calculate the mass percent of aluminum, $\%_{Al}$, in the alloy sample:

_____ $\%_{mass}$ Al

Experiment 11

The Atomic Spectrum of Hydrogen

When electrons in the atoms of an element are excited, by either an electric discharge or very high temperatures, they give off <u>light</u> (defined here as electromagnetic radiation of any wavelength, not just visible light). The light is emitted only at certain wavelengths, characteristic of the element. These wavelengths make up what is called the <u>atomic spectrum</u> of the element, and they reveal detailed information regarding the electronic structure of atoms.

Atomic spectra are interpreted in terms of <u>quantum theory</u>. According to this theory, atoms can only exist in certain allowed states, each of which has a fixed amount of energy associated with it. When an atom changes state, it must absorb or emit an amount of energy equal to the difference between the energies of the initial and final states. This energy may be absorbed or emitted in the form of light. The <u>emission spectrum</u> of an atom is obtained when excited electrons transition from higher to lower energy levels. Because there are many allowed energy levels, the atomic spectrum of most elements is very complex.

Light is absorbed or emitted by atoms in the form of <u>photons</u> (packets of light energy), each of which has a specific amount of energy, ϵ_{photon}. This energy is related to the wavelength of the light, λ, by the equation

$$\epsilon_{photon} = \frac{hc}{\lambda} \tag{1}$$

where h is Planck's constant, 6.62608×10^{-34} joule seconds; c is the speed of light (in vacuum), 2.997925×10^{8} meters per second; and λ is the wavelength of the light, in meters. The energy ϵ_{photon} is therefore in joules, the SI base unit for energy (see Appendix D). Because energy is <u>conserved</u> in these transitions (the total amount of energy does not change, because energy is not created or destroyed), the change in energy of the atom, $\Delta\epsilon_{atom}$, must equal the energy of the photon emitted or absorbed:

$$\Delta\epsilon_{atom} = \pm\epsilon_{photon} \tag{2}$$

For emission the energy of the emitted photon is equal to $-\Delta\epsilon_{atom}$, which is the energy of the higher (and initial) allowed energy level minus the energy of the lower (and final) one. Combining Equations 1 and 2 yields the relationship between the change in energy of an atom and the wavelength of light associated with that change:

$$-\Delta\epsilon_{atom} = \epsilon_{higher} - \epsilon_{lower} = \epsilon_{photon} = \frac{hc}{\lambda} \tag{3}$$

The energy of the photon associated with an atom transitioning from one electronic level to another is very small: on the order of 1×10^{-19} joules. This is not surprising, because atoms are very small! To avoid working with such small numbers, Equation 3 is often multiplied by the number of atoms in a mole, N:

$$-N\Delta\epsilon = -\Delta E = N\epsilon_{higher} - N\epsilon_{lower} = E_{higher} - E_{lower} = \frac{Nhc}{\lambda}$$

Substituting in values for N, h, and c, and expressing the wavelength in nanometers (nm) rather than meters (<u>1 meter</u> $= 1 \times 10^{9}$ nanometers), an equation is obtained relating energy change in kilojoules per mole of atoms, $-\Delta E$, to the wavelength of photons associated with such a change, λ:

$$-\Delta E = \frac{6.02214 \times 10^{23} \times 6.62608 \times 10^{-34} \text{ J sec} \times 2.997925 \times 10^{8} \text{ m/sec}}{\lambda \text{ \{nm\}}} \times \frac{1 \times 10^{9} \text{ nm}}{1 \text{ m}} \times \frac{1 \text{ kJ}}{1000 \text{ J}}$$

$$-\Delta E \text{ \{kJ/mol\}} = E_{higher} - E_{lower} = \frac{1.19627 \times 10^{5}}{\lambda \text{ \{nm\}}} \quad \text{or} \quad \lambda \text{ \{nm\}} = \frac{1.19627 \times 10^{5}}{-\Delta E \text{ \{kJ/mol\}}} \tag{4}$$

Equation 4 is useful in interpreting atomic spectra. In a study of the atomic spectrum of sodium, the wavelength of its strong yellow emission line is found to be 589.16 nm. This line is known to result

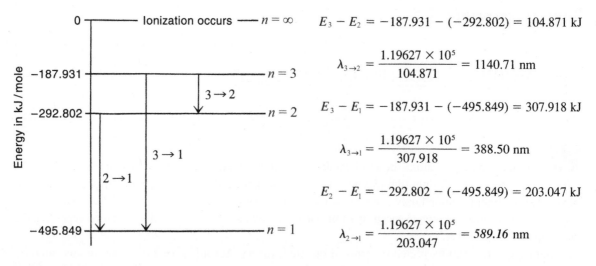

Figure 11.1 Spectral line wavelengths are calculated from the electronic energy levels of the sodium atom.

from a transition between a pair of the three lowest allowed electronic states (or levels) in the atom. The energies of these allowed levels are shown in Figure 11.1. To determine which pair of levels give rise to the 589.16-nm line, note that there are three possible transitions, shown by downward arrows in the figure. The wavelengths associated with those transitions are found by first calculating the change in the energy of the atom, $-\Delta E = E_{higher} - E_{lower}$, for each transition. Knowing $-\Delta E$ allows the calculation of λ by Equation 4. In this case the $2 \rightarrow 1$ transition is the source of the bright yellow line in the sodium emission spectrum.

The simplest of all atomic spectra is that of the hydrogen atom. In 1913 Niels Bohr proposed a theoretical explanation for this spectrum with his model of the hydrogen atom. According to Bohr's model, the energies allowed to the electron in a hydrogen atom are given by the equation

$$\epsilon_n = \frac{-B}{n^2} \tag{5}$$

where B is a constant predicted by the theory and n is a counting number (1, 2, 3, ...) called a <u>quantum number</u>. It has been found that all the lines in the atomic spectrum of hydrogen can be associated with energy levels in the atom that are predicted with great accuracy by Equation 5. Substituting a numerical value for B into Equation 5 and scaling it up to a mole of H atoms, it becomes

$$E_n = \frac{-1312.04 \text{ kJ/mol}}{n^2} \text{ where } n = 1, 2, 3, \ldots \tag{6}$$

This equation accurately calculates the allowed electronic energy levels for hydrogen. Transitions between these levels give rise to the wavelengths of light observed in the atomic spectrum of hydrogen. These wavelengths are known very accurately. Given both the energy levels and the wavelengths, it is possible to determine the actual quantum numbers of the electronic states (levels) associated with each wavelength. In this experiment your goal will be to do this for the observed wavelengths in the atomic spectrum of hydrogen listed in Table 11.1.

Table 11.1

Some Wavelengths in the Atomic Spectrum of Hydrogen, as Measured in a Vacuum					
Observed wavelength, λ {nm}	Assignment $n_{high} \rightarrow n_{low}$	Observed wavelength, λ {nm}	Assignment $n_{high} \rightarrow n_{low}$	Observed wavelength, λ {nm}	Assignment $n_{high} \rightarrow n_{low}$
97.25	_____	410.29	_____	1005.2	_____
102.57	_____	434.17	_____	1094.1	_____
121.57	_____	486.27	_____	1282.2	_____
389.02	_____	656.47	_____	1875.6	_____
397.12	_____	954.86	_____	4052.3	_____

In this experiment your instructor may allow you to work without safety glasses.

Experimental Procedure

There are several ways to analyze an atomic spectrum, given the energy levels of the atom involved. A simple and effective method is to calculate the wavelengths of some specific lines arising from transitions between some of the lower allowed energy levels, and see if they match those that are observed. You will use this method in this experiment. All the data are reliable to at least five significant figures, so you will be able to make very precise determinations. A spreadsheet can be used to good advantage in this experiment. Your instructor may give you some suggestions on how you might proceed.

A. Calculating the electronic energy levels in a hydrogen atom

Given the expression for E_n in Equation 6, it is possible to calculate the energy for each of the allowed electronic levels of the H atom, starting with $n = 1$. Calculate the energy in kJ/mole of each of the 10 lowest levels of the H atom. Note that the energies are all negative, so the *smallest* quantum number will be associated with the most negative energy value. Enter these values in Table 11.2, on the first report page. On the energy level diagram provided on the last report page, just before the Advance Study Assignment (ASA), plot each of the six lowest energies, as was done for the three lowest levels of sodium in Figure 11.1, by drawing a horizontal line at the allowed energy level and writing the numerical energy value to the left of this horizontal line, near the y-axis. Write the quantum number associated with the level to the right of the line.

B. Calculating the wavelengths of the lines in the atomic spectrum of hydrogen

The lines in the hydrogen spectrum all arise from transitions made by electrons from one allowed energy level to another. The wavelengths of these lines (in nm) can be calculated by Equation 4, where $-\Delta E$ is the difference in energy (in kJ/mole) between any two allowed levels. For example, to find the wavelength of the spectral line associated with a transition from the $n = 2$ level to the $n = 1$ level, calculate the difference, $-\Delta E$, between the energies of those two levels. Then substitute $-\Delta E$ into Equation 4 to obtain the wavelength, in nanometers.

Using this procedure, calculate the wavelengths of all the lines indicated in Table 11.3. That is, calculate the wavelengths of all the lines that can arise from transitions between any combination of two of the six lowest levels of the H atom, recording these values in Table 11.3.

C. Assigning observed lines in the atomic spectrum of hydrogen

If you have made your calculations correctly, some of the wavelengths in Table 11.3 should match up, within the error of your calculation, with some of the observed wavelengths in Table 11.1. On the line to the right of each of those wavelengths in Table 11.1, write the quantum numbers of the upper and lower states associated with that wavelength, n_{high} and n_{low}. On the energy level diagram just before the ASA, draw a vertical arrow pointing down (light is emitted, meaning energy leaves the atom and ΔE_{atom} is negative) between those pairs of levels that you associate with any of the observed wavelengths, as was done in Figure 11.1. Next to each arrow write the wavelength of the emission line resulting from that transition.

There are a few wavelengths in Table 11.1 that have not yet been calculated. Enter those wavelengths in Table 11.4. By considering assignments already made and by examining the transitions you have marked on the energy level diagram, figure out what quantum states are likely to be associated with the as-yet-unassigned lines. A good approach is to first calculate the energy change associated with each wavelength, $-\Delta E$, then find two values of E_n from Table 11.2 whose difference is equal to $-\Delta E$. The quantum numbers for the two E_n states whose energy difference is $-\Delta E$ will be the ones to assign to the given wavelength. When you have found n_{high} and n_{low} for a wavelength, write them in Table 11.1 and Table 11.4; continue until all the lines in both tables have been assigned.

D. Visible lines in the atomic spectrum of hydrogen

Not all light is visible to the human eye; in fact, we can only see in a small part of the electromagnetic spectrum, called the visible spectrum, which runs from about 400 nm in the blue (shorter wavelength) direction to 750 nm in the red (longer wavelength) direction. Outside this range we humans quickly lose the ability to differentiate color, and almost as quickly to even detect any light at all, especially as we get older.

Most of the lines in the atomic spectrum of hydrogen can not be seen directly with the human eye, but there are four bright lines in it that humans *can* reliably see. It is not surprising, then, that these lines were the first to be discovered and mathematically described.

Your instructor may have a hydrogen source tube and a spectroscope with which you may be able to observe these lines yourself. Whether or not that is the case, there are some questions you should answer relating to these lines on the report page.

E. The ionization energy of a hydrogen atom

As the quantum number of the electron in a hydrogen atom gets larger and its energy becomes less negative (as the electron gains more energy and its E_n approaches zero), the electron gets closer and closer to being able to overcome its attraction to the positively charged nucleus of the atom and fly away, leaving behind a hydrogen ion, H^+. The ionization energy of the hydrogen atom is the amount of energy required to make this happen, and you will calculate it in the final section of the report.

Experiment 11

Data and Calculations: The Atomic Spectrum of Hydrogen

A. Calculating the electronic energy levels in a hydrogen atom

Energies are to be calculated using Equation 6, for the 10 lowest energy states (smallest allowed values of n).

Table 11.2

Quantum number, n	Energy, E_n {kJ/mol}	Quantum number, n	Energy, E_n {kJ/mol}
_____	_____	_____	_____
_____	_____	_____	_____
_____	_____	_____	_____
_____	_____	_____	_____
_____	_____	_____	_____

B. Calculating the wavelengths of the lines in the atomic spectrum of hydrogen

In the upper half of each box write $-\Delta E$, the difference in energy (in kJ/mole) between $E_{n_{high}}$ and $E_{n_{low}}$. In the lower half of the box, write λ, the wavelength (in nm) associated with that value of $-\Delta E$.

Table 11.3

n_{high}	6	5	4	3	2	1
n_{low}						
1						
2						
3						
4						
5						

$$-\Delta E = E_{n_{high}} - E_{n_{low}}$$

$$\lambda \text{ \{nm\}} = \frac{1.19627 \times 10^5}{-\Delta E \text{ \{kJ/mol\}}}$$

(continued on following page)

C. Assigning observed lines in the atomic spectrum of hydrogen

1. As directed in the procedure, assign n_{high} and n_{low} for each wavelength in Table 11.1 that corresponds to a wavelength calculated in Table 11.3.

2. List all the wavelengths from Table 11.1 that you cannot yet assign in Table 11.4. Then assign these, with the help of Table 11.4:

Table 11.4

Observed wavelength, λ {nm}	Energy of transition, $-\Delta E$ {kJ/mol}	Probable transition, $n_{high} \rightarrow n_{low}$	Calculated wavelength, λ {nm} (Equation 4)
_____	_____	_____	_____
_____	_____	_____	_____
_____	_____	_____	_____
_____	_____	_____	_____

D. Visible lines in the atomic spectrum of hydrogen

1. The first lines in the atomic spectrum of hydrogen to be observed, studied, and quantified were the four lines easily observed with the human eye: those with wavelengths between 400 nm and 750 nm. Based on your completed Table 11.1, what do these four lines have in common?

2. You will find two lines in Table 11.1 just barely outside the 400 nm to 750 nm window. They are part of the same <u>series</u> as the four lines you identified above, because they share the same characteristic you identified above. A young human eye might spot them, under the right conditions (for example, in a very dark room). What is the shortest possible wavelength that a line in the same series can have? (*Hint*: What is the largest possible value of $-\Delta E$ that can be associated with a line in this series?)

$\lambda = $ _____ nm

E. The ionization energy of the hydrogen atom

1. In a typical hydrogen atom at room temperature, the electron is in its lowest energy state, $n = 1$, which is called the (electronic) <u>ground state</u> of the atom. The maximum electronic energy that a hydrogen atom can have is 0 kJ/mole, at which point the electron is essentially removed from the atom, leaving behind an H^+ ion. How much energy, in kilojoules per mole, does it take to <u>ionize</u> (remove the electron from) a typical hydrogen atom at room temperature? (This is called the <u>ionization energy</u> of hydrogen.)

_____ kJ/mole

2. The ionization energy of hydrogen is often expressed in units other than kJ/mole. What would it be in joules per atom? In electron volts per atom? ($\underline{1}$ eV $= 1.602 \times 10^{-19}$ J)

_____ J/atom; _____ eV/atom

(The energy level diagram to be completed in Part A of the instructions is on the following page.)

Experiment 11

Data and Calculations: The Atomic Spectrum of Hydrogen

Energy Level Diagram

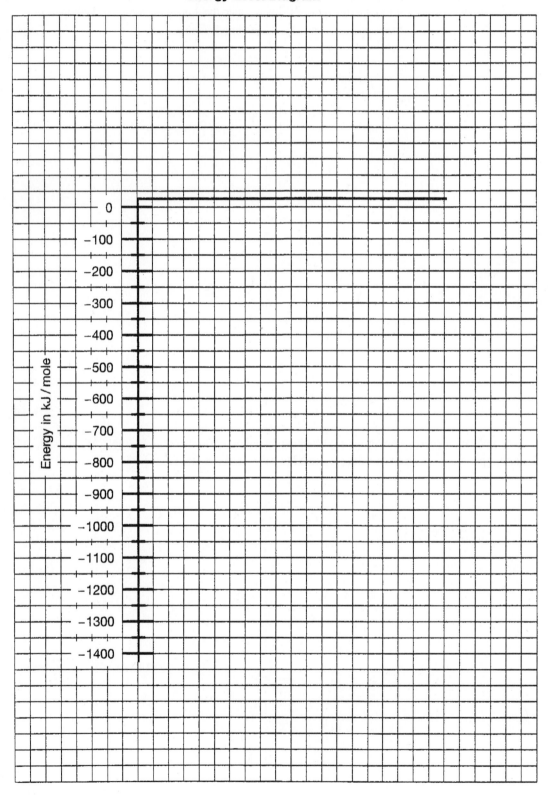

Experiment 11

Advance Study Assignment: The Atomic Spectrum of Hydrogen

1. Found in the gas phase, the boron tetracation (quadruply charged positive, or +4, ion), B^{4+}, has an energy level formula similar to that of the hydrogen atom—because both have only one electron. The electronic energy levels of the B^{4+} ion are given by the equation

$$E_n = -\frac{32815}{n^2} \text{ kJ/mole where } n = 1, 2, 3, \dots$$

a. Calculate the energies in kJ/mole for the four lowest energy levels of the B^{4+} ion.

$E_1 = $ _____ kJ/mole

$E_2 = $ _____ kJ/mole

$E_3 = $ _____ kJ/mole

$E_4 = $ _____ kJ/mole

b. One of the most important transitions for the B^{4+} ion involves a fall from the $n = 3$ to the $n = 1$ level. $-\Delta E$ for this transition equals $E_3 - E_1$, energies you obtained in Part (a). Find the value of $-\Delta E$, in kJ/mole, for this transition. Then find the wavelength, in nanometers, of the light emitted when this transition occurs, using Equation 4.

$-\Delta E = $ _____ kJ/mole; $\lambda = $_____ nm

(continued on following page)

c. Three of the stronger lines in the B^{4+} ion spectrum are observed at the following wavelengths: (1) 74.993 nm; (2) 19.443 nm; (3) 3.8885 nm. Find the quantum numbers of the initial and final states for the transitions that give rise to these three lines. Do this by calculating, using Equation 4, the wavelengths of lines that can originate from transitions involving any pair of the four lowest energy levels. You calculated one such wavelength in Part (b). Make similar calculations with the other possible pairs of levels. When a calculated wavelength matches an observed one, write down n_{high} and n_{low} for that line. Continue until you have assigned all three of the lines. Show your work below the answer blanks.

(1) _____ → _____   (2) _____ → _____   (3) _____ → _____

Experiment 12

Periodic Trends in the Alkaline Earth Metals (Group 2) and the Halogens (Group 17)

The Periodic Table arranges the elements in order of increasing atomic number, in horizontal rows of such length that elements with similar properties recur periodically: that is, they fall directly beneath each other in the table. The elements in a given vertical column are referred to as a family, or group. The physical and chemical properties of the elements in a given family often change gradually, going down a column from one element to the next. By observing the trends in properties, the elements can be arranged in the order in which they appear in the Periodic Table. In this experiment you will study the properties of the elements in two families in the Periodic Table: the alkaline earths (Group 2) and the halogens (Group 17).

The alkaline earth elements are all moderately reactive metals and include barium (Ba), beryllium (Be), calcium (Ca), magnesium (Mg), radium (Ra), and strontium (Sr). (However, because beryllium compounds are rarely encountered and are often very poisonous, and radium compounds are highly radioactive, you will not work with these elements in this experiment!) All the alkaline earths exist in chemical compounds and in solution as M^{2+} cations (ions with a +2 charge: Mg^{2+}, Ca^{2+}, and so on). If a solution containing one of these cations is mixed with one containing an anion (negatively charged ions such as CO_3^{2-}, SO_4^{2-}, IO_3^{-}, and so on), an alkaline earth salt will precipitate if the compound containing those two ions is insoluble. For example:

$$M^{2+}(aq) + SO_4^{2-}(aq) \rightarrow MSO_4(s) \quad \text{if } MSO_4 \text{ is insoluble} \tag{1a}$$

$$M^{2+}(aq) + 2\,IO_3^{-}(aq) \rightarrow M(IO_3)_2(s) \quad \text{if } M(IO_3)_2 \text{ is insoluble} \tag{1b}$$

The salts composed of a given anion and the alkaline earth cations generally show a smooth trend in solubility, consistent with the order of the cations' corresponding elements in the Periodic Table. That is, going down the Group 2 column of the Periodic Table, the solubilities of, say, the sulfate (SO_4^{2-}) salts either gradually increase or decrease. Similar trends exist for the carbonates (CO_3^{2-} salts), oxalates ($C_2O_4^{2-}$ salts), and iodates (IO_3^{-} salts) formed by those cations. By determining such trends in this experiment, you will be able to confirm the order in which the alkaline earth elements appear in the Periodic Table.

The elemental halogens of Group 17 are also relatively reactive. They include astatine (At), bromine (Br), chlorine (Cl), fluorine (F), iodine (I), and the recently discovered element tennessine (Ts). (You will study only bromine, chlorine, and iodine in this experiment. Astatine is radioactive, while fluorine is too reactive to be safely used in general chemistry labs. Tennessine is extremely unstable and decays to lighter elements so quickly that nobody has really had a chance to study it!) Unlike the alkaline earths, the halogens tend to gain electrons, forming X^{-} anions (ions with a single negative charge: Cl^{-}, Br^{-}, and so on), called halides. Because of this, the halogens are oxidizing agents: substances that tend to oxidize (remove electrons from) other substances, being reduced (gaining an electron themselves) to become halides in the process. An interesting example of the sort of reaction that may occur arises when a solution containing a halogen (Cl_2, Br_2, or I_2) is mixed with a solution containing the halide ion (Cl^{-}, Br^{-}, or I^{-}) of a different element. Taking X_2 to be the halogen, and Y^{-} to be the halide ion, the following reaction may occur, in which the halogen Y_2 is formed:

$$X_2(aq) + 2\,Y^{-}(aq) \rightarrow 2\,X^{-}(aq) + Y_2(aq) \tag{2}$$

Reaction 2 will occur if X_2 is a stronger oxidizing agent than Y_2, since then X_2 can produce Y_2 by oxidizing (removing electrons from) the Y^{-} ions. If Y_2 is a stronger oxidizing agent than X_2, Reaction 2 will not proceed, but the reverse reaction (in which Y_2 oxidizes X^{-}, taking its electron away) will.

In this experiment you will mix solutions of halogens and halide ions to determine the relative oxidizing strengths of the halogens. These strengths show a smooth variation as one goes from one halogen to the next down Group 17 of the Periodic Table. You will be able to tell if a reaction occurs by the colors you observe. In water, and particularly in some organic solvents, the halogens have characteristic colors. The halide ions are colorless when dissolved in water and insoluble in organic solvents. Bromine (Br_2) is orange when dissolved in the organic solvent heptane, C_7H_{16} (HEP for short), while Cl_2 and I_2 in that solvent have quite different colors (which you will discover when you do this experiment).

If a solution of the halogen bromine, Br_2, in water is mixed with a little heptane (HEP), most of the bromine will move from the water into the HEP, where it is much more soluble. HEP is insoluble in water and does not mix with it; instead, HEP forms a distinct layer: and because HEP is not as dense as water, the HEP layer floats on top of the water. With Br_2 in the HEP, that top layer is orange. If a solution containing a halide ion, say Cl^- ion, is added to that mixture and everything is shaken together, the Br_2 will have the opportunity to oxidize (take electrons from) the Cl^-. The reaction would be

$$Br_2(aq) + 2\,Cl^-(aq) \rightarrow 2\,Br^-(aq) + Cl_2(aq) \tag{3}$$

but it will only actually happen if Br_2 is a stronger oxidizing agent (better at taking electrons) than Cl_2. If Reaction 3 reaction occurs, the color of the HEP layer will change because Br_2 will be turned into colorless Br^-, and (differently colored) Cl_2 will form. The color of the HEP layer will go from orange to that of a solution of Cl_2 in HEP. If the reaction does *not* occur, the color of the HEP layer will *remain* orange. Using this line of reasoning, and by working with the possible mixtures of halogens and halide ions, you will be able to arrange the halogens in order of increasing oxidizing strength, which should correspond (either up or down the Group 17 column) to their order in the Periodic Table.

One difficulty that you may have in this experiment involves terminology. You must learn to distinguish the halogen *elements* from the halide *ions*—since the two are very different things, even though their names are similar:

Elemental halogens	**Halide ions**
Bromine, Br_2	Bromide ion, Br^-
Chlorine, Cl_2	Chloride ion, Cl^-
Iodine, I_2	Iodide ion, I^-

The *halogens* are molecular substances and oxidizing agents, and all have odors. They are only slightly soluble in water and are much more soluble in HEP, where they have distinct colors. The *halide ions* exist in solution only in water, have no color or odor, and are *not* oxidizing agents (they already have the only extra electron they can grab). They do not dissolve in HEP.

Once you know the solubility properties of the alkaline earth cations, and the relative oxidizing strength of the halogens in this experiment, it is possible to develop a systematic procedure for determining which pair of any Group 2 cation and any Group 17 anion from this experiment is present in a solution. In the last part of this experiment you will be asked to set up such a procedure and use it to identify an unknown solution containing a single alkaline earth halide.

Wear your safety glasses while performing this experiment.

Experimental Procedure

A. Determining relative solubilities of some salts of the alkaline earths

To each of four small (13 × 100 mm) test tubes add about one milliliter (in these test tubes, that is a depth of about one centimeter) of 1.0 M sodium sulfate, Na_2SO_4. Then add about one milliliter (an additional centimeter of depth) of 0.10 M solutions of the nitrate salts of barium, calcium, magnesium, and strontium to those tubes, one solution to a tube. Swirl each test tube around thoroughly to mix the contents. Record your results on the solubilities of the sulfates of the alkaline earths in the table on the report page, noting whether a precipitate forms, and any characteristics (such as color, amount, size of particles, and settling tendencies) that might distinguish it.

Rinse out the test tubes, and to each add about one milliliter of 1.0 M sodium carbonate, Na_2CO_3. Then add about one milliliter of the solutions of the alkaline earth nitrates again, one solution to a tube, as before. Record your observations on the solubility properties of the carbonates of the alkaline earth cations. Rinse out the tubes, and test for the solubilities of the oxalates of these cations, using about one milliliter of 0.25 M ammonium oxalate, $(NH_4)_2C_2O_4$, as the precipitating agent. Finally, determine the relative solubilities of the iodates of the alkaline earths, using about one milliliter of 0.10 M potassium iodate, KIO_3, as the precipitating agent.

B. Relative oxidizing strengths of the halogens

Place a few milliliters of bromine-saturated water in a small test tube, then add about one milliliter of heptane, HEP, to it. Mix the two layers well [stopper the tube and shake it, or suck the lower (water) layer into a Pasteur pipet and squirt it back into the tube through the upper (HEP) layer], until the bromine color is mostly in the (upper) HEP layer. CAUTION: **Avoid breathing halogen fumes (gas). Do not use your finger to stopper the tube, or otherwise touch the solutions, because halogens can give you bad chemical burns.** Repeat this process with chlorine water, and then with iodine water, each time with fresh HEP, noting any color changes as the bromine, chlorine, and iodine are extracted from the water layer into the HEP layer. Record your results on the report page.

Now, to each of three small test tubes add about one milliliter of bromine water and about one milliliter of HEP. Then add about one milliliter of 0.10 M NaCl to the first test tube, about one milliliter of 0.10 M NaBr to the second, and about one milliliter of 0.10 M NaI to the third. Mix the two layers in each tube well. Note the color of the HEP phase above each solution. If the color is not that of Br_2 in HEP, a chemical reaction has occurred in which Br_2 oxidized the halide added to that tube, producing its corresponding halogen. In such a case, Br_2 is a stronger oxidizing agent than the halogen that was produced.

Rinse out the tubes, and this time add about one milliliter of chlorine water and about one milliliter of HEP to each tube. Then add about one milliliter of the 0.10 M solutions of the sodium halide salts, one solution to a tube, as before. Mix the two layers well and then note the color of the HEP layer. Depending on whether the color is that of Cl_2 in HEP or not, decide whether Cl_2 is a better oxidizing agent than Br_2 and/or I_2. Rinse out the tubes again and then add about one milliliter of iodine water and about one milliliter of HEP to each. Test each tube with about one milliliter of a sodium halide salt solution, and determine whether I_2 is able to oxidize Cl^- or Br^- ions. Record all your observations in the table on the report page.

C. Identifying an alkaline earth halide

Your observations on the solubility properties of the alkaline earth cations from Part A (or D) should allow you to develop an efficient method for determining which of those cations is present in a solution containing one of those cations and no others. The method will involve testing samples of the solution with one or more of the precipitating agents you used in Part A or D. You will not need to repeat everything you did in Part A or D; try to only carry out steps that *quickly narrow down* the identity of the unknown. Indicate on the report page how you would proceed.

In a similar way you can determine which halide ion is present in a solution containing only one such anion and no others. For this you will need to test a solution of an oxidizing halogen with your unknown to see how the halide ion is affected. From the behavior of the halogen-halide ion mixtures you studied in Part B, you should be able to efficiently identify which halide is present. Describe your method on the report page.

Obtain an unknown solution of an alkaline earth halide, and then use your procedures to determine the (one) alkaline earth cation and the (one) halide anion that it contains.

Optional   D. Microscale procedure for determining relative solubilities of some salts of the alkaline earths

Your instructor may have you carry out Part A of this experiment using a microscale approach, which requires smaller amounts of the chemicals. Individual cells in a single plastic well plate replace the test tubes.

Start by adding 4 drops of 0.10 M barium nitrate, $Ba(NO_3)_2$, to wells A1–A4, 4 drops in each well. Similarly, add 4 drops of 0.10 M calcium nitrate, $Ca(NO_3)_2$, to wells B1–B4; 4 drops of 0.10 M magnesium nitrate, $Mg(NO_3)_2$, to wells C1–C4; and 4 drops of 0.10 M strontium nitrate, $Sr(NO_3)_2$, to wells D1–D4. Each row of your well plate now contains a different alkaline earth metal cation, consistent with Figure 12.1.

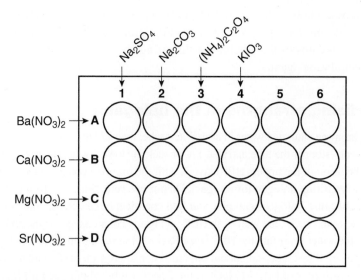

Figure 12.1   In the microscale version of this experiment (Part D), each row contains a solution of a different alkaline earth metal nitrate salt, and each column contains a different precipitating agent (a solution containing a precipitating anion).

Now you will add a different precipitating anion solution to each column in the well plate. First, add 4 drops of 1.0 M sodium sulfate, Na_2SO_4, to wells A1–D1. Next, add 4 drops of 1.0 M sodium carbonate, Na_2CO_3, to wells A2–D2. Do the same with 0.25 M ammonium oxalate, $(NH_4)_2C_2O_4$, in wells A3–D3, and finally with 0.10 M potassium iodate, KIO_3, in wells A4–D4. Record your results on the solubilities of these alkaline earth salts in the table on the report page. Note whether a precipitate formed, along with any characteristics (such as the amount, the size of the particles, and cloudiness) that might help distinguish each one. ▨

Take it further (optional): Dissolve a sample of limestone in acid, and determine whether it contains Mg^{2+} as well as Ca^{2+} ions. If both are present, estimate the relative concentrations of the two ions.

> **Disposal of reaction products:** Dispose of the reaction products from this experiment as directed by your instructor.

Experiment 12

Observations and Analysis: Periodic Trends in the Alkaline Earth Metals
(Group 2) and the Halogens (Group 17)

A or D. Determining relative solubilities of some salts of the alkaline earths

	1.0 M Na_2SO_4	1.0 M Na_2CO_3	0.25 M $(NH_4)_2C_2O_4$	0.10 M KIO_3
$Ba(NO_3)_2$				
$Ca(NO_3)_2$				
$Mg(NO_3)_2$				
$Sr(NO_3)_2$				

Key: P = precipitate forms; S = no precipitate forms.

Note any distinguishing characteristics of each precipitate, such as its color, the amount of precipitate formed, and the degree of cloudiness observed in the precipitating solution.

Based on your results above, list the four alkaline earth cations in order of increasing tendency to form insoluble salts:

_____ _____ _____ _____

forms the most   forms the least
soluble salts   soluble salts

Is this the same order in which the corresponding elements appear in the Periodic Table?

Does solubility increase or decrease going down a group, at least in this case?

B. Relative oxidizing strengths of the halogens

1. Colors of the halogens in solution:

	Br_2	Cl_2	I_2
in heptane (HEP)	_____	_____	_____
in water	_____	_____	_____

(continued on following page)

2. Reactions between halogens and halides:

Example based on ASA

	Br^-	Cl^-	I^-
Br_2 (orange in HEP)			
Cl_2			
I_2			

	A^-	B^-
A_2 (green)		
B_2 (blue)		
C_2 (red)	initial: red final: green A_2: R	initial: red final: red C_2: NR

Key: R = reaction occurs; NR = no reaction occurs.

Record your observations of each mixture in the table above; record the initial and final colors of the HEP layer, which halogen ends up in the HEP layer, and whether or not a reaction occurred.

Rank the halogens in order of increasing strength as oxidizing agents:

_____ < _____ < _____

Weakest oxidizing agent   Strongest oxidizing agent
(Least able to pull electrons away)   (Most able to pull electrons away)

Is this the order in which they appear in the Periodic Table?

C. Identifying an alkaline earth halide

Procedure for identifying the Group 2 cation:

Procedure for identifying the Group 17 anion:

Observations on unknown alkaline earth halide solution:

Cation present _____ Anion present _____ Unknown ID code _____

Experiment 12

Advance Study Assignment: Periodic Trends in the Alkaline Earth Metals
(Group 2) and the Halogens (Group 17)

1. As pure elements, all of the halogens are diatomic molecular substances. Their melting and boiling points are as listed in Table 12.1.

Table 12.1 Elemental Forms and Melting and Boiling Points of the Elemental Halogens

Element	Elemental form	Melting point {K}	Boiling point {K}
F	F_2	54	85
Cl	Cl_2	172	239
Br	Br_2	266	332
I	I_2	387	457

Based on the Periodic Table, predict the elemental form of pure astatine, At, a radioactive halogen. Also predict whether it will be a solid, liquid, or gas at room temperature. Explain your reasoning.

Elemental form of At: _____ Phase at room temperature: _____

2. Substances A_2, B_2, and C_2 can each act as oxidizing agents. In solution, A_2 is green, B_2 is blue, and C_2 is red. In the reactions in which they participate, they are reduced to A^-, B^-, and C^- ions, all of which are colorless. When a solution of C_2 is mixed with one containing A^- ions, the color changes from red to green.

Which substance is oxidized? _____

Which substance is reduced? _____

Is C_2 a stronger oxidizing agent than A_2? _____

When a solution of C_2 is mixed with one containing B^- ions, the color remains red.

Is C_2 a stronger oxidizing agent than B_2? _____

Arrange A_2, B_2, and C_2 in order of increasing strength as oxidizing agents:

_____ < _____ < _____

Weakest oxidizing agent   Strongest oxidizing agent
(Least able to pull electrons away)   (Most able to pull electrons away)

(continued on following page)

3. You are given a colorless unknown solution that contains one of the following salts: NaA, NaB, or NaC. In solution, each salt dissociates completely into the Na^+ ion and the anion A^-, B^-, or C^-, whose properties are as described in Problem 2. The Na^+ ion is effectively inert. Given the availability of solutions of A_2, B_2, and C_2, develop an efficient procedure for determining which salt is present in your unknown. Try to avoid carrying out steps that do not narrow down the possible identity of the salt, and ideally have your procedure arrive at the identity of the unknown in as few steps as possible. (That is what "efficient" means in this context!) Describe your procedure in detail here:

Experiment 13

The Geometric Structure of Molecules—
An Experiment Using Molecular Models

Many years ago it was observed that the carbon atom typically forms four chemical linkages to other atoms. As early as 1870, structural formulas of carbon compounds were drawn as shown here:

$$
\begin{array}{cc}
\begin{array}{c}
\text{H} \\
| \\
\text{H}-\text{C}-\text{H} \\
| \\
\text{H} \\
\text{methane}
\end{array}
&
\begin{array}{c}
\text{H}\quad\text{H} \\
|\quad\ | \\
\text{C}=\text{C} \\
|\quad\ | \\
\text{H}\quad\text{H} \\
\text{ethylene}
\end{array}
\end{array}
$$

Drawings such as these imply that the atom-atom linkages (chemical bonds), indicated by the lines connecting the atoms, lie in a plane. But chemical evidence, for example the existence of only one substance with the chemical formula CH_2Cl_2, requires that the linkages be arranged in three dimensions (coming out of and going behind the plane of the page). Were the molecules truly flat, two distinct (geometric) isomers of CH_2Cl_2 would exist, with the following structures:

$$
\begin{array}{cc}
\begin{array}{c}
\text{Cl} \\
| \\
\text{H}-\text{C}-\text{Cl} \\
| \\
\text{H}
\end{array}
&
\begin{array}{c}
\text{H} \\
| \\
\text{Cl}-\text{C}-\text{Cl} \\
| \\
\text{H}
\end{array}
\end{array}
$$

Instead, the hydrogen atoms in methane, CH_4, are positioned at the tips of a <u>tetrahedron</u>, which is a four-sided three-dimensional shape: a pyramid with a triangular base, as shown in Figure 13.1(a). In a <u>regular tetrahedron</u>, the four faces are identical equilateral triangles. Methane's carbon atom is located at the center of a regular tetrahedron, such that the hydrogen atoms bound to it are all the same distance away from the carbon and positioned as far away from each other (in three dimensional space) as possible, as shown in Figure 13.1(b). When the atoms bound to a central atom are not all identical, as in CH_2Cl_2, the molecular frame is no longer a regular tetrahedron. The basic shape, however, remains that of a tetrahedron, and the molecule is still said to have <u>tetrahedral geometry</u> or to "be <u>tetrahedral</u>".

The physical significance of the chemical linkages between atoms, indicated by the lines in molecular structure diagrams, became evident soon after the discovery of the electron. In a classic paper, G. N. Lewis suggested, on the basis of chemical evidence, that the single bonds in structural formulas involve two electrons and that atoms tend to hold eight electrons in their outermost, or <u>valence</u>, shells.

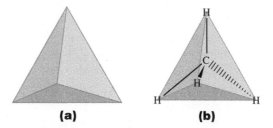

(a)   **(b)**

Figure 13.1 (a) A tetrahedron is a four-sided three-dimensional shape. A regular tetrahedron has four identical sides: all equilateral triangles of the same size. (b) In the methane molecule, CH_4, the hydrogen atoms are located at the tips of a regular tetrahedron, and the carbon atom is located at its center.

Lewis's proposal that atoms generally have eight (valence) electrons in their outer shells proved extremely useful, and has come to be known as the "octet rule". It can be applied to many atoms, but is most reliable for covalent compounds of elements in the second row of the Periodic Table. For atoms such as carbon, oxygen, nitrogen, and fluorine, the eight valence electrons occur in pairs that occupy tetrahedral positions around the central atom core. Some of the electron pairs do not participate directly in chemical bonding and are called "unshared electrons" or "nonbonding pairs"; however, the structures of compounds containing such unshared pairs reflect the tetrahedral arrangement of the four pairs of valence-shell electrons, whether they are involved in bonding or not. In the H_2O molecule, which obeys the octet rule, the four pairs of electrons around the central oxygen atom occupy essentially tetrahedral positions; there are two unshared nonbonding pairs, and two bonding pairs shared by the O atom and the two H atoms. The H—O—H bond angle is nearly but not exactly tetrahedral, because the properties of shared and unshared pairs of electrons differ.

$$\overset{\cdot\cdot}{\underset{H \qquad H}{\overset{/ \quad \backslash}{O}}}$$

Most molecules obey the octet rule. Essentially, all organic molecules obey the rule, and so do most inorganic molecules and ions. For substances that obey the octet rule it is possible to predict electron-dot structures, or Lewis structures. The previous drawing of the H_2O molecule is an example of a Lewis structure. Here are several others:

$$
\begin{array}{ccccc}
:\overset{\cdot\cdot}{\underset{}{Cl}}: & & & & \\
| & & & H\;\;H & \\
H-C-H & H-\overset{\cdot\cdot}{N}-H & \left[H-\overset{\cdot\cdot}{\underset{\cdot\cdot}{O}}:\right]^- & H-\overset{|}{\underset{|}{C}}-\overset{|}{\underset{|}{C}}-H & H-C=C-H \\
| & | & & H\;\;H & |\;\;| \\
:\underset{\cdot\cdot}{Cl}: & H & & & H\;\;H \\
\\
CH_2Cl_2 & NH_3 & OH^- & C_2H_6 & C_2H_4
\end{array}
$$

In each of these structures there are eight electrons around each atom, except for around the H atoms, which always have two electrons around them. There are two electrons in each bond. When counting electrons in these structures, consider *both* of the shared electrons in a bond between two atoms as belonging to (being shared by) *each* of the atoms participating in the bond. In the CH_2Cl_2 molecule, for example, each Cl atom has eight electrons, including the two in the single bond to the C atom. The C atom also has eight electrons, two from each of the four bonds to it. The bonding and nonbonding electrons in Lewis structures are all from the *outermost* occupied electron shells of the atoms involved and are called the "valence electrons" of those atoms. For the main group elements, the number of valence electrons in an atom is equal to the last digit in the element's group number in the Periodic Table. Atoms of carbon, in Group 14, have four valence electrons each; atoms of hydrogen, in Group 1, have one; atoms of chlorine, in Group 17, each have seven valence electrons. In an octet rule structure the valence electrons from all the atoms are arranged so that each atom can "claim" eight electrons (except hydrogen, which "claims" two).

The first step in constructing an octet rule structure for a molecule or ion is to determine the number of valence electrons available. This is done by adding up the valence electrons provided by each atom, then for ions adding one more electron for each negative charge or subtracting one for each positive charge. For example, because an oxygen atom has six valence electrons (Group 16) and a hydrogen atom has one (Group 1), one O and one H atom together have a total of seven valence electrons; in the OH^- ion, which has a -1 charge (because it contains one extra electron), we add one more to this, for a total of eight valence electrons. The octet rule structure for OH^- places eight electrons around the oxygen atom, two electrons around the hydrogen atom, and connects all of the atoms together by at least one bond. Structures like that of OH^-, involving only single bonds and nonbonding electron pairs, are common. Sometimes, however, there is a "shortage" of electrons; that is, it is not possible to construct an octet rule structure in which all the electron pairs are either in single bonds or are nonbonding. C_2H_4 is a typical example: in such cases, octet rule structures can often be constructed in which two atoms are bonded by two pairs, rather than one pair, of electrons. The two pairs of electrons form a double bond. In the C_2H_4 molecule the C atoms each get four of their electrons from the double bond. The assumption that electrons behave this way is supported by the fact that the $C=C$ double bond is both shorter and stronger than the $C-C$ single bond in the C_2H_6 molecule. Double bonds, and triple bonds, occur in many molecules, usually between C, O, N, and/or S atoms.

Singly-bonded Lewis structures can be used to predict molecular and ionic geometries by assuming that the four pairs of electrons around each atom are arranged tetrahedrally, or at least nearly so. Applying this principle to the CH_2Cl_2 molecule predicts it to be (at least roughly) tetrahedral. Molecular geometries are assigned based on the positions of the *atoms* in a molecule; while the nonbonding pairs influence those positions, the positions of the nonbonding pairs themselves are not considered when assigning molecular geometries. Taking the case of NH_3 as an example, the four pairs of valence electrons around the central N atom are arranged tetrahedrally, but one of them is a nonbonding pair. The three H atoms and the N atom form a triangular pyramid, and this is the molecular geometry assigned to NH_3. Similarly, in H_2O there are four tetrahedrally arranged pairs of valence electrons around the O atom, but two of these are nonbonding pairs. The two H atoms and the O atom are arranged in the shape of a boomerang, a molecular geometry called "bent".

Lewis structures are also useful in predicting the <u>polarity</u> of molecules: the extent to which their internal charge distribution is asymmetric, and thus how strongly they interact with electric fields. Bonding electrons are only truly shared equally between two atoms of the same element. In all other cases, the electrons in a bond spend more of their time close to one end of the bond (which then has a slight negative charge) than to the other end (which is left with a slight positive charge). This makes the bond itself <u>polar</u>: it has a positive and a negative end. Thus, two-atom molecules in which the two atoms differ (technical term: <u>heteronuclear diatomics</u>), such as HCl, are polar because they contain a polar bond, while two-atom molecules composed of two atoms of the same element (technical term: <u>homonuclear diatomics</u>), such as Cl_2, are nonpolar because they contain a nonpolar bond. For molecules and ions made up of more than two atoms, the molecule or ion as a whole is said to be (at least slightly) polar unless any polarity in the bonds is cancelled out by symmetry. This will only happen when the polar bonds are all identical *and* symmetrically arranged around the *molecule's center*. For example, the C-Cl bonds in CCl_4 are each polar, but they are arranged symmetrically around the carbon atom such that the CCl_4 molecule as a whole is nonpolar. It might be tempting to think that one of the Lewis structures of CH_2Cl_2 would be nonpolar, but CH_2Cl_2 is not actually flat, as drawn on the page: it is tetrahedral. Therefore there is only one isomer, and the Cl atoms can néver actually be on opposite sides of the central carbon atom from each other, meaning the two polar bonds are never arranged symmetrically. It is important to consider the three-dimensional shape of a molecule when assessing its polarity!

For some molecules with a given molecular formula, it is possible to satisfy the octet rule with multiple atomic arrangements. An example of this is C_2H_6O:

ethanol (ethyl alcohol)   dimethyl ether
C_2H_5OH CH_3OCH_3

These two molecules are called <u>isomers</u> of one another, and the phenomenon is called <u>isomerism</u>. Although the chemical formulas of both substances are the same, C_2H_6O, these are very different molecules.

Isomerism is very common, particularly in organic chemistry; when double bonds are present, isomerism can occur in very small molecules, such as $C_2H_2Cl_2$:

In the course of this experiment you will experience how double bonds prevent rotation, which is what causes the first two of these isomers to be distinct, despite each having one Cl and one H on each C atom.

With certain molecules it is possible to satisfy the octet rule with more than one bonding arrangement *without moving any of the atoms*. The classic example is benzene, whose molecular formula is C_6H_6:

These two individual bonding depictions are called <u>resonance structures</u>, and molecules such as benzene, which have two or more resonance structures, are said to exhibit <u>resonance</u>. The actual bonding in such molecules is best thought of as an average of the bonding in the resonance structures. The double-headed arrow between resonance structures is *not* meant to indicate that the molecule spends some of its time in each structure, but rather that the one actual structure of the molecule is a (sometimes equal, sometimes unequal) *mix* of the resonance structures. Thus, the bond between carbon atoms in benzene does not ever have the characteristics of a single bond, or those of a double bond, but instead it lies somewhere near the middle: slightly stronger, it turns out, than would be expected for a "one-and-a-half" bond. The stability of molecules exhibiting resonance is found to be higher than what would be expected for any single one of their resonance structures.

Although the conclusions drawn here regarding molecular geometry and polarity can be obtained from Lewis structures alone (and some fancy three-dimensional thinking), it is much easier to draw such conclusions from physical, three-dimensional models of molecules and ions. The rules cited here for octet rule structures transfer well to models. In many ways the models are easier to construct than are drawings of Lewis structures on paper. In addition, because the models are three-dimensional, they make it easier to determine both geometry and polarity. In this experiment you will assemble models for a number of common molecules and ions and interpret their geometry, polarity, and isomerism.

Experimental Procedure

In this experiment your instructor may allow you to work without safety glasses.

The models you will use consist of balls with holes in them, short sticks, and springs: or something equivalent. The balls represent atomic nuclei surrounded by their inner electron shells. The sticks and springs represent valence electron pairs, and fit in the holes in the balls. The model (molecule or ion) consists of balls (atoms) connected by sticks or springs (chemical bonds). Some sticks may be connected to only one atom (these represent nonbonding pairs).

This experiment will deal with atoms that obey the octet rule; such atoms have four electron pairs around the central core and will be represented by balls with four tetrahedral holes, each accepting one stick or spring. The only exception will be hydrogen atoms, which share in only two electrons and will be represented by balls with a single hole.

A systematic approach to assembling a molecular model is recommended, as demonstrated here for CH_2O:

1. Determine the total number of valence electrons. The number of valence electrons on an atom of a main group element is equal to the last digit of the number of the group to which the element belongs in the Periodic Table. For CH_2O:

    C   Group 14   H   Group 1   O   Group 16

    Therefore each carbon atom in a molecule or ion contributes four valence electrons, each hydrogen atom contributes one valence electron, and each oxygen atom contributes six valence electrons. The total number of valence electrons equals the sum of the valence electrons on all of the atoms in the substance being studied. For CH_2O this total is $4 + (2 \times 1) + 6$, or 12 valence electrons. If working with an ion, add one valence electron for each negative charge or subtract one for each positive charge on the ion.

2. Select balls and sticks to represent the atoms and electron pairs in the molecule. You should use four-holed balls for the carbon atom and the oxygen atom, and one-holed balls to represent the hydrogen atoms. Because

there are 12 valence electrons in CH_2O and electrons occur in pairs, six sticks are needed to represent the electron pairs. The sticks are used to represent both bonds between atoms and nonbonding electron pairs.

3. Connect the balls with some of the sticks. (Assemble a skeleton structure for the molecule, joining atoms by single bonds.) In some cases this can only be done in one way. Usually, however, there are various possibilities, some of which are more reasonable than others. In CH_2O the model can be assembled by connecting the two H atom balls to the C atom ball with two of the available sticks, and then using a third stick to connect the C atom and O atom balls. It is generally the case that atoms with valence electron counts closest to four have the most other atoms connected to them, which is why a good starting point is to connect the H atoms to the C atom rather than to the O atom.

4. The next step is to use the sticks that are left over to fill all the remaining holes in the balls. (Distribute the electron pairs so as to give each atom eight electrons and so satisfy the octet rule.) In the CH_2O structure assembled so far, one unfilled hole remains in the C atom ball, three unfilled holes remain in the O atom ball, and three sticks remain unused. The way to fill all the holes is to use two sticks to fill two of the holes in the O atom ball, and then replace the last two sticks with (equivalent, but bendable) springs and use those to connect the C atom and O atom balls. The completed model is shown in Figure 13.2.

5. Interpret the model in terms of the atoms and bonds represented. The sticks and spatial arrangement of the balls will closely correspond to the electronic and atomic arrangement in the molecule. Given the model for CH_2O, the molecular geometry is planar, with single bonds between the C and H atoms and a double bond between the C and O atoms. There are two nonbonding electron pairs on the O atom. Because there are polar bonds in CH_2O and molecular symmetry does not cancel the polarity out, the molecule is polar. The Lewis structure of CH_2O is shown here:

$$\ddot{\ddot{\text{O}}}\!=\!\text{C}\!\!\overset{\displaystyle \diagup\ \text{H}}{\underset{\displaystyle \diagdown\ \text{H}}{}}$$

(The compound having molecules with the formula CH_2O is well known, and is called formaldehyde. The bonding and structure in formaldehyde are indeed as indicated by this model.)

6. Investigate the possibility of the existence of isomers or resonance structures. It turns out that in the case of CH_2O an isomeric form that obeys the octet rule can be constructed in which the central atom

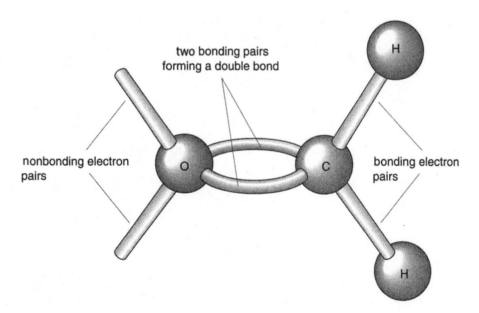

Figure 13.2 Formaldehyde, CH_2O, looks like this when constructed using a molecular model kit of the type typically used in this experiment.

is oxygen rather than carbon. This isomeric form of CH_2O is not found in nature. As a general rule, carbon atoms almost always form a total of four bonds; put another way, nonbonding electron pairs on carbon atoms are very rare (and usually very unstable). Another useful rule of a similar nature is that if a substance contains several atoms of one kind and one of another, the atoms of the same kind will usually assume equivalent positions in the structure. In the sulfate ion, SO_4^{2-}, for example, the four O atoms are all equivalent, and are bonded to the S atom and not to one another. Resonance structures are reasonably common. For resonance to occur, however, the atomic arrangement must remain fixed for two or more possible electronic structures. (The sticks and springs can move, but the balls cannot.) For CH_2O there are no resonance structures.

You are now ready to try this on your own!

A. Using the procedure outlined above, construct and report on models of the molecules and ions listed here and/or other substances assigned by your instructor. Draw the complete Lewis structure for each molecule, showing nonbonding as well as bonding electrons. Given the structure you come up with, describe the geometry of the molecule or ion, and state whether the substance is polar. Finally, draw the Lewis structures of any likely isomers or resonance forms.

CH_4	H_3O^+	N_2	C_2H_2	SCN^-
CH_2Cl_2	HF	P_4	SO_2	NO_3^-
CH_4O	NH_3	C_2H_4	SO_4^{2-}	HNO_3
H_2O	H_2O_2	$C_2H_2Br_2$	CO_2	$C_2H_4Cl_2$

B. Assuming that stability requires that each non-hydrogen atom obey the octet rule, predict the stability of the following substances. (Can you build a structure that obeys the octet rule?)

$$PCl_3 \quad H_3O \quad CH_2 \quad CO$$

C. When you have completed parts A and B, consult with your laboratory instructor, who will check your results and assign you a personal set of substances to create. Assemble models for each of these substances as you did in the previous section, and report on the geometry and bonding in each, on the basis of the model you construct. Also consider and report the polarity, as well as the Lewis structures of any isomers and/or resonance forms.

Experiment 13

Observations and Analysis: The Geometric Structure of Molecules

A. Substance	Lewis structure	Molecular geometry	Polar?	Isomers or resonance structures
CH_4				
CH_2Cl_2				
CH_4O				
H_2O				
H_3O^+				
HF				

(continued on following page)

Substance	Lewis structure	Molecular geometry	Polar?	Isomers or resonance structures
NH_3				
H_2O_2				
N_2				
P_4				
C_2H_4				
$C_2H_2Br_2$				
C_2H_2				

Substance	Lewis structure	Molecular geometry	Polar?	Isomers or resonance structures
SO_2				
SO_4^{2-}				
CO_2				
SCN^-				
NO_3^-				
HNO_3				
$C_2H_4Cl_2$				

B. Stability prediction for PCl_3 _____ H_3O _____ CH_2 _____ CO _____ (Yes/No)

C. Personal set

Substance	Lewis structure	Molecular geometry	Polar?	Isomers or resonance structures
_____				
_____				
_____				
_____				
_____				

Experiment 13

Advance Study Assignment: The Geometric Structure of Molecules

You are asked by your instructor to construct a model of the NH_2F molecule. You proceed as directed in the Experimental Procedure section.

1. First you need to find the number of valence electrons in NH_2F. For counting purposes with Lewis structures, the number of valence electrons in an atom of a main group element is equal to the last digit in the group number of that element in the Periodic Table.

   N is in Group _____   H is in Group _____   F is in Group _____

   In NH_2F there are a total of _____ valence electrons.

2. The model consists of balls and sticks.

   a. How many holes should be in the ball you select for the N atom? _____

   b. How many holes should be in the balls you select for the H atoms? _____

   c. How many holes should be in the ball you select for the F atom? _____

   The electrons in the molecule are paired, and each stick represents a valence electron pair.

   d. How many sticks do you need? _____

3. You assemble a skeleton structure for the molecule, connecting the balls and sticks to make one unit. Follow the rule that N atoms form three bonds but F and H atoms form only one. Draw a sketch of the skeleton here:

4. a. How many sticks did you need to make the skeleton structure? _____

   b. How many sticks are left over? _____

(continued on following page)

If your model is to obey the octet rule, each ball must have four sticks in it (except for hydrogen atom balls, which need and can only have one). (Each non-hydrogen atom in an octet rule substance is surrounded by four pairs of electrons.)

c. How many holes remain to be filled? _____

Fill them with the remaining sticks, which represent nonbonding electron pairs. Draw the complete Lewis structure for NH_2F, using lines for bonds and pairs of dots for nonbonding electrons.

5. Describe the geometry of the model, which is that of NH_2F. _____

Is the NH_2F molecule polar? _____ Why?

Do you expect NH_2F to have any isomeric forms? _____ Explain your reasoning.

6. Does NH_2F have any resonance structures? _____ If so, draw them here:

Experiment 14

Heat Effects and Calorimetry

Heat is a transfer of random kinetic energy (random motion of particles), often called thermal energy: the transfer occurs automatically (technical term: spontaneously) from any object at a higher temperature to any object at a lower temperature. If two otherwise isolated objects are in contact, they will, given sufficient time, both reach the same temperature.

Heat is ordinarily measured in a device called an adiabatic calorimeter: a container with insulating walls, made so that essentially no thermal energy is exchanged between its contents and the surroundings. Within the calorimeter chemical reactions may occur or thermal energy may transfer from one part of the contents to another, but (ideally) no thermal energy flows into or out of the calorimeter from or to the surroundings.

Temperature is a measure of average thermal energy. Throughout this experiment, you should report actual temperatures in °C but temperature *differences* in C°. This is because a temperature difference of 4 C° is the same as a temperature difference of 4 K°, but the temperature 4°C is very different from the temperature 4 K!

Specific Heat Capacity

When thermal energy flows into a substance, the temperature of that substance increases. The heat (quantity of thermal energy transfer), q, required to cause a temperature change in any pure substance increases with the mass m of the substance and the size of the temperature change, Δt, as shown in Equation 1. How much thermal energy must transfer also depends on the substance; that dependence is quantified with a value called the specific heat capacity, abbreviated c, of that substance:

$$\text{heat} = q = (\text{specific heat capacity}) \times (\text{mass}) \times (\text{change in temperature}) = c \times m \times \Delta t \tag{1}$$

The specific heat capacity of a pure substance is the amount of thermal energy required to raise the temperature of one gram of the substance by $\underline{1}$ C°. (If you make m and Δt in Equation 1 both equal to $\underline{1}$, then q will equal c.) Thermal energy is measured in either joules or calories. To raise the temperature of $\underline{1}$ g of water by $\underline{1}$ C°, 4.18 joules of thermal energy must be transferred into the water. The specific heat capacity of water is therefore 4.18 joules/(g·C°). Because 4.18 joules equals $\underline{1}$ calorie, the specific heat capacity of water is also $\underline{1}$ calorie/(g·C°). Ordinarily the heat into or out of a substance is determined by the effect that it has on a known amount of water. Because water plays such an important role in these measurements, the calorie was actually defined based on the specific heat capacity of water.

The specific heat capacity of a metal can be determined with an adiabatic calorimeter, which thermally insulates its contents from the outside world. A weighed amount of metal is heated to some known temperature and then quickly transferred into a calorimeter containing a measured amount of water at a known temperature. Thermal energy is transferred from the metal to the water, and the two equilibrate at some temperature between the initial temperatures of the metal and the water.

Assuming that no thermal energy transfers between the calorimeter and the surroundings, and that the amount of thermal energy absorbed by the calorimeter walls is small enough that it can be ignored, the amount of thermal energy that flows from the metal as it cools is equal to the amount absorbed by the water. In thermodynamic terms, the heating of the metal is equal in magnitude but opposite in direction, and therefore in sign, to that of the water:

$$q_{H_2O} = -q_{metal} \tag{2}$$

If each heat is then expressed in terms of Equation 1 (for both the water, H_2O, and the metal, M), Equation 2 becomes

$$q_{H_2O} = c_{H_2O} \times m_{H_2O} \times \Delta t_{H_2O} = -c_M \times m_M \times \Delta t_M = -q_M \tag{3}$$

In this experiment you will measure the masses of water and metal and their initial and final temperatures. (Note that $\Delta t_M < 0$ and $\Delta t_{H_2O} > 0$ because $\Delta t = t_{final} - t_{initial}$, and the water heats up as the metal cools down.) Given the specific heat capacity of water, the specific heat capacity of the metal can be determined using Equation 3. You will use this procedure to determine the specific heat capacity of an unknown metal.

It turns out the specific heat capacity of a metal is generally, if somewhat roughly, related to its molar mass. French physicists Dulong and Petit discovered many years ago that about twenty-five joules of thermal energy are required to raise the temperature of one mole of many metals by 1 C°. This relation, shown in Equation 4, is known as the law of Dulong and Petit:

$$MM_{metal} \{g/mol\} \sim \frac{25}{c_{metal} \{J/(g \cdot C°)\}} \tag{4}$$

where MM_{metal} is the molar mass of the metal and c_{metal} is its specific heat capacity. Once the specific heat capacity of a metal is known, its molar mass can be approximated using Equation 4. The law of Dulong and Petit was one of the few rules available to early chemists in their studies of molar masses.

Enthalpy of Reaction

When a chemical reaction occurs in a solution having water as its solvent, the situation is similar to that of a hot metal sample placed in water. There is an exchange of thermal energy between the reaction mixture and the solvent, water. As in the specific heat capacity of a metal experiment, the heating of the reaction mixture is equal in magnitude but opposite in sign to that of the water. Under these conditions the heat, q, associated with the reaction mixture is also equal to the <u>enthalpy change</u>, ΔH, for the reaction:

$$q_{reaction} = \Delta H_{reaction} = -q_{H_2O} \tag{5}$$

By measuring the mass of the water used as the solvent, and the temperature change the water undergoes, you can find q_{H_2O} by Equation 1 and ΔH by Equation 5. If the temperature of the water goes up, energy has left the reaction mixture as thermal energy and the reaction is <u>exothermic</u>: q_{H_2O} is *positive* and ΔH is *negative*. If the temperature of the water goes down, the reaction mixture has gained thermal energy from the water and the reaction is <u>endothermic</u>: q_{H_2O} is *negative* and ΔH is *positive*. Both exothermic and endothermic reactions exist.

One reaction that can be studied in solution occurs when a solid is dissolved in water. As an example of such a reaction, consider the dissolution of solid sodium hydroxide, $NaOH(s)$, in water:

$$NaOH(s) \rightarrow Na^+(aq) + OH^-(aq); \quad \Delta H = \Delta H_{solution} \tag{6}$$

When this reaction occurs, the temperature of the solution becomes much higher than that of the NaOH and water that were used. If you dissolve a known amount of NaOH in a measured amount of water in a calorimeter and measure the temperature change that occurs, you can use Equation 1 to find q_{H_2O} for the reaction and Equation 5 to obtain ΔH. Because ΔH increases with the amount of NaOH used, it makes sense to report $\Delta H_{solution}$ per gram or mole of NaOH. In the second part of this experiment you will measure $\Delta H_{solution}$ for an unknown ionic solid.

Chemical reactions often occur when solutions are mixed. For example, a precipitate may form, in a reaction opposite in direction to Equation 6. Another common reaction is <u>neutralization</u>, which occurs when an acid is mixed with a base. In the last part of this experiment you will measure the heat effect when a solution of hydrochloric acid, HCl, is mixed with one containing sodium hydroxide, NaOH, which is a base. The heat effect is quite large, and results from the reaction between H^+ ions from the HCl solution with OH^- ions from the NaOH solution to form water:

$$H^+(aq) + OH^-(aq) \rightarrow H_2O(\ell); \quad \Delta H = \Delta H_{neutralization} \tag{7}$$

Enthalpies of reaction and solution are often expressed per gram and given the symbol $\Delta \hat{H}$, or per mole and given the symbol $\Delta \tilde{H}$.

Experimental Procedure

A. Specific heat capacity

Obtain a calorimeter, a sensitive thermometer, a special glass stirring rod with a perpendicular loop on one end, a sample of metal in a large stoppered test tube, and a sample of unknown solid.

The calorimeter consists of two foam coffee cups, one inside the other, with a foam cover. There are two holes in the cover: one for the thermometer and the other for the special stirring rod. Assemble the experimental set-up as shown in Figure 14.1.

Fill a 400-mL or larger beaker about two-thirds full of water and begin heating it to the boiling point. While the water is heating, weigh your sample of unknown metal in the large stoppered test tube to ± 0.1 g and record the result on the data page. Pour the metal into a dry container or onto a sheet of paper and weigh the empty test tube and stopper. Replace the metal in the test tube and put the *loosely* stoppered tube into the hot water in the beaker. The water level in the beaker must be high enough that the top of the metal is below the water surface, but water should not go inside the test tube. Continue heating the test tube filled with metal in the water for at least 10 minutes after the water begins to boil, to ensure that the metal reaches the temperature of the boiling water. Add small amounts of water to the beaker as necessary to maintain the water level.

While the water is boiling, weigh the empty calorimeter to ± 0.1 g. Put about forty milliliters of room temperature water into the calorimeter and weigh it again. Insert the stirrer and thermometer into the cover and put it on the calorimeter. If a glass thermometer is used, insert it so that it almost touches the bottom of the calorimeter.

Measure the temperature of the water in the calorimeter to ± 0.1°C. Take the test tube out of the beaker of boiling water, remove the stopper, and pour the hot, dry metal into the water in the calorimeter. Make sure that no water from the outside of the test tube runs into the calorimeter when you pour out the metal. Replace the calorimeter cover and agitate the water as best you can with the glass stirrer. Record to ± 0.1°C the maximum temperature reached by the water. Repeat the entire process, but this time use about fifty milliliters of water in the calorimeter. Be sure to dry your metal before reusing it; this can be done by heating the metal briefly in the test tube in boiling water and then pouring the metal out onto a paper towel. You can dry the hot test tube with a little compressed air.

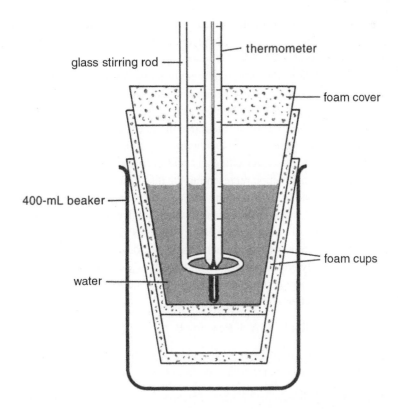

Figure 14.1 The calorimeter employed in this experiment is assumed to be <u>adiabatic</u>: it does not allow thermal energy exchange between its contents and the surroundings.

B. Enthalpy of solution

Place about fifty milliliters of room-temperature deionized water in the calorimeter and weigh it again, as in the previous procedure. Measure the temperature of the water in the calorimeter to $\pm 0.1°C$. In a small beaker, weigh out about five grams of the solid compound assigned to you. Tare the empty beaker and add solid to it, recording the amount added to ± 0.1 g. Then add the solid to the calorimeter. Stirring continuously and occasionally swirling the calorimeter, determine to $\pm 0.1°C$ the maximum or minimum temperature reached as the solid dissolves. After this extreme is reached, quickly open the lid and check to make sure that *all* the solid dissolved. A temperature change of at least 5 C° should be obtained in this experiment. If necessary, repeat the experiment, increasing the amount of solid used if the temperature change was too small, or decreasing it if not all the solid dissolved.

Disposal of reaction products: Dispose of the solution from Part B as directed by your instructor.

C. Enthalpy of neutralization

Rinse out your calorimeter with deionized water, pouring the rinse water into the sink. With a graduated cylinder, measure out 25.0 mL of 2.00 M HCl; pour that solution into the calorimeter. Rinse out the cylinder with deionized water, then measure out 25.0 mL of 2.00 M NaOH. Measure the temperature of the acid and of the base (still in the cylinder) to $\pm 0.1°C$, making sure to rinse and dry your thermometer before immersing it in each solution. Put the thermometer back in the calorimeter cover. Pour all of the NaOH solution into the HCl solution and put the cover back on the calorimeter. Stir the reaction mixture and swirl the calorimeter occasionally; record the maximum temperature reached by the neutralized solution.

Calculate the enthalpy of neutralization. The densities of the solutions are a little higher than that of pure water; assume a value of 1.02 g/mL for all solutions, such that the mass of solution is obtained by multiplying 50.0 mL by 1.02 g/mL. Use the specific heat capacity of water, 4.18 J/(g·C°), as the specific heat capacity of the solutions.

Optional   D. Hess's law

Using the same procedure as in Part C, measure the enthalpy of neutralization of 2.00 M acetic acid, CH_3COOH. You should use 25.0 mL of the acetic acid solution in place of the 2.00 M HCl used in Part C. Calculate the molar enthalpy of dissociation of acetic acid, "HAc" for short, again using 1.02 g/mL as the density for all solutions. Record your data and calculations on a separate sheet of paper. Write out the net ionic equation for the strong acid–strong base neutralization as well as the net ionic equation for the acetic acid–strong base neutralization reaction. (Acetic acid is a weak acid; use HAc to represent undissociated acetic acid and Ac^- for the acetate ion, CH_3COO^-.) In each case, show the value of the measured molar enthalpy of reaction on the right side of the equation. Remember that the enthalpy change for an exothermic reaction is negative. Combine the two net ionic equations (add the reactions, or their reverse, cancelling any substance that appears on opposite sides of the sum) in such a way that you end up with an equation representing the dissociation of the weak acid HAc into $H^+(aq)$ and $Ac^-(aq)$. If you need to reverse one of the equations, its enthalpy change value will change sign. As you add the two equations you should add the two enthalpy change values, to give the enthalpy change for the dissociation reaction. Just as the dissociation of water to form $H^+(aq)$ and $OH^-(aq)$ is expected to be endothermic, the corresponding dissociation reaction for a weak acid, such as acetic acid, is also expected to be endothermic.

If you did this exercise correctly, you have just illustrated Hess's law, which states that the value of ΔH for a reaction is the same whether it occurs directly, or in a series of steps. Thus, if a thermochemical equation can be expressed as the sum of two chemical reactions,

$$\text{If Reaction C = Reaction A + Reaction B, then } \Delta H_C = \Delta H_A + \Delta H_B$$

When you have completed these experiments, you may pour the neutralized solutions down the sink. Rinse the calorimeter, stirrer, and thermometer with water and return them, along with the metal sample.

Take it further (optional): Measure the heat of neutralization of household vinegar in its reaction with NaOH solution. Assuming that household vinegar is 5.0%$_{mass}$ acetic acid (or, better, using the actual concentration indicated on the vinegar, if so labeled), find the molar heat of neutralization of acetic acid.

Name _____ Section _____

Experiment 14

Data and Calculations: Heat Effects and Calorimetry

A. Specific heat capacity

	Trial 1	Trial 2
Mass of stoppered test tube plus dry metal	_____ g →	_____ g
Mass of empty test tube and stopper	_____ g →	_____ g
Mass of empty calorimeter	_____ g →	_____ g
Mass of calorimeter and water	_____ g	_____ g
Mass of dry metal	_____ g →	_____ g
Mass of water in calorimeter	_____ g	_____ g
Initial temperature of water in calorimeter	_____ °C	_____ °C
Initial temperature of metal (assume 100°C unless directed to do otherwise)	_____ °C	_____ °C
Equilibrium temperature of metal and water in calorimeter	_____ °C	_____ °C
Δt_{water} ($t_{final} - t_{initial}$)	_____ C°	_____ C°
Δt_{metal} ($t_{final} - t_{initial}$)	_____ C°	_____ C°
q_{H_2O} (Equation 3)	_____ J	_____ J
Specific heat capacity of the metal (Equation 3)	_____ J/(g C°)	_____ J/(g·C°)
Approximate molar mass of the metal (Equation 4)	_____ g/mol	_____ g/mol
Unknown metal ID code	_____	

(continued on following page)

B. Enthalpy of solution

Mass of calorimeter plus water   _____ g

Initial temperature of the water, $t_{initial}$ _____ °C

Mass of solid, m_s _____ g

Final temperature of the solution, t_{final} _____ °C

Δt_{water} $(t_{final} - t_{initial})$ _____ C°

Mass of water, m_{H_2O} _____ g

q_{H_2O} for the reaction (Equation 1) $\left(c_{H_2O} = \dfrac{4.18\ J}{g \cdot C°} \right)$ _____ joules

ΔH for the reaction (Equation 5)   _____ joules

The quantity you have just calculated is approximately* equal to the enthalpy of solution for the mass of sample you used, $\Delta H_{solution}$. Calculate the enthalpy of solution per gram of solid sample, $\Delta\hat{H}_{solution}$.

$$\Delta\hat{H}_{solution} = \underline{\hspace{2cm}}\ joules/g$$

The dissolution reaction is endothermic/exothermic. (Circle the correct answer.) Explain your reasoning.

Solid unknown ID code _____

Optional   Formula of compound used (if furnished) _____   Molar mass _____ g/mol

Enthalpy of solution per mole of compound, $\Delta\tilde{H}_{solution}$ _____ kJ/mol

C. Enthalpy of neutralization

Initial temperature of HCl solution   _____ °C

Initial temperature of NaOH solution   _____ °C

Final temperature of neutralized mixture   _____ °C

Change in temperature, Δt [Use the average (arithmetic mean) of the initial temperatures of HCl and NaOH as the initial temperature.]   _____ C°

q_{H_2O} (Assume 50.0 mL of solution and a solution density of 1.02 g/mL.)   _____ J

Total ΔH for the neutralization reaction, $\Delta H_{reaction}$ _____ J

ΔH per mole of H^+ and OH^- ions reacting, $\Delta\tilde{H}_{reaction}$ _____ kJ/mol

*Your ΔH value will be approximate because some thermal energy released by the reaction actually does go into and through the walls of the calorimeter. Some also goes into heating up the dissolved salt. But this analysis assumes it *all* goes into heating up the water.

Experiment 14

Advance Study Assignment: Heat Effects and Calorimetry

1. A metal sample weighing 147.90 g and at a temperature of 99.5°C was placed in 49.73 g of water in an adiabatic calorimeter at 21.0°C. At equilibrium, the temperature of the water and metal was 51.8°C.

   a. What was Δt for the water? ($\Delta t = t_{final} - t_{initial}$)

   _____ C°

   b. What was Δt for the metal? ($\Delta t = t_{final} - t_{initial}$)

   _____ C°

   c. How much thermal energy flowed into the water? Use 4.18 J/(g·C°) as the specific heat capacity of water.

   _____ joules

   d. Calculate the specific heat capacity of the metal, c_{metal}, using Equation 3.

   _____ joules/(g·C°)

   e. What is the approximate molar mass of the metal? (Use Equation 4.)

   _____ g/mol

2. When 4.89 g of solid NaOH was dissolved in 47.92 g of water in an adiabatic calorimeter at 24.7°C, the temperature of the solution went up to 50.1°C.

   a. Is this dissolution reaction exothermic? _____ Why?

(continued on following page)

b. Calculate q_{H_2O}, using Equation 1.

_____ joules

c. Use Equation 5 to calculate ΔH for the dissolution reaction as it
    occurred in the calorimeter, $\Delta H_{solution}$.

$\Delta H_{solution} =$ _____ joules

d. Find ΔH for the dissolution of 1.00 g of solid NaOH in water, $\Delta \hat{H}_{solution}$.

$\Delta \hat{H}_{solution} =$ _____ kJ/g

e. Find ΔH for the dissolution of 1 mole of solid NaOH in water, $\Delta \tilde{H}_{solution}$.

$\Delta \tilde{H}_{solution} =$ _____ kJ/mol

f. Given that NaOH exists as Na^+ and OH^- ions in solution, write the equation for the reaction that
    occurs when solid NaOH is dissolved in water.

g. Given the heats of formation, $\Delta \tilde{H}_f$, listed in Table 14.1, calculate $\Delta \tilde{H}_{reaction}$ for the dissolution reaction
    in Part (f). Compare your answer with the result you obtained in Part (e).

Table 14.1 Selected Molar Enthalpies of Formation, $\Delta \tilde{H}_f$

Substance	$\Delta \tilde{H}_f$ {kJ/mol}
NaOH(s)	−425.6
Na^+(aq)	−240.1
OH^-(aq)	−230.0

$\Delta \tilde{H}_{reaction} =$ _____ kJ/mol

Experiment 15

Vapor Pressure and Enthalpy of Vaporization*

If a pure liquid is poured into an empty container and the container is then sealed, some of the liquid will evaporate into the air in the container. After a short time, the <u>vapor</u> (evaporated liquid) will establish a <u>partial pressure</u> (a contribution to the total pressure due to a specific gas), which remains constant so long as the temperature does not change. That pressure is called the <u>vapor pressure</u> of the liquid, and at a given temperature it remains the same, regardless of what other gases are present in the container and what the total pressure in the container may be.

In the air we breathe, there is usually some water vapor, but its partial pressure remains well below the vapor pressure of water (unless it is foggy or raining). The <u>relative humidity</u> is defined as the partial pressure of water vapor in a given air sample, divided by the vapor pressure of water at that temperature. On a hot, sticky day the relative humidity is high, perhaps 80 or 90%, and the air is nearly saturated with water vapor; most people (but not all!) feel uncomfortable under such conditions.

If the temperature of a liquid is increased, its vapor pressure increases. At the "normal boiling point" of the liquid, its vapor pressure becomes equal to 1 atm. (This is the pressure at 100°C in a closed container in which there is just water and its vapor, with no air or other gases present.)

At high temperatures, vapor pressures can become very large: reaching about 5 atm for water at 150°C and 15 atm at 200°C. A liquid that boils at a low temperature, say −50°C, will already have a huge vapor pressure at room temperature, and confining it at 25°C in a closed container may well cause the container to explode.

There is an equation relating the vapor pressure of a liquid, VP, to the Kelvin temperature, T, and the enthalpy of vaporization of the liquid, ΔH_{vap}. It is called the <u>Clausius-Clapeyron equation</u> and has the form

$$\ln VP = \frac{-\Delta H_{vap}}{RT} + \text{a constant} \tag{1}$$

Equation 1 indicates that if the natural log of the vapor pressure, $\ln VP$, is plotted against the reciprocal of the Kelvin temperature, $1/T$, a straight line with slope $-\Delta H_{vap}/R$ should be obtained. In using this equation, the units of ΔH_{vap} and R must match. VP may be in any pressure unit, such as atm, mm Hg, or pascals.

There are many methods for measuring vapor pressure, but doing so directly (by trying to measure the pressure in a volume containing only the liquid and vapor of interest, no air) can be challenging. Instead, you will measure vapor pressure indirectly, by monitoring an air bubble surrounded by the liquid whose vapor pressure you wish to measure. A measured sample of air, free of the liquid under study, will be injected into a graduated pipet filled with the liquid under study. You will measure the volume of the air–vapor bubble that forms, which will be larger than that of the injected air because some of the liquid will vaporize into the air bubble. As you raise the temperature, the volume of the bubble will increase: in part because the air in the bubble expands as the temperature rises, but also because more and more liquid vaporizes as the temperature increases. (Recall that vapor pressure increases with temperature!) The total pressure in the bubble will remain equal to the ambient pressure, so by calculating the partial pressure of air in the bubble, and realizing that the bubble's total pressure is that of the atmosphere, you can find the vapor pressure of the liquid at that temperature by subtraction.

Wear your safety glasses while performing this experiment.

Experimental Procedure

Obtain a "device" (a half-sealed 0.5-mL graduated pipet segment: see Figure 15.1), a Pasteur pipet, a digital thermometer, and a sample of a liquid unknown, whose density will be given to you.

*The general method used in this experiment is one described by DeMuro, Margarian, Mkhikian, No, and Peterson, *J Chem Ed* 76:1113–1116, 1999.

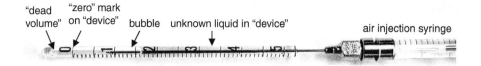

Figure 15.1 The "device" used in this experiment will have an air bubble of known volume injected into it by your instructor. *(Robert Rossi)*

A. Finding the vapor pressure of the liquid at room temperature

The accuracy of your results in this experiment hinges strongly on the quality of your mass measurements—so make them very carefully! Determine the mass of the empty, dry device to ± 0.0001 g and record the result on the report page. Then place the device, open end up, in a small (25- or 50-mL) Erlenmeyer flask (so you do not warm it by holding it with your hand while it is being filled). Using your Pasteur pipet, carefully fill the device completely with your unknown liquid. Dry off the outside of the device and accurately determine its mass again. Be careful around the top of the device—dry only up to the top edge, and do not remove liquid from inside the device by drying the top face. You want the liquid in the device even with the rim, but the outside of the device must be dry. The surface tension of the liquid will keep it from escaping the device, even when it is set on its side, so weigh *only* the device, setting it directly on the balance—do not weigh your Erlenmeyer flask.

Now request your instructor's help. Using a syringe, your instructor will inject 0.200 mL of ambient air at the closed end of the device (holding it horizontally, or with the sealed end tilted slightly up), forming an air bubble. (Surface tension will again keep the liquid in the device, even if it is inverted.) After the syringe is removed, the device will no longer be completely full at the open end, so gently return the device to the Erlenmeyer flask, open end up, and add more unknown with your Pasteur pipet, until the liquid level is back up to the rim of the device. Do not worry: the bubble should not escape so long as the device does not experience any sudden motion. Do not heat up the bubble by holding the part of the device containing it with your hand. You need to keep the bubble at room temperature in Part A to get good results!

Carefully dry and reweigh your device on the same balance you used before. Be sure that the liquid level in the device is again even with the rim, as close as possible to the way it was when you first weighed the device (without the bubble in it). This is your last crucial mass measurement. Record it, and return the device to the Erlenmeyer flask.

Read and record the ambient air temperature, t_{air}, and the atmospheric (ambient) pressure, P_{atm}.

Now fill a medium (18 × 150 mm) test tube to within about 3 cm of the top with your unknown liquid, and also add a bit more to the top of the device, so that the liquid is at or above its rim. (It is no longer important that the outside of the device be dry.) Tilt the test tube to a 45° angle and gently drop in your device, *open end down and sealed end up,* so that it slides down the wall of the liquid-filled test tube, to the bottom. Swirl your test tube and the device inside until the bubble moves all the way to the very top of the device. Read and record the reading on the device as accurately as possible. *Note*: This is a crucial step! Take your time, and read to the nearest 0.002 mL. Note that the numbers on the pipet are in 0.1-mL increments, so each line (technical term: <u>graduation</u>) indicates 0.01 mL.

Next, insert the temperature probe of the digital thermometer into the unknown liquid in the test tube, holding it in position with a stopper with a large enough hole that **the test tube is not tightly sealed,** and record the unknown liquid's (and the bubble's) temperature, t_{bubble}. Do not hold the liquid-filled part of the test tube with your hand while taking this measurement, or you will warm it up!

From the data you have obtained so far, you can calculate the vapor pressure of your unknown at the ambient temperature in the lab. Using the mass measurements, you can find the mass of the liquid in the full device and the mass of liquid driven out by the bubble. The volume of the bubble is equal to the mass of liquid driven out divided by the liquid's density. This volume is larger than that of the air injected because some of the unknown liquid vaporizes into the bubble, increasing the amount of gas present. You may safely assume the bubble always remains at atmospheric pressure: the pressure on it due to the mass of the liquid is tiny. At room temperature the partial pressure of the air in the bubble is equal to the atmospheric pressure multiplied by the volume of air injected, divided by the actual volume of the bubble. The vapor pressure of the liquid is obtained by subtracting the partial pressure of the air from the atmospheric pressure. You can (and should) perform these calculations while waiting for water bath temperatures throughout the rest of the experiment. *Note*: If the initial volume of the bubble at room temperature calculates to less than 0.210 mL, consult with your instructor. This value is critical, so do not leave lab (or even return your device) before calculating it.

B. Finding the vapor pressure of the liquid at other temperatures

You will now use a water bath to study how the vapor pressure of your unknown liquid changes with temperature. Set up a water bath as shown in Figure 15.2. Use a 1000-mL beaker that you can heat on a hotplate. Fill the beaker to near the top with warm water from the tap (between 40° and 45 °C). Clamp the test tube with the device and digital thermometer in it as shown in Figure 15.2, so that the water level in the bath is at least as high as the liquid level in the test tube. Place your lab thermometer in the water in the beaker.

The temperature of the water bath should be at 40 ± 3 °C for the first measurement. Heat or cool the bath if necessary, while stirring, to get into that range. When the temperature on the digital thermometer holds steady for at least 30 seconds, compare it against that of the water bath; the two temperatures should be close to one another when steady. Record the reading on the device and the temperature of the unknown liquid, as indicated by the digital thermometer.

Now heat the water in the large beaker while stirring. When the temperature of the bath is roughly 5 C° above the previous value, stop heating. The temperature will continue to rise for a few minutes as you stir, but will finally level off. When the temperatures on the digital thermometer and the bath thermometer become about equal, and the temperature on the digital thermometer has remained constant for at least 30 seconds, record the reading on the device and the temperature of the unknown liquid (and of the bubble), as indicated by the digital thermometer.

Repeat the measurements you have just made at two higher temperatures, each time heating to about five degrees higher than the previous temperature actually measured by the digital thermometer. Each time stop heating at that point, let the temperature level off, and when the bath and device temperatures are similar and both the device temperature and reading are steady for at least 30 seconds, record them. At a high enough temperature, the bubble volume will exceed the volume of the device and the bubble will escape. If this appears likely to happen, because you mistakenly overheat the bath, *immediately* remove the test tube from the bath, then add some cold water to the bath and stir. Put the test tube back in position, and then proceed to make a measurement.

The data you have now obtained will allow you to calculate the vapor pressure of the liquid at each temperature you used. It is important to realize that there is a unique "dead volume" above the zero mark in each device (see Figure 15.1) that the volume reading on the device does not take into account. (Making a device that measures the bubble volume without this complication would require custom glassware and would be very expensive.) You calculated this "dead volume" in Part A, and on the report page you need to add this value to the device readings taken in Part B to obtain the actual volume of the bubble at each temperature. You must also calculate the volume that the air sample alone would occupy at each temperature and atmospheric pressure (use the ideal gas law; the

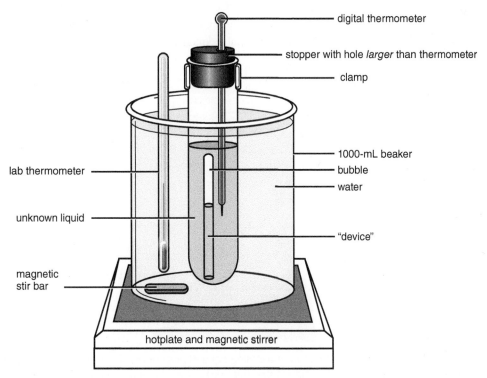

Figure 15.2 After taking a reading at room temperature, the device is heated to progressively higher temperatures in this apparatus, and readings are taken once the temperature stabilizes.

moles of *air* present in the bubble remains constant, so the air volume effectively increases with the ratio of the Kelvin temperatures). The partial pressure of air in the bubble will equal the atmospheric pressure multiplied by the ratio of the air volume over the total bubble volume at each temperature. Calculate the partial pressure of air and the liquid vapor pressure at each temperature as you wait for boiling during the next part of this experiment.

C. Measuring the boiling point directly

In the last part of this experiment you will measure the boiling point of your liquid at atmospheric pressure. Pour the liquid sample from your test tube into a larger test tube. Determine the boiling point of the liquid using the apparatus shown in Figure 15.3. The thermometer bulb should be just *above* the liquid surface. Heat the water in the bath until the liquid in the tube boils gently, with the vapor condensing *at least 5 cm below* the top of the test tube, but well above the thermometer bulb. Boiling chips may help to keep the liquid boiling smoothly. As the boiling proceeds, liquid will condense onto the thermometer bulb, and droplets will begin falling from it. Shortly after that the temperature reading should become reasonably steady. Record that temperature. The liquid you are using may be flammable and toxic, so you should not inhale the vapor unnecessarily. *Do not* heat the water bath so strongly that vapor condenses only at the top of the test tube.

Calculating the Enthalpy of Vaporization of the Liquid

When you have completed the calculations described in the previous sections, you should have a table of the vapor pressure of the liquid at each of the temperatures you used. To find the molar enthalpy of vaporization of the liquid, use Equation 1. First make a table listing ln VP (the natural logarithm of the vapor pressure), the temperature T in Kelvin, and $1/T$, as found from your data. Make a graph of ln VP vs $1/T$, using the graph paper provided in the report pages (see Appendix E) or spreadsheet software such as Excel or Google Sheets (see Appendix G1 or G2). The slope of the line, which should be straight, is equal to $-\Delta H_{vap}/R$: the negative of the molar enthalpy of vaporization, ΔH_{vap}, divided by the gas constant, R. If you want to express ΔH_{vap} in joules/mole, R will equal 8.3145 J/(mol·K).

Compare the value of the boiling point you measured directly with that which you can calculate from your graph. The value of $1/T$ at the boiling point will occur where ln VP equals ln P_{atm}.

Optional   Finally, make a graph of the vapor pressure as a function of the temperature in °C. Use a spreadsheet or the other sheet of graph paper in the report pages. Connect the points with a smooth curve. Comment on how the vapor pressure, VP, varies with the Celsius temperature, t. ▪

Include your graph(s) with your report. When you are finished with the experiment, pour the liquid from the test tube back into its container and return it along with the digital thermometer, the device, and the Pasteur pipet.

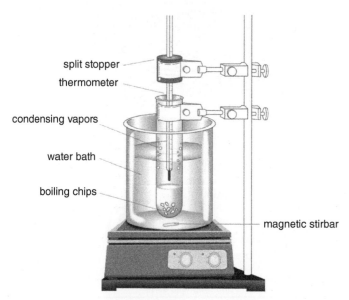

Figure 15.3 Use this apparatus to determine the boiling point of your unknown. Note that the stopper **does not seal the test tube**, and that the bulb of the thermometer is *above* the unknown liquid itself. The temperature of the liquid's vapor condensing onto the thermometer bulb provides a better boiling point value than the temperature of the boiling liquid itself.

Experiment 15

Data and Calculations: Vapor Pressure and Enthalpy of Vaporization

A. Finding the vapor pressure of the liquid at room temperature

Mass of empty, dry device   _____ g

Mass of device, completely filled with unknown liquid   _____ g

Mass of device containing liquid and bubble   _____ g

Atmospheric (ambient) pressure, P_{atm} _____ mm Hg

Ambient air temperature, t_{air} _____ °C

Volume reading on device when at room temperature   _____ mL

Bubble (liquid) temperature for unheated measurement, t_{bubble} _____ °C

Ambient air temperature in absolute units, T_{air} _____ K

Mass of liquid driven out by adding the bubble   _____ g

Volume of liquid driven out by adding the bubble (use the density of the liquid)
> (*Note*: This is the volume of the air bubble after it has been saturated with the liquid's vapor, which will be referred to from here on as the "volume of the bubble", V_{bubble}.)

_____ mL

Partial pressure of air in the bubble at room temperature, $P_{air} = P_{atm}\left(\dfrac{0.200 \text{ mL}}{\text{volume of the bubble}}\right)$

_____ mm Hg

Vapor pressure of the liquid (at room temperature), $VP = P_{atm} - P_{air}$

_____ mm Hg

Difference between volume reading on device and volume of the bubble (based on readings at room temperature—it will be unique to your device, but the same at all temperatures: this is the "dead volume")

_____ mL

(continued on following page)

B. Finding the vapor pressure of the liquid at other temperatures

Device (and bubble) temperature (t_{bubble}) {°C}	Volume reading on device {mL}	Dead volume** {mL}	Actual volume of the bubble (V_{bubble}) {mL}

**The dead volume is specific to your device and is the same for all readings. You calculated it at the bottom of the previous page. You must add it to each volume reading to get the actual bubble volume.

Calculations

Absolute bubble temperature (T_{bubble}) {K}	Actual volume of the bubble (V_{bubble}) {mL}	Volume that just the air in the bubble would occupy at this temperature (V_{air}) {mL}	Partial pressure of air in the bubble {mm Hg} (V_{air}/V_{bubble}) $\times$ P_{atm}	Partial pressure of vapor in the bubble (VP) {mm Hg}

Absolute bubble temperature (T_{bubble}) {K}	Partial pressure of vapor in the bubble (VP) {mm Hg}	Reciprocal of absolute bubble temperature ($1/T_{bubble}$) {K^{-1}}	Natural logarithm of vapor pressure (ln VP) {dimensionless}

Calculated enthalpy of vaporization of unknown liquid (attach your plot and show your work here!)

_____ kJ/mol

C. Measuring the boiling point directly

Observed boiling point of unknown liquid　　　　　　　　　　_____ °C

Unknown ID code　　　　　　　　　　_____

Predicted boiling point of unknown liquid based on enthalpy of vaporization graph _____ °C

Experiment 15

Data and Calculations: Vapor Pressure and Enthalpy of Vaporization

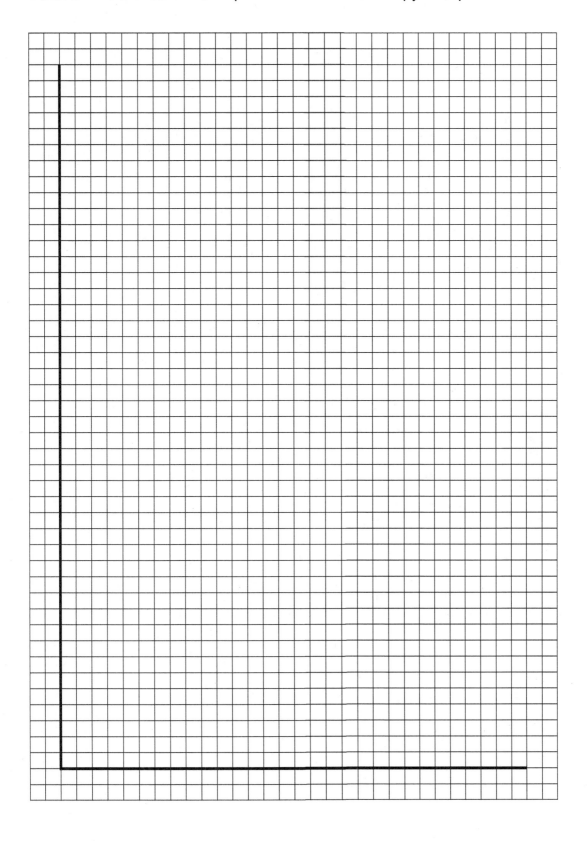

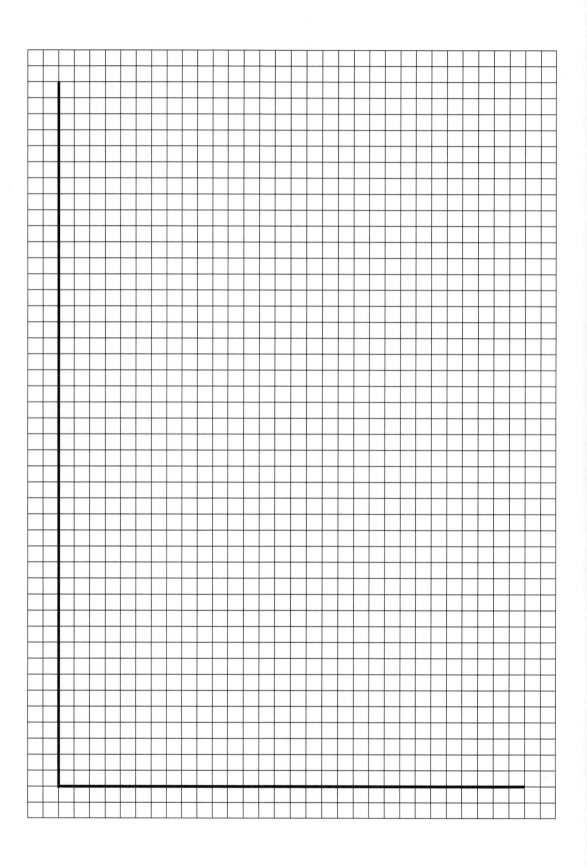

Experiment 15

Advance Study Assignment: Vapor Pressure and Enthalpy of Vaporization

1. In an experiment to measure the vapor pressure of cyclohexane, the following data were obtained:

Mass of empty device	2.4033 g	$P_{atm} = 731$ mm Hg
Mass of device full of cyclohexane	2.7870 g	$t_{air} = 23.1\,°C$
Mass after injecting 0.200 mL of air	2.6093 g	density of cyclohexane $= 0.778$ g/mL
Volume reading on device	0.182 mL	

a. How many grams of cyclohexane were in the full device?

_____ g

b. How many grams were driven out by the air?

_____ g

c. What is the volume of the cyclohexane driven out, and thus the actual volume of the bubble (calculate this using the density)?

_____ mL

d. What is the "dead volume" for this device (see Figure 15.1)? This is the difference between the volume reading on the device and the actual volume of the bubble, which you just calculated.

_____ mL

e. Find the partial pressure of air in the bubble.
   [*Note*: The mass and moles of air in the bubble remain the same, but the bubble's volume grows larger than 0.200 mL because some vapor enters the bubble. The partial pressure of air in the bubble is equal to the fraction of the bubble's volume that the air is responsible for (0.200 mL divided by the volume you just calculated) multiplied by the total pressure of the bubble (which is still the atmospheric pressure, P_{atm}).]

_____ mm Hg

f. What is the partial pressure of vapor in the bubble? The partial pressure of the cyclohexane and the partial pressure of the air must add up to the total pressure of the bubble, which is the atmospheric pressure. (This result is the vapor pressure of cyclohexane at 23.1 °C!)

_____ mm Hg

(continued on following page)

2. When the device is heated to 46.4 °C, the volume reading on the device increases to 0.271 mL.

   a.   What is the new volume of the bubble? Note that the *difference* in device readings gives you the *difference* in the actual volumes of the two bubbles. Adding the "dead volume" to the device readings gives actual bubble volumes.

   _____ mL

   b.   Why does the volume of the bubble increase? (There are two reasons!)

   c.   What volume would the 0.200 mL of air in the bubble expand to at 46.4 °C? The pressure and the moles of air remain the same, but you must use the ratio of the two *absolute* temperatures.

   _____ mL

   d.   What is the partial pressure of air in the bubble at 46.4 °C? The partial pressure of air in the bubble is again equal to the fraction of the bubble's volume that the air is responsible for, multiplied by the total pressure of the bubble (which is still the atmospheric pressure, P_{atm}).

   _____ mm Hg

   e.   What is the vapor pressure of cyclohexane at 46.4 °C? (The partial pressure of the air, P_{air}, and the vapor pressure of the cyclohexane, VP, add up to the total pressure: which is that of the atmosphere, P_{atm}.)

   _____ mm Hg

Experiment 16

The Structure of Crystals— An Experiment Using Models

Under a magnifying glass or microscope, an ordinary substance such as table salt is seen to be composed of many small particles, most with remarkably similar shapes. In the specific case of table salt, NaCl, many cubic particles, with surfaces at right angles to one another, are observed. A similar situation arises with many common solids. The observed consistency in particle shape reflects a deeper regularity in the arrangement of atoms or ions in the solid. Indeed, when crystals are probed with X-rays to determine their atomic construction (technical term: X-ray diffraction), the atomic nuclei are found to be arranged in remarkably symmetric patterns that continue in three dimensions for many millions of units. Substances having a regular arrangement of atom-size particles as solids are called crystalline, and the solid material is said to consist of crystals. This experiment deals with some of the simpler ways in which atoms or ions are arranged in crystals, and what these indicate about atomic and ionic size, the density of solids, and efficient packing of particles.

Many crystals are complex; this analysis will be limited to the simplest crystals: those with cubic structures containing only one or two elements. These relatively simple crystal structures still exhibit many of the interesting properties of more complicated structures. Substances composed of two elements are called binary compounds: NaCl, KBr, H_2O, and NH_3 are examples. All of the binary crystals in this experiment are those of binary salts, meaning they are composed of the ions of two different elements: NaCl and KBr are examples, but H_2O and NH_3 are not because they exist as molecules rather than ionic solids.

The atoms in crystals occur in planes. Often a crystal will preferentially break (technical term: cleave) along such planes. This is the reason the cubic structure of salt at the atomic level shows up in so many of its crystals that are much, much larger. As a first step into the study of crystals, consider a two-dimensional crystal in which all of the atoms lie in a square array (multidimensional pattern) in a plane, as shown here:

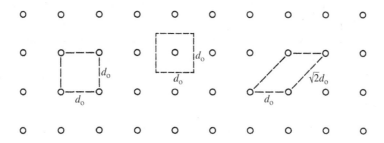

Although only a small section can be drawn here, the array goes on essentially forever—in both directions. This is called the (in this case, two-dimensional) crystal lattice, or just the lattice. In this array o is always the same kind of atom, so all the atoms are equivalent. The distance, d_o, between atoms in the vertical direction is the same as in the horizontal, so knowing this one distance would allow you to generate the array completely. Crystal studies focus on a small section of the array, called the unit cell, that represents the array in the sense that by moving the unit cell repeatedly in either the x- or y-direction, a distance d_o at a time, you could generate the entire array. In the sketch above, the dotted lines indicate several possible unit cells. All the cells have the same area, and making a pattern of any one of them in all directions will locate all the atoms in the array. The unit cell on the left includes four of the atoms and has its edges along the natural axes, x and y, for the array, but that does not necessarily make it the best unit cell. The middle cell clearly indicates that the number of o atoms in the cell is equal to 1. The number of atoms in whatever cell we choose must also equal 1, but that is not so easy to see in the cell on the left. However, once you realize that only one-fourth of each atom on the corners of the leftmost cell actually belongs to that cell—because each atom is shared by three other cells—the number of atoms in the whole cell becomes equal to $4 \times \frac{1}{4}$, or 1, which is the same result obtained by analyzing the center unit cell.

When the array is extended to three dimensions, the same ideas regarding unit cells apply. The unit cell is the smallest portion of the array that could be used to generate the array. With cubic cells the unit cell is usually chosen to have edges parallel to the x-, y-, and z-axes.

Experimental Procedure

In this experiment your instructor may allow you to work without safety glasses.

In this experiment the discussion and the experimental procedure will be combined—because one supports and illustrates the other. You will start with the simplest possible cubic crystal, deal with its properties, and then go on to the next more complex example, and the next, and so on. Each kind of crystal will be related in some ways to the earlier ones, but will have its own properties as well.

Be sure to complete each section before going on to the next, unless your instructor directs otherwise, because they build on one another. Start by opening your model kit and looking over the contents. Your instructor may provide some additional advice about what pieces you should use in this experiment.

A. Simple cubic (SC) crystals

The crystal structure called simple cubic (<u>SC</u>) is shown on the report page for this experiment. The SC structure only contains atoms (no ions), all of the same element, and is the three-dimensional analog of the two-dimensional array presented in the introduction.

The SC unit cell is a cube with an edge length equal to the distance from the center of one atom to the center of the next. The length of the unit cell edge is usually given the symbol d_o. The volume of the unit cell is therefore d_o^3, which is very small because d_o is about half a nanometer. Using X-rays, d_o can be measured to four significant figures. The number of atoms in the unit cell of a simple cubic crystal is equal to $\underline{8} \times \underline{⅛}$, or $\underline{1}$, because only one-eighth of each corner atom actually belongs to the cell; each corner atom is shared equally by eight cells.

Using your model set, assemble three attached unit cells having the SC structure shown in the image on the report page. Use the short (#6) bonds and the gray atoms. Each bond goes into a *square* face on the atom. An actual crystal with this structure would have many such cells, extending in all three dimensions.

If you extended the model you have made further, in the other dimensions, you would find each atom to be connected to six others. The <u>coordination number</u> of the atoms in this structure is <u>6</u>. Only atoms or ions *bonded* to each other, by covalent or ionic bonds, are counted in arriving at a coordination number.

Although the model has an open structure to make it easier to see relationships between atoms, in an actual crystal the atoms that are closest to one another are usually touching. It is possible to determine atomic and ionic radii if this is assumed to be the case. In this SC crystal, if d_o is known, the atomic radius r of the atoms can be calculated: if the atoms are touching, d_o must equal $\underline{2}r$. Knowing d_o also allows the density to be calculated, if the chemical identity of the atoms in the crystal is known. The volume V of the unit cell can be calculated from d_o. Because there is one atom per unit cell, a mole will occupy the volume of 6.02×10^{23} cells. Knowing the molar mass of the element, the mass m of a single unit cell can be calculated. The density is m/V. With more complex crystals the same calculations can be made, but they must take account of the fact that the number of atoms or ions per unit cell may not equal one.

Almost no elements crystallize in an SC structure: the radioactive metal polonium is one of the very few that do (if not the only one). This is because SC packing is inefficient, in that the atoms are farther apart than they need be. There is a big cubic <u>hole</u> (technical term: <u>void</u> or <u>interstitial site</u>) in the middle of each unit cell—big enough for an additional atom or ion to squeeze into—and generally, the closer together things get in a crystal, the more stable (energetically favorable) it becomes. That situation will be considered further in Part B. For now, calculate the fraction of the volume of the unit cell that is actually occupied by atoms in the SC structure: do so on the report page, which will lead you through how to do so.

That fraction is small; only 52% of the cell volume is occupied. Most of the empty space is in the hole at the center of the SC unit cell. Geometric calculations indicate that an atom or ion having a radius r_{void} equal to 73% of that of the atom or ion on a corner would fit into the hole. Or, putting it another way, an atom bigger than that would have to push the corner atoms apart to fit in. If the corners of a cubic unit cell are occupied by <u>anions</u> (negatively charged ions), and a <u>cation</u> (a positively charged ion), were placed in the hole, the cation would push the anions apart, decreasing the repulsion between negatively charged anions and maximizing the attraction between negative anions and positive cations. The key takeaway here, which will become important in Part E when analyzing binary salts, is that if $r_+/r_- > 0.732$ a cation will not fit in the cubic hole formed by anions on the corners of a cubic cell.

B. Body-centered cubic (BCC) crystals

In a body-centered cubic (BCC) crystal, the unit cell still contains the corner atoms present in the SC structure, but in the center of the cell there is another atom of the same kind. The unit cell is shown on the report page.

Using your model set, assemble a BCC crystal. Use the blue balls. Put a short (#6) bond in each of the eight holes in the *triangular* faces on one blue ball. Attach a blue ball to the other end of each of these bonds, again using the holes in *triangular* faces. These eight blue balls define the unit cell in this structure. Add as many atoms to the structure as are available, so that you can see how the atoms are arranged. In a BCC crystal each atom is bonded to eight others, so the coordination number is $\underline{8}$. Verify this with your model. (There are no bonds along the edges of the unit cell, because the corner atoms are not as close to one another as they are to the central atom.)

The BCC lattice (crystal structure) is much more stable than SC, partly because of the higher coordination number. Many metals crystallize in a BCC lattice at room temperature—including sodium, chromium, tungsten, and iron.

There are several properties of the BCC structure that you should note. The number of atoms per unit cell is two: one from the corner atoms ($\underline{8} \times \frac{1}{8}$) and one from the atom in the middle of the cell, which is unshared. As with SC cells, there is a relationship between the unit cell size and the atom radius. Given that in sodium metal the unit cell edge is 0.429 nm, calculate the radius of a sodium atom on the report page. Once you have done that, calculate the density of sodium metal.

The fraction of the volume that is occupied by atoms in BCC crystals is quite a bit larger than in SC crystals: in BCC crystals, 68% of the cell volume is occupied, as opposed to 52% in SC.

C. Close-packed structures

Although many elements form BCC crystals, still more prefer structures in which the atoms in a layer are *close-packed*: touching as many nearest neighbors as possible—which is six, as shown here:

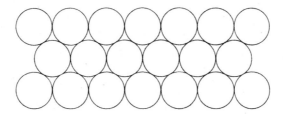

This is the way that balls of the same size naturally arrange themselves when pushed together, and the way the cells are arranged in a honeycomb. It is the most efficient way to pack spheres, with 74% of the volume in a close-packed structure filled with atoms. It turns out that in three dimensions, more than one close-packed crystal structure exists. The layers all have the same structure, but they can be stacked on top of each other in two different ways. This is certainly not apparent at first sight! These two possibilities are discussed further in the next section.

D. Face-centered cubic (FCC) crystals

In the face-centered cubic (FCC) unit cell there are atoms on the corners and an atom at the center of each face; all are of the same element. There is no atom in the center of the cell (consult the image on the first report page; the bonding is shown on only the three exposed faces).

Constructing an FCC unit cell, using bonds between closest atoms, is not as easy as it might seem. Start with the large gray balls, the ones with the most holes. Make the bottom face of the unit cell by attaching four of these larger gray balls to a center ball of the same kind, using short (#6) bonds; use the holes in the *rectangular* faces. You should get the structure shown on the left in the following figure:

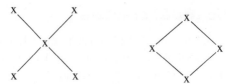

To make the middle layer of atoms, connect four of the same type of balls in a square, again using short bonds and the holes in *rectangular* faces, as in the sketch on the right. The top layer is made the same way as the bottom layer, and again looks like the sketch on the left. Connect these layers to one another, again using holes in the *rectangular* faces. *All* of the rectangular holes that face each other are the same distance apart, so they should *all* be connected by a short (#6) bond. There are quite a few bonds in the final cell: <u>32</u> in all. Put them all in. You should then have an FCC unit cell.

The FCC, or cubic close-packed, structure is a common one. Among the metals with this type of crystal structure at room temperature are copper, silver, nickel, and calcium.

In close-packed structures the coordination number has the largest possible value. It can be found by looking at a face-centered atom, say on the top face of the cell. That atom has eight bonds to it and would also have bonds to atoms in the cell immediately above: four of them. So its coordination number, like that of every atom in the cell, is <u>12</u>.

Having seen how to deal with other structures, you should now be able to find the number of atoms in the FCC unit cell and the radius of an atom as related to d_o. From these relationships and the measured density of a chosen metal, you can calculate how many atoms are in a mole. On the report page, find these quantities using copper metal, which has an FCC unit cell edge d_o equal to 0.361 nm and a density of 8.92 g/cm^3.

In the center of the FCC unit cell there is a hole. It is smaller than a cubic hole, but of significant size nonetheless. It is called an <u>octahedral hole</u>, because the six atoms around it define an octahedron (an eight-sided polygon). In an ionic crystal, the anions (negative ions) often occupy the sites of the atoms in an FCC unit cell, and there is a cation (positively charged ion) in the octahedral hole. On the report page, calculate the maximum radius, r_+, of the cation that would just fit in an octahedral hole surrounded by anions of radius r_-. You should find that the r_+/r_- ratio turns out to be 0.414.

There is another kind of hole in the FCC lattice that is important. See if you can find it; there are eight of them in the unit cell. If you look at an atom on the corner of the cell, you can see that it is at one corner of a tetrahedron (a triangular pyramid) of atoms. In the center of the tetrahedron there is a <u>tetrahedral hole</u>, which is small compared to an octahedral or cubic hole. In some ionic crystals small cations are found in some or all of the tetrahedral holes, again in a close-packed anion lattice. The cation–anion radius ratio at which cations just fit into a tetrahedral hole is challenging to calculate: r_+/r_- turns out to be 0.225.

The close-packed layers of atoms in the FCC structure are not parallel to the unit cell faces, but instead perpendicular to the cell diagonal. If you look down the cell diagonal, you see six atoms in a close-packed triangle in the layer immediately below the corner atom, and another layer of close-packed atoms below that, followed by another corner atom. The layers are indeed close-packed, and, as you go down the diagonal of this and succeeding cells, the layers repeat their positions in the order ABCABC...—meaning that atoms in every fourth layer lie directly below one another.

However, there is one other way such layers can stack and still be close-packed. The first and second layers remain in the same relative positions, but the third layer can be positioned directly below the first one if it is shifted properly. This results in a close-packed structure in which the arrangement of the layers is ABABAB...—with every other layer directly above the other. The crystal obtained from this arrangement of layers is not cubic, but hexagonal. It too is a common structure for metals. Cadmium, zinc, and manganese adopt this crystal structure at room temperature. The stability of this structure is very similar to that of FCC crystals. Just changing the temperature often converts a metal from one of these crystal structures to the other. Calcium, for example, is FCC (cubic close-packed) at room temperature, but if heated to 450 °C it converts to hexagonal close-packed.

E. Crystal structures of some common binary salts

All of the possible cubic crystal structures for metals (single-element unit cells) have now been discussed. The structures of many binary ionic compounds are closely related to these metal structures. In many ionic crystals the anions (negatively charged ions)—which are larger than cations (positively charged ions)—are essentially in contact with each other, arranged as in either an SC or FCC structure. The cations go into cubic, octahedral, or tetrahedral holes, depending on the cation–anion radius ratios previously calculated. The cation will tend to go into the hole type in which it will not *quite* fit. This increases the unit cell size from the value it would have if the anions were touching, reducing the repulsion energy due to anion–anion interactions and maximizing the cation–anion attraction energy, thereby producing the most stable possible crystal structure. According to the so-called <u>radius-ratio rule</u>, large cations go into cubic holes, smaller ones into octahedral holes, and the smallest ones into tetrahedral holes.

The deciding factor for which hole is favored is given by the radius ratio:

If $r_+/r_- > 0.732$	cations go into cubic holes
If $0.732 > r_+/r_- > 0.414$	cations go into octahedral holes
If $0.414 > r_+/r_- > 0.225$	cations go into tetrahedral holes

1. The sodium chloride (NaCl) crystal

To apply the radius-ratio rule to sodium chloride, NaCl, first calculate the r_+/r_- ratio using the data in Table 16.1, on the first report page. Because $r_{Na^+} = 0.095$ nm, and $r_{Cl^-} = 0.181$ nm, the radius ratio is $0.095/0.181$, or 0.525. That value is less than 0.732 and greater than 0.414, so the sodium ions should go into octahedral holes. We saw the octahedral hole in the center of the FCC unit cell, so Na^+ ions go there, and the Cl^- ions have the FCC structure. Actually, there are 12 other octahedral holes associated with the cell, one on each edge, which would have been apparent if you had been able to make more cells. An Na^+ ion goes into each of these holes, giving the NaCl structure shown on the report page. (Only the ions and bonds on the exposed faces are shown.)

Make a model of the unit cell for NaCl, using the large gray balls with lots of holes for the Cl^- ions and the smaller blue balls for Na^+ ions. While the Na^+ ion at the center of the cell is clearly in an octahedral hole, so are all of the other sodium ions: if you extend the lattice (the crystal structure), every Na^+ ion will be surrounded by six Cl^- ions. The coordination number of each Na^+ ion in NaCl is six. The coordination number of the Cl^- ions in NaCl is also six. The NaCl crystal is FCC in Cl^-, but also in Na^+ because you could exchange the positions of the Na^+ and Cl^- ions, putting the Na^+ ions on the corners of the unit cell, and maintain the same structure. The unit cell extends from the center of one Cl^- ion to the center of the next Cl^- along the cell edge—or, from the center of one Na^+ ion to the center of the next Na^+ ion.

On the report page, calculate the number of Na^+ and of Cl^- ions in the unit cell.

2. The cesium chloride (CsCl) crystal

Like NaCl, cesium chloride, CsCl, is a binary salt with a 1:1 cation:anion ratio. The Cs^+ ion, however, is larger than Na^+, having a radius of 0.169 nm. This makes r_+/r_- equal to 0.933. Because this value is greater than 0.732, the Cs^+ ions fill cubic holes in CsCl. The structure of CsCl looks like that of the BCC unit cell you made earlier, except that there is a Cs^+ ion in the center and Cl^- ions on the corners. If you put small gray balls in the center of each BCC unit cell you made from the blue balls, you would have the CsCl structure. This structure is *not* BCC: because the corner and center atoms are not the same. It consists of two interpenetrating simple cubic (SC) lattices, one made from Cl^- ions and the other from Cs^+ ions.

3. The zinc sulfide (ZnS) crystal

Zinc sulfide is another 1:1 binary salt, but its crystal structure is not that of either NaCl or CsCl. In ZnS, the Zn^{2+} ions have a radius of 0.074 nm and the S^{2-} ions a radius of 0.184 nm, making r_+/r_- equal to 0.402. The radius-ratio rule predicts ZnS should have close-packed S^{2-} ions, with the Zn^{2+} ions in tetrahedral holes. This proves to be correct: in the ZnS unit cell the S^{2-} ions are FCC, and alternate tetrahedral holes in the unit cell are occupied by Zn^{2+} ions—which themselves form a tetrahedron.

To make a model for ZnS, first assemble a (large) SC unit cell, using blue balls and long (#17) bonds placed in their *square* holes. Using small gray balls (g) for the zinc ions, assemble the unit shown in the following figure, using short (#6) bonds and placing them in the *triangular* holes on either side of the same square hole in two gray balls and one new (additional) blue ball (b):

Attach this unit to the *triangular* holes on the corner atoms of a diagonal of the bottom face of the large SC unit cell, such that the gray atoms are inside the unit cell and the blue one is positioned at the center of the bottom face (making that an FCC face). Make another copy of the same unit and attach it to the two corner atoms in the top face of the large SC unit cell, on the face diagonal that is *not* parallel to the bottom face diagonal you chose. The gray balls will again be inside the unit cell, and the blue ball will end up positioned at the center of the *top* face, making it FCC. The gray balls should then form a tetrahedron. Now attach four more blue balls to the gray ones, using two short (#6) bonds for each, placed in *triangular* holes on opposite sides of the same square hole on the blue balls and inserted into *triangular* holes on the gray balls such that each new blue ball is positioned at the center of one of the unit cell faces. When you are done, the blue atoms will form a complete FCC lattice. In this structure the coordination number is 4 (the long bonds do not count: they just keep the unit cell from falling apart). If the gray balls in the unit cell are replaced with blue ones, so that all atoms are of the same element, the lattice obtained is called the diamond crystal structure. (And yes, it is the crystal structure of diamond, which is made of pure carbon!)

You are now familiar with the three common cubic structures of 1:1 binary compounds. The radius-ratio rule predicts which structure a given compound will have. It does not always work, but it is correct most of the time. On the report page, use the radius-ratio rule to predict the cubic structures that crystals of the following substances will have: KI, CuBr, and TlBr.

4. The calcium fluoride (CaF$_2$) crystal

Calcium fluoride, CaF_2, is a 1:2 binary compound, so it cannot have the structure of any crystal discussed so far. The radii of Ca^{2+} and F^- are 0.099 and 0.136 nm, respectively, so r_+/r_- is 0.727. This places CaF_2 on the boundary between compounds with cations in cubic holes and those with cations in octahedral holes. It turns out that in CaF_2, the F^- ions have a simple cubic structure, with half of the cubic holes filled by Ca^{2+} ions. This produces a crystal in which the Ca^{2+} ions lie in an FCC lattice, with 8 F^- ions in a cube inside the unit cell.

Use your model set to make a unit cell for CaF_2, with small gray balls (g) representing the Ca^{2+} ions and red balls (r) representing F^-. Start by building a single unit as shown in the following diagram, again using short (#6) bonds to connect *triangular* holes on opposite sides of the same square hole:

To complete the bottom face of the unit cell, attach two red and two small gray balls to the center gray ball of this unit, forming an identical unit that is perpendicular to the first and has the center gray ball in common. If done correctly, the four red balls will form a square and the gray ones an FCC face. The top face of the unit cell is made the same way. Attach the two faces through four small gray balls, which lie at the centers of the side faces and have four short bonds in them, one in each *triangular* hole around the same square hole. When you are done, the red balls should form a cube inside an FCC unit cell made of gray balls. Extend this model to one more unit cell to show that every other cube of F^- ions has a Ca^{2+} ion at its center.

When you have completed this part of the experiment, your instructor may assign you a crystal structure of your own to work on. Report your results as directed. Then disassemble the models you made and pack the components back up the way they came to you.

Experiment 16

Data and Calculations: The Structure of Crystals

Table 16.1

Atom	Molar mass {g/mol}	Atomic radius {nm}	Ionic radius {nm}
Na	22.99	calculated on	0.095
Cu	63.55	following page	0.096
Cl	35.45	0.099	0.181
Cs	132.9	0.262	0.169
Zn	65.38	0.133	0.074
S	32.06	0.104	0.184
K	39.10	0.231	0.133
I	126.9	0.133	0.216
Br	79.90	0.114	0.195
Tl	204.4	0.171	0.147

Some cubic unit cells

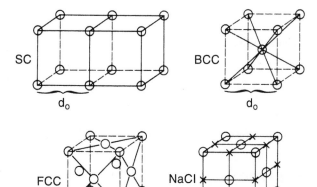

A. Simple cubic (SC) crystals

Fraction of volume of unit cell occupied by atoms

Volume of unit cell in terms of d_o _____

Number of atoms per unit cell ===============

Radius r of atom in terms of d_o _____

Volume of atom in terms of d_o $\left(V = \dfrac{4\pi}{3} r^3\right)$ _____

(Volume of atom)/(volume of unit cell) _____ = _____

            (fraction)   (decimal equivalent)

(continued on following page)

B. Body-centered cubic (BCC) crystals

Radius of a sodium atom, Na

In BCC, atoms touch along the cube diagonal (consult your model and the Advance Study Assignment/ASA)
Length of cube diagonal if d_o = 0.429 nm

_____ nm

$\underline{4} \times r_{Na}$ = length of cube diagonal

r_{Na} = _____ nm

Density of sodium metal, Na(s)

Unit cell side length, d_o, in cm (1 cm = $\underline{10^7}$ nm)   _____cm

Volume of unit cell, V _____cm^3

Number of atoms per unit cell   ═══════════

Number of unit cells per mole of Na   _____ mol^{-1}

Mass of a single unit cell, m _____ g

Calculated density of sodium metal, m/V (Observed value = 0.97 g/cm^3)   _____ g/cm^3

D. Face-centered cubic (FCC) crystals

Number of atoms per unit cell

Number of atoms on corners _____, shared by _____ cell(s) → _____ × _____ = _____
                                                                                          (fraction)

Number of atoms on faces _____, shared by _____ cell(s) → _____ × _____ = _____
                                                                                          (fraction)

Total atoms per unit cell _____

Radius of a copper atom, Cu

In FCC, atoms touch along the face diagonal (see your model and the Advance Study Assignment/ASA)
Length of face diagonal in Cu, where d_o = 0.361 nm

_____ nm

Number of Cu atom radii on face diagonal _____
Radius of a copper atom, r_{Cu}

r_{Cu} = _____ nm

(continued on following page)

The number of atoms in a mole, N

Length of cell edge, d_o, in cm, in Cu metal   _____ cm

Volume of a unit cell in Cu metal   _____ cm^3

Volume of a mole of Cu metal (V = mass/density)   _____ cm^3

Number of unit cells per mole Cu   _____ mol^{-1}

Number of atoms per unit cell   _____

Number of atoms per mole, $N_{calculated}$ _____ mol^{-1}

Size of an octahedral hole

This is a sketch of one face of NaCl's unit cell. The ions are drawn so that they are just touching their nearest neighbors. The anions are FCC. The cations, shaded in gray, are in octahedral holes.

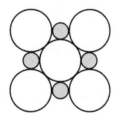

Show the length of d_o on the sketch. (Your model can help you identify the unit cell edges!)

What is the relationship between r_- and d_o?

$$r_- = \underline{\hspace{2cm}} \times d_o = \underline{\hspace{2cm}} \times d_o$$
$$\text{(fraction)} \qquad \text{(decimal equivalent)}$$

What is the equation relating r_+, r_-, and d_o?

$$d_o = \underline{\hspace{2cm}}$$

What is the relationship between r_+ and d_o?

$$r_+ = \underline{\hspace{2cm}} \times d_o$$

What is the value of the radius ratio r_+/r_-? Express the ratio to three significant figures.

$$r_+/r_- = \underline{\hspace{2cm}}$$

(continued on following page)

E. Crystal structures of some common binary compounds

1. The sodium chloride (NaCl) crystal

Number of Cl^- ions in the unit cell (FCC)   _____

Number of Na^+ ions on edges of cell _____, shared by _____ cell(s)

Number of Na^+ ions in center of cell _____, shared by _____ cell(s)

Total number of Na^+ ions in unit cell   _____

3. The zinc sulfide (ZnS) crystal

Applying the radius-ratio rule (necessary data is in Table 16.1, at the start of the report pages):

KI r_+/r_- _____ Predicted structure _____ Observed structure = NaCl

CuBr r_+/r_- _____ Predicted structure _____ Observed structure = ZnS

TlBr r_+/r_- _____ Predicted structure _____ Observed structure = CsCl

Report on individual crystal structure (if assigned)

Experiment 16

Advance Study Assignment: The Structure of Crystals

1. Many substances crystallize in a cubic structure. The unit cell for such crystals is a cube having an edge with a length equal to d_o.

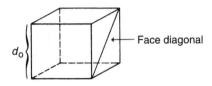

   a. What is the length, in terms of d_o, of the face diagonal, which runs diagonally across one face of the cube? (*Hint*: Use the Pythagorean theorem.)

   _____

   b. What is the length, again in terms of d_o, of the cube diagonal, which runs from one corner, through the center of the cube, to the opposite corner? (*Hint*: Make a right triangle having a face diagonal and an edge of the cube as its sides, with the hypotenuse equal to the cube diagonal, then use the Pythagorean theorem again.)

   _____

2. In an FCC structure, the centers of the atoms are found on the corners of the cubic unit cell and at the center of each face. The unit cell has an edge whose length is the distance from the center of one corner atom to the center of another corner atom on the same edge. The atoms on the diagonal of any face are touching. One of the faces of the unit cell is shown here:

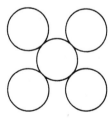

   a. Show the distance d_o on the sketch. Draw in the boundaries of the unit cell.

   b. What is the relationship between the length of the face diagonal and the radius of the atoms, r?

   Face diagonal = _____

(continued on following page)

c.   How is the radius of the atoms related to d_o?

$r =$ _____ $\times d_o$

d.   Platinum metal crystals have an FCC structure. The unit cell edge in platinum is 0.3932 nm long. What is the radius of a platinum atom, r_{Pt}?

$r_{Pt} =$ _____nm

Experiment 17

Classifying Chemical Substances

Depending on the kind of bonding present in a chemical substance, the substance may be called ionic, molecular, or metallic.

Ionic compounds are composed of ions: the large electrostatic forces between the positively and negatively charged ions are responsible for the ionic bonding that holds them together.

In a molecular substance the atoms are bound to each other by sharing electrons in covalent bonds. When the number of atoms connected to each other by covalent bonds is relatively small, the covalently bound units are called molecules. If huge numbers of atoms are interconnected by covalent bonds, the substance is called macromolecular.

Metals are characterized by their own special kind of bonding, in which the electrons are much freer to move than in other kinds of substances. Metallic bonding might be best characterized as relatively stationary atomic nuclei in a gas of free-flowing electrons attracted to the positively charged nuclei.

The terms "ionic", "molecular", "macromolecular", and "metallic" are generalizations, and some substances have properties that would place them in a borderline category, somewhere between one group and another. It is useful, however, to consider some of the general characteristics of typical ionic, molecular, macromolecular, and metallic substances, because many common substances can be clearly assigned to one of these categories.

Ionic Substances

Most ionic substances are solids at room temperature. They are all salts, consisting of positive and negative ions. They are typically crystalline but may exist as fine powders as well as clearly defined crystals. While many ionic substances are stable up to their melting points, some decompose when heated. It is common for an ionic crystal to release loosely bound waters of hydration at temperatures below 200 °C. Anhydrous (dehydrated) ionic salts typically have high melting points, usually above 300 °C but below 1000 °C. They are not easily turned into gases (technical term: volatilized) and boil only at very high temperatures (see Table 17.1).

When melted, ionic compounds will conduct an alternating (back and forth) electric current. In the solid state they do not conduct electricity. Their conductivity when melted is due to the ions' freedom to move.

Ionic substances are frequently, but not always, appreciably soluble in water. The solutions produced conduct alternating electric current well. The conductivity of a solution of even a slightly soluble ionic substance is often several times that of the solvent water. Ionic substances are usually not nearly so soluble in other liquids as they are in water. For a liquid to be a good solvent for ionic compounds it must be highly polar, containing molecules with well-defined positive and negative regions with which the ions can interact.

Molecular Substances

All gases and almost all liquids at room temperature are molecular in nature. If the molar mass of a molecular substance is greater than about one hundred grams per mole, it may be a solid at room temperature. The melting points of molecular substances are usually below 300 °C; these substances are relatively easy to turn into gases, but quite a few will decompose before they boil. Most molecular substances do not conduct electric current as either solids or liquids.

Organic compounds contain primarily carbon and hydrogen, often in combination with other nonmetals, and are essentially molecular in nature. Because there are a great many organic substances, it is true that most substances are molecular. If an organic compound decomposes on heating, what is left behind is frequently a black carbonaceous (carbon-containing) material. Reasonably large numbers of inorganic substances are also molecular; those that are solids at room temperature include some of the two-element (technical term: binary) compounds of elements from Groups 14, 15, 16, and 17.

Table 17.1

Physical Properties of Some Representative Chemical Substances						
	Melting point, M.P. {°C}	Boiling point, B.P. {°C}	Solubility in		Electrical conductivity	Classification
Substance			Water	Hexane		
NaCl	801	1413	Sol	Insol	High for melted solid and in soln	Ionic
MgO	2800	—	Sl sol	Insol	Low in sat'd soln	Ionic
CoCl$_2$	Sublimes	1049	Sol	Insol	High in soln	Ionic
CoCl$_2 \cdot 6H_2O$	86	Dec	Sol	Insol	High in soln	Ionic hydrate, loses H$_2$O at 110 °C
C$_{10}$H$_8$	70	255	Insol	Sol	Zero for melted solid	Molecular
C$_6$H$_5$COOH	122	249	Sl sol	Sl sol	Low in sat'd soln	Molecular-ionic
FeCl$_3$	282	315	Sol	Sl sol	High in soln and in melted solid	Molecular-ionic
SnI$_4$	144	341	Dec	Sol	~Zero in melted solid	Molecular
SiO$_2$	1600	2590	Insol	Insol	Zero in solid	Macromolecular
Fe	1535	3000	Insol	Insol	High in solid	Metallic

Key: "Sol" = soluble, having a solubility of at least 0.1 mole/L; "Sl sol" = slightly soluble, having significant solubility but less than 0.1 mole/L; "Insol" = (essentially) insoluble; "Dec" = decomposes (changes chemical form); "sat'd" = saturated; "soln" = solution

Molecular substances are usually soluble in at least a few organic solvents, with the solubility being enhanced if the substance and the solvent have similar molecular structures.

Some molecular compounds are quite polar, which tends to increase their solubility in water and other polar solvents. Such substances may ionize appreciably in water, or even as melted solids, so that they become conductors of electricity. Usually the conductivity of a molecular material is considerably lower than that of an ionic material, however. Most polar molecular compounds in this category are organic, but a few, including some transition metal compounds, are inorganic.

Macromolecular Substances

Macromolecular substances are solids at room temperature, and include diamond, silicon, glass, silicon carbide, proteins, and plastics. They are extremely difficult, if not impossible, to get into the gas phase. Inorganic macromolecular materials have very high melting points, usually above 1000 °C, and typically resist decomposition at high temperatures, while organic macromolecular substances often melt at much lower temperatures and decompose before or shortly after doing so. Macromolecular materials generally do not conduct electric current and are not easily dissolved in water or any organic solvents. They are frequently chemically inert, and inorganic macromolecular substances may be used as abrasives (very hard materials used for grinding other things) or refractories (materials that resist extremely high temperatures).

Metallic Substances

The unique properties of metals mostly come from the freedom of movement of their bonding electrons. Metals are good electrical conductors as both solids and liquids, reflect light, and are malleable (they reshape, rather than crack or break apart, when hit or bent as solids), at least to some extent. Most metals are solid at room temperature, but their melting points range widely, from below 0 °C to over 2000 °C. They are not soluble in water or organic solvents. Some metals are prepared as gray or black powders, and these may not appear to be electrical conductors. However, if you measure their ability to conduct electricity with the powder under pressure, such that many of the particles touch one another, their metallic character is revealed.

Experimental Procedure

In this experiment you will investigate the properties of several substances with the purpose of determining whether they are molecular, ionic, macromolecular, or metallic. In some cases, the classification will be clear. In others, you may find that the substance exhibits characteristics associated with more than one class.

As the earlier discussion indicates, there are several properties that can be used to find out which class a substance belongs to. In this experiment you will use the melting point; the solubility in water and organic solvents; and the electrical conductivity of the <u>aqueous</u> (water-as-the-solvent) solution, the solid, and the melted solid.

The substances to be studied in the first part of this experiment are on the laboratory tables along with two organic solvents: one polar and one nonpolar. You need only carry out enough tests on each substance to establish the class to which it belongs, so you will not need to perform every test on every substance. You may, however, carry out any extra tests you wish. Follow the directions for each test, recording your findings in Table 17.2 on the report page.

A. Melting point

Approximate melting points of substances can be determined by heating a small amount of the substance (a sample the size of a pea) in a test tube. Substances with low melting points, less than 100 °C, will easily melt when warmed gently over a laboratory burner flame in a small (13 × 100 mm) test tube, or when the tube is immersed in boiling water. If the sample melts between 100° and 300 °C, it will require more than gentle warming over a laboratory burner flame, but it will melt before the test tube imparts a yellow-orange color to the flame. Above 300 °C, the flame will take on an increasingly yellow-orange color; up to 500 °C you can still use a (true, laboratory-grade) test tube and a strong burner flame, but at 550 °C even the special borosilicate glass that laboratory test tubes are made of will begin to soften. In this experiment you should not attempt to measure any melting points above 500 °C. Stop heating as soon as the test tube itself begins to glow (you will need to take it out of the flame periodically to check; even the slightest glow from the glass indicates you are over 500 °C).

While heating a sample, keep the tube **loosely** stoppered with a cork. **Do not breathe** any gases that are given off, and **do not continue to heat** a sample after it has melted. Keep the tube tilted, with only the bottom in the flame, so that the upper parts of the tube remain cool. As you heat the sample, watch for and document any evidence of decomposition, sublimation (transition into the gas phase without melting first; a transition back to the solid form on the cooler parts of the test tube is the giveaway), or the release of water (indicated by liquid droplets condensing on the cool parts of the test tube).

B. Solubility and conductivity of solutions

In testing for solubility, again use a sample about the size of a pea, this time in a medium (18 mm × 150 mm) test tube. Use about two milliliters of solvent, enough to fill the tube to a depth of a bit over one centimeter from the bottom of the tube. Stir well, using a clean stirring rod. Some samples will dissolve completely almost immediately, some are only slightly soluble and may produce a cloudy suspension, and others are completely insoluble. Test solubility in deionized water and the two organic solvents provided, and record your results. Use fresh deionized water in your wash bottle for the aqueous solubility tests.

You should measure the electrical conductivity of the aqueous solutions using a portable ohmmeter. Your instructor may take the measurements for you, or may have you take them. An <u>ohmmeter</u> measures the electrical resistance of a sample in ohms, Ω. A solution with a high resistance has low electrical conductivity. Some of your solutions will have a low resistance, on the order of 1000 Ω or less; these are good conductors. Deionized water has a high resistance; with your meter it will probably have a resistance of 50,000 Ω or greater. Be warned that small amounts of contaminants can lower the resistance of a solution well below where it "should" be for a given <u>solute</u> (dissolved substance), so be sure to rinse your stirring rod, the ohmmeter probes, and anything else that moves from one solution to the next.

Measure the resistance of any of the aqueous solutions that contain soluble or slightly soluble substances. Between tests, rinse the electrodes in a beaker filled with deionized water. You should consider a solution with a resistance of less than 2000 Ω to be a good conductor, denoted "G". Between 2000 and 20,000 Ω it is a weak conductor, denoted "W". Above 20,000 Ω, or if the ohmmeter reads "OL" (overload), consider the solution to be essentially nonconducting, and denote it with an "N". Record the resistances you observe, in ohms (Ω). Then note, with a G, W, or N, whether the solution is a good, weak, or poor conductor.

C. Electrical conductivity of solids and melted solids

Some substances conduct electricity in the solid state. If a sample contains large crystals, you can test their conductivity by selecting a crystal, putting it on the lab bench, and touching it in two different places with the tips of the ohmmeter probes. (The crystal must be large enough for you to do this without the probes touching one another.) Metals have a very low resistance. In powder form, however, most substances—including metals—appear to have essentially infinite resistance. The difference is that when pressed together, metal powders—unlike those of other substances—show good conductivity. To test a powder for conductivity, put a penny on the lab bench. Then place a small rubber washer on it. Fill the hole in the washer with the powder, and put another penny on top of the washer. Put the whole sandwich between the jaws of a pair of pliers with insulated jaws. Touch the electrodes from the ohmmeter probe to the pennies, one electrode to each penny, and squeeze the pliers. If the powder is a metal, the resistance will fall gradually, from overload to a small value, as you squeeze the pliers. Make sure any drop in resistance is not caused by the pennies touching each other (which results in a sudden drop to nearly zero resistance). Record your results.

To check the conductivity of a melted solid, put a pea-size sample in a dry, medium test tube and melt it. Heat the electrodes on the probe for a few seconds in the burner flame and touch them to the melted solid. Heat gently to ensure that no solid has formed on the electrodes. Many melted solids are good conductors. After testing a melted solid, clean the electrodes by washing them with water or an organic solvent, or, if necessary, scraping them off with a spatula.

Having made the tests just described, you should be able to assign each substance to its class, or, possibly, to one or both of two classes. Make this classification for each substance, and give your reasons for doing so.

When you have classified each substance, report to your laboratory instructor, who will assign you two unknowns for characterization.

Disposal of reaction products: All materials from your tests should be discarded in a waste container unless your instructor directs otherwise.

Experiment 17

Observations and Analysis: Classifying Chemical Substances

Table 17.2

Substance ID code	Approximate melting point {°C} (< 100, 100–300, 300–500, > 500)	Solubility, by solvent			Electrical resistance {Ω}			Classification and reason
		H₂O	Nonpolar organic	Polar organic	Solution in H₂O	Solid	Melted solid	
A								
B								
C								
D								

Key: Sol = soluble; Sl sol = slightly soluble; Insol = insoluble; G = good electrical conductor, $R < 2000\ \Omega$; W = weak electrical conductor, $2000\ \Omega < R < 20{,}000\ \Omega$; N = nonconductor, $R > 20{,}000\ \Omega$.

Table 17.2 (*continued*)

Substance ID code	Approximate melting point {°C} (< 100, 100–300, 300–500, > 500)	Solubility, by solvent			Electrical resistance {Ω}			Classification and reason
		H₂O	Nonpolar organic	Polar organic	Solution in H₂O	Solid	Melted solid	
E								
F								
Unknown ID code								
_____								
_____								

Key: Sol = soluble; Sl sol = slightly soluble; Insol = insoluble; G = good electrical conductor, $R < 2000\ \Omega$; W = weak electrical conductor, $2000 < R < 20{,}000\ \Omega$; N = nonconductor, $R > 20{,}000\ \Omega$.

Experiment 17

Advance Study Assignment: Classifying Chemical Substances

1. List the properties of a substance that would strongly indicate it to be a macromolecular material.

2. If substances are classified as ionic, molecular, macromolecular, or metallic, in which (if any) categories are the members generally

    a. soluble in water?

    b. insoluble in all common solvents?

    c. solids at room temperature?

    d. electrical conductors as melted solids?

3. A given substance is a white, granular solid at 25 °C that does not conduct electricity. It melts at 250 °C, then slowly chars to a black solid above 390 °C, never entering the gas phase. When melted it does not conduct electricity. It does not easily dissolve in any solvent. What would the proper classification of the substance be, based on this information? Explain your logic.

4. A gray-white solid melts at a bit above 1650 °C. It is electrically conductive as both a solid and a liquid, but not soluble in either water or any organic solvent. Classify the substance as best you can from these properties, explaining your logic.

Experiment 18

Some Nonmetals and Their Compounds— Preparations and Properties

Some of the most commonly encountered chemical substances are nonmetallic elements or their two-element (technical term: <u>binary</u>) compounds. O_2 and N_2 in the air, CO_2 exhaled by animals, and H_2O in rivers, lakes, and air are typical of such substances. Substances containing nonmetallic atoms are all composed of molecules (technical term: <u>molecular</u>), reflecting the covalent bonding that holds their atoms together. They are often gases at room temperature, due to weak intermolecular forces. However, with high molecular masses or hydrogen bonding they can be liquids, such as H_2O and Br_2, or solids, such as I_2 and graphite.

Several of the common nonmetallic elements and some of their gaseous compounds can be prepared by uncomplicated reactions. In this experiment you will prepare some typical examples of such substances and examine a few of their characteristic properties.

Experimental Procedure

Wear your safety glasses while performing this experiment.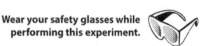

In several of these experiments you will prepare gases. Prepare for this carefully and report what you actually observe, as there may be a few surprises!

There are several tests you will conduct on the gases you prepare, and the way each of these should be carried out is explained in the following instructions.

Test for odor

To determine the odor of a gas, first try moving your hand, with your palm facing your nose and your fingers horizontal, toward you over the top of the test tube with the gas in it (technical term: waft). If you do not detect an odor after doing that a few times, sniff near the end of the tube, first at some distance and then gradually closer. **Do not just put the tube under your nose immediately and inhale deeply!** Some of the gases you will make have no odor, and some will have very potent ones. Some of the gases are quite toxic, and even though you will be making only small amounts, caution in testing for their odor is very important.

CAUTION: The following gases prepared in this experiment are toxic: Br_2, SO_2, NO_2, NH_3, and H_2S. **Do not inhale these gases unnecessarily when testing their odors. One small sniff will be sufficient and will not be harmful. It is good for you to know these odors, in case you encounter them in the future.**

Test for support of combustion

A few gases will support combustion, but most will not. To test for this, ignite a wood splint, blow out the flame, and put the glowing—but not burning—splint into the gas in the test tube. If the gas supports combustion, the splint will glow more brightly, or it may make a small popping noise. If the gas does not support combustion, the splint will go out almost instantly. You can use the same splint for multiple combustion tests.

Test for acid-base properties

Many gases are acids. This means that if the gas is dissolved in water it will increase the concentration of H^+ ions. A few gases are bases; in <u>aqueous</u> (water-as-the-solvent) solution such gases decrease the concentration of H^+ ions and increase the concentration of OH^- ions. Other gases do not react with water, and are neutral. The acidic or basic nature of a gas can be determined using a chemical indicator. One of the most common acid-base indicators is litmus, which is red in acidic solution and blue in basic solution. To test whether a gas is an acid, moisten a piece of blue litmus paper with water from your wash bottle and lower the paper down into the test tube in which the gas is present. If the gas is an acid, the paper will turn red. Similarly, to test if a gas is a base, moisten a piece of red litmus paper and lower it into the tube. A color change to blue will occur if the gas is basic. Because you may have used an acid or a base in making the gas, do not touch the walls of the test tube with the paper. The color change will occur fairly quickly and evenly over the surface of the paper if the gas is acidic or basic. It is not necessary to use a new piece of litmus for each test. Start with a piece of blue and a piece of red litmus. If you need to (re-)generate blue litmus paper, hold moist red paper over an open bottle of 6 M NH_3, a gas that is basic. If you need to (re-)generate red litmus, hold moist blue paper in the acidic gas above an open bottle of 6 M acetic acid.

A. Preparation and properties of nonmetallic elements: O_2, N_2, Br_2, I_2

1. Oxygen, O_2

Oxygen can be conveniently prepared by the decomposition of hydrogen peroxide, H_2O_2, in aqueous solution. Hydrogen peroxide is chemically unstable and will break down (technical term: <u>decompose</u>) into water and oxygen gas upon adding a suitable <u>catalyst</u> (a substance that increases the rate of a chemical reaction without being chemically altered itself), such as the Fe^{3+} ion:

$$2\ H_2O_2(aq) \xrightarrow{\ Fe^{3+}\ } O_2(g) + 2\ H_2O(\ell) \tag{1}$$

Add 5 mL of $3\%_{mass}$ aqueous H_2O_2 to a small (13×100 mm) test tube. (This will be a depth of about five centimeters in such a tube.) Add 5 drops of concentrated $Fe(NO_3)_3$ solution to the test tube and swirl. You may wish to hold your finger lightly over the end of the tube to help confine the O_2 that is generated. Test the product gas for odor. Test the gas for any acid-base properties, using moist blue and red litmus paper. Test the gas for support of combustion. Record your observations.

2. Nitrogen, N_2

Sodium sulfamate in aqueous solution, $NH_2SO_3^-(aq)$, will react with nitrite ion, $NO_2^-(aq)$, to produce nitrogen gas:

$$NO_2^-(aq) + NH_2SO_3^-(aq) \rightarrow N_2(g) + SO_4^{2-}(aq) + H_2O(\ell) \tag{2}$$

Add about one milliliter of 1.0 M potassium nitrite, KNO_2, to a small test tube. Then add about eleven drops of 0.5 M sodium sulfamate, $NaNH_2SO_3$, and place the test tube in a hot-water bath made from a 250-mL beaker half full of water. Bubbles of nitrogen should form within a few moments. Confine the gas for a few seconds with a stopper. Test the product gas for odor. Carry out the tests for acid-base properties and for support of combustion. If you need to generate more N_2 to complete the tests, add sodium sulfamate solution as necessary, about seven drops at a time. Report your observations.

3. Iodine, I_2

The halogen elements are most easily prepared from their sodium or potassium salts. The reaction involves an <u>oxidizing agent</u> that removes electrons from the halide ions, generating the halogen. The reaction that occurs when a solution of potassium iodide, KI, is treated with 6 M HCl and a little manganese dioxide, MnO_2 (the oxidizing agent), is typical:

$$2\ I^-(aq) + 4\ H^+(aq) + MnO_2(s) \rightarrow I_2(aq) + Mn^{2+}(aq) + 2\ H_2O(\ell) \tag{3}$$

At 25 °C, I_2 is a solid, with relatively low solubility in water and a tendency to enter the gas phase. I_2 can be extracted from aqueous solution into organic solvents, particularly heptane, C_7H_{16} ("HEP" for short). The solid, its gas, and solutions of iodine in both water and HEP all have characteristic colors.

Put 2 drops of 1.0 M KI into a *medium* (18×150 mm) test tube. Add 6 drops of 6 M HCl and a small amount of solid manganese dioxide, $MnO_2(s)$—about the size of a small pea. Swirl the mixture and note any changes that occur. Put the test tube into a hot-water bath. After a minute or two, a noticeable amount of I_2

gas should be visible above the liquid. Remove the test tube from the water bath and add 10. mL of deionized water. Stopper the tube and shake. Note the color of I_2 in the aqueous solution. **Cautiously** test for the odor of I_2. Pour off the liquid into another medium test tube, leaving the solids behind, and add 3 mL of heptane, C_7H_{16}, to the liquid. Stopper and shake the tube. Observe the color of the HEP layer (on top) and the relative solubility of I_2 in water (the bottom layer) and HEP (on top). (Where is the color?) Record your observations.

4. Bromine, Br_2

Bromine can be made by the same reaction used to make iodine, substituting bromide ion for iodide. Bromine at 25 °C is a liquid. Br_2 can also be extracted from aqueous solution into heptane. Bromine's liquid, gas, and solutions in water and HEP are colored.

Put 3 drops of 1.0 M NaBr in a *medium* (18 × 150 mm) test tube and add 6 drops of 6 M HCl and a small amount of solid MnO_2—about the size of a small pea. Swirl to mix the contents of the tube, and observe any changes. Heat the tube in a hot-water bath for a minute or two. Try to detect Br_2 gas above the liquid by observing its color. Add 10. mL of water to the tube, stopper, and shake. Note the color of Br_2 in the aqueous solution. **Cautiously** test for the odor of Br_2. Pour off the liquid into another medium test tube, leaving the solid behind. Add 3 mL of HEP, stopper, and shake. Note the color of Br_2 in HEP (the top layer) and in which solvent Br_2 appears to be more soluble. Record your observations.

B. Preparation and properties of some nonmetallic oxides: CO_2, SO_2, NO, and NO_2

1. Carbon dioxide, CO_2

Carbon dioxide, like several other nonmetallic oxides, can be made by treating an oxygen-enriched ionic version of it (technical term: oxyanion) with an acid. With carbon dioxide the ion is carbonate, CO_3^{2-}, which is present in solutions of carbonate salts such as sodium carbonate, Na_2CO_3. The reaction is

$$CO_3^{2-}(aq) + 2\,H^+(aq) \rightarrow H_2CO_3(aq) \rightarrow CO_2(g) + H_2O(\ell) \qquad \textbf{(4)}$$

Carbon dioxide is not very soluble in water, and upon being made acidic, carbonate solutions will tend to bubble (technical term: effervesce) as CO_2 is released into the gas phase. Nonmetallic oxides in solution are often acidic, but never basic.

To about one milliliter of 1.0 M Na_2CO_3 in a small test tube add 6 drops of 3 M sulfuric acid, H_2SO_4. Test the gas for odor, acidic properties, and ability to support combustion.

2. Sulfur dioxide, SO_2

Sulfur dioxide can be prepared by making a solution of sodium sulfite, containing sulfite ion, SO_3^{2-}, acidic. The reaction that occurs is very similar to Reaction 4:

$$SO_3^{2-}(aq) + 2\,H^+(aq) \rightarrow H_2SO_3(aq) \rightarrow SO_2(g) + H_2O(\ell) \qquad \textbf{(5)}$$

Sulfur dioxide is considerably more soluble in water than carbon dioxide is. This means that it is much more likely for all of the SO_2 produced in this reaction to dissolve into the solution; however, some SO_2 gas may be seen bubbling out of concentrated sulfite solutions when they are made acidic, and this bubbling will increase if the solution is heated.

To about one milliliter of 1.0 M Na_2SO_3 in a small test tube add 6 drops of 3 M sulfuric acid, H_2SO_4. **Cautiously** test the product gas for odor. Test its acidic properties and its ability to support combustion. Put the test tube into a hot-water bath for a few seconds to see if bubbling is observed when the solution is hot.

3. Nitrogen dioxide, NO_2, and nitric oxide, NO

If a solution containing nitrite ion, NO_2^-, is treated with acid, two oxides are produced: NO_2 and NO. In solution these gases are combined in the form of N_2O_3, which is colored. When these gases come out of solution, they do so as a mixture of NO and NO_2. NO_2 is colored, while NO is colorless but easily reacts with oxygen in the air to form NO_2. The preparation reaction is

$$2 \, NO_2^-(aq) + 2 \, H^+(aq) \rightarrow N_2O_3(aq) + H_2O(\ell) \rightarrow NO(g) + NO_2(g) + H_2O(\ell) \qquad \textbf{(6)}$$

To about one milliliter of 1.0 M KNO_2 in a small test tube, add 6 drops of 3 M sulfuric acid, H_2SO_4. Swirl the mixture for a few seconds and note the color of the solution. Warm the tube in a hot-water bath for a few seconds to increase the rate of gas release. Note the color of the gas that is given off. **Cautiously** test the odor of the gas. Test its acidic properties and its ability to support combustion. Record your observations.

C. Preparation and properties of some nonmetallic hydrides: NH₃, H₂S

1. Ammonia, NH₃

A solution of 6 M ammonia, NH_3, in water is a common laboratory chemical. You may have already used it in this experiment to make your litmus paper turn blue. Ammonia gas can be released from solution by heating a strong solution of it, such as 6 M $NH_3(aq)$. It can also be prepared by treating a solution of an ammonium salt, such as NH_4Cl, which contains the ammonium ion, NH_4^+, with base. On treatment with a source of OH^- ion, such as a solution of NaOH, the following reaction occurs:

$$NH_4^+(aq) + OH^-(aq) \rightarrow NH_3(aq) + H_2O(\ell) \rightarrow NH_3(g) + H_2O(\ell) \qquad \textbf{(7)}$$

The odor of NH_3 is distinctive, and strong. NH_3 is very soluble in water, so bubbling is not observed, even on heating concentrated solutions.

Add about one milliliter of 1.0 M NH_4Cl to a small test tube. Add about one milliliter of 6 M sodium hydroxide, NaOH. Swirl the mixture and **cautiously** test the odor of the product gas. **WARNING: Inhaling too much ammonia at once can knock you unconscious!** Put the test tube into a hot-water bath for a few moments to increase the amount of NH_3 in the gas phase, then test the gas with moistened blue and red litmus paper, and for support of combustion.

2. Hydrogen sulfide, H₂S

Hydrogen sulfide can be made by treating some solid sulfides, particularly iron(II) sulfide, FeS, with an acid such as hydrochloric acid, HCl, or sulfuric acid, H_2SO_4. With FeS the reaction is

$$FeS(s) + 2 \, H^+(aq) \rightarrow H_2S(g) + Fe^{2+}(aq) \qquad \textbf{(8)}$$

This reaction was used for many years to make H_2S in laboratory in courses in qualitative analysis. In this experiment you will instead use the method currently used in such courses for H_2S generation. It involves the decomposition (chemical breaking apart) of thioacetamide, CH_3CSNH_2, which occurs in hot acid. The reaction is

$$CH_3CSNH_2(aq) + 2 \, H_2O(\ell) \rightarrow H_2S(g) + CH_3COO^-(aq) + NH_4^+(aq) \qquad \textbf{(9)}$$

The odor of H_2S is notorious, and very strong. H_2S is moderately soluble in water, and because it is produced reasonably slowly in Reaction 9 you will observe little—if any—bubbling.

To about one milliliter of 1.0 M thioacetamide in a small test tube add 6 drops of 3 M sulfuric acid, H_2SO_4. Put the test tube in a boiling-water bath for about one minute. You may observe some cloudiness due to the formation of solid elemental sulfur in a side reaction. **Carefully** smell the gas in the tube; H_2S is toxic! Test the gas for acidic and basic properties and for the ability to support combustion.

Optional D. Identifying an unknown solution

In this experiment you prepared nine different substances composed of nonmetallic elements. In each case the source of the substance was in solution. In this part of the experiment you will be given an unknown solution that can be used to make one of the nine substances you studied in this experiment. It will be a solution used in preparing one of them, but it will not be an acid or a base. Identify by suitable tests the substance that can be made from your unknown and the chemical that is present in it. ▪

Disposal of reaction products: Most of the chemicals used in this experiment can be discarded down the sink drain. Pour the solutions from Sections 3 and 4 of Part A, containing I_2 and Br_2 in heptane (HEP), into a waste container, unless your instructor directs otherwise.

Name _____   **Section** _____

Experiment 18

Observations and Analysis: Some Nonmetals and Their Compounds

A. Preparation and properties of nonmetallic elements: O_2, N_2, Br_2, I_2

Element prepared	Amount of bubbling	Odor	Acid–base tests	Supports combustion
1. O_2	_____	_____		_____
2. N_2	_____	_____		_____

		Color	
	Odor	As a gas	In H_2O	In HEP
3. I_2	_____	_____	_____
4. Br_2	_____	_____	_____

B. Preparation and properties of some nonmetallic oxides: CO₂, SO₂, NO, and NO₂

	Oxide prepared	Amount of bubbling	Odor	Acid test	Color	Supports combustion
1.	CO₂	_____	_____	_____	_____	_____
2.	SO₂	_____	_____	_____	_____	_____
3.	NO₂ + NO	_____	solution: _____	_____		
			as a gas: _____			

C. Preparation and properties of some nonmetallic hydrides: NH₃, H₂S

	Hydride prepared	Amount of bubbling	Odor	Acid–base tests	Supports combustion
1.	NH₃	_____	_____	_____	_____
2.	H₂S	_____	_____	_____	_____

Optional **D. Identifying an unknown solution**

Unknown ID code _____

Observations:

Nonmetal substance that can be made from unknown _____   Identity of unknown solution _____

Experiment 18

Advance Study Assignment: Some Nonmetals and Their Compounds

1. In this experiment nine substances are prepared and studied. For each one, list the chemicals used in its preparation and the chemical reaction that produces the substance.

	Chemicals used	**Chemical reaction**

a. O_2

b. N_2

c. I_2

d. Br_2

Chemicals used   **Chemical reaction**

e. CO_2

f. SO_2

g. $NO_2 + NO$

h. NH_3

i. H_2S

Experiment 19

Determining Molar Mass by Freezing Point Depression

The most common liquid we encounter in our daily lives is water. In this experiment you will study the equilibria that can exist between pure water and its aqueous (water-as-the-solvent) solutions, and ice, the solid form of water. (Water is one of a very few substances for which we have a separate name for the solid!)

If some ice cubes are put into a glass of water, the water temperature will fall and some of the ice will melt. This occurs because thermal energy always tends to flow from higher to lower temperatures. Thermal energy from the (warmer) water flows into the (colder) ice, until they reach the same temperature. It takes the flow of a certain amount of thermal energy, called the enthalpy of fusion, to melt a given amount of ice. If there is enough ice present, the water temperature will ultimately fall to the freezing point of pure water, $T_f°$, which is $0°C$, and stay there. At that point, ice and water are in equilibrium: at the freezing point of water. At the freezing point the vapor pressures of ice and water are equal: it is that requirement that sets the freezing point. (See Figure 19.1.)

Now consider what happens if a soluble liquid or solid (a solute) is added to the equilibrium mixture of ice and water—for example, if a small amount of ethanol or table salt is added. Perhaps surprisingly, the temperature of the ice and the solution falls as equilibrium is reestablished. This is because in the solution the vapor pressure of water at $0°C$ is lower than that of pure water. At equilibrium, the vapor pressures of the solid and liquid phases must be equal. Because the vapor pressure of ice at $0°C$ is unchanged, and now higher than that of the solution, some ice melts. Melting ice requires thermal energy, which is taken from the solution and from the ice, cooling them both to a lower temperature. The vapor pressures of the ice and the solution both fall, but that of the ice falls faster, until at some temperature below $0°C$ the two vapor pressures become equal and equilibrium is reestablished at the new freezing point, T_f. This situation is shown in Figure 19.1.

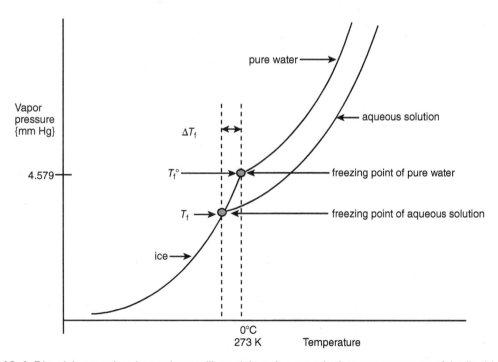

Figure 19.1 Dissolving a solute in a solvent will result in a decrease in the vapor pressure of the liquid phase; this causes the freezing point of the solution, T_f, to be lower than that of the pure solvent, $T_f°$.

The change in freezing point observed in this situation is called the freezing point depression, ΔT_f, and is equal to $T_f° - T_f$. It is observed with solutions of any solvent. Freezing point depression is one of the <u>colligative properties</u> of solutions. Others are boiling point elevation (which requires a solute that will not enter the gas phase), osmotic pressure, and vapor pressure lowering. The colligative properties of solutions depend on the number of (independently moving) solute particles present in a given amount of solvent, but not on the kinds of particles dissolved: they can be molecules, atoms, or ions.

When working with colligative properties it is convenient to express the solute concentration in terms of its <u>molality</u>, b, as defined by this equation:

$$\text{molality of A} = b_A := \frac{\text{moles of A dissolved}}{\text{kg of solvent in the solution}} \tag{1}$$

For this unit of concentration, the boiling point elevation, $T_b - T_b°$, or ΔT_b, and the freezing point depression, $T_f° - T_f$, or ΔT_f, in C° at very low concentrations are estimated reasonably well by these equations:

$$\Delta T_b = k_b b \quad \text{and} \quad \Delta T_f = k_f b \tag{2}$$

where k_b and k_f depend on the solvent used. For water, $k_b = 0.52$ (kg·C°)/mol and $k_f = 1.86$ (kg·C°)/mol. For benzene, $k_b = 2.53$ (kg·C°)/mol and $k_f = 5.10$ (kg·C°)/mol. In this experiment you will assume that Equation 2 is valid, even though your solutions will be moderately concentrated. Note that you should write temperature *differences* in C°, while temperatures are in °C. This is because a temperature difference of 4 C° is the same as a temperature difference of 4 K°, but a temperature of 4 °C is very different from a temperature of 4 K!

Colligative properties offered a way for early chemists to determine molar masses. With organic molecules, molar masses indicated by freezing point depression agreed with those found by other methods. However, with ionic salts, such as NaCl, molar masses suggested by freezing point depression measurements were much lower than the actual values. On the basis of such experiments, Arrhenius suggested that ionic substances exist as separated ions in aqueous solution, consistent with the observation that such solutions conduct an electric current. Arrhenius's general idea turned out to be correct and is now a basic part of modern chemical theory.

The <u>van't Hoff factor</u>, i, provides insight into how an ionic solute dissociates, and any solid dissolves (becomes solvated by water molecules), in a liquid at a given concentration. The van't Hoff factor is defined as

$$i := \frac{\text{moles of independent particles released into a solution}}{\text{moles of solute that actually dissolve in that solution}} = \frac{\text{actual molar mass}}{\text{apparent (measured) molar mass}} \tag{3}$$

Introductory chemistry teaches that most ionic solids dissociate completely in aqueous solution. For example:

$$NaCl(s) \xrightarrow{\text{H}_2\text{O}(\ell)} Na^+(aq) + Cl^-(aq) \tag{4}$$

This chemical equation suggests that i should be exactly 2 for sodium chloride. Reaction 4 indicates that every mole of sodium chloride that dissolves provides two moles of independent particles (ions) in solution. It may surprise you to learn that the van't Hoff factor for these salts only equals 2.00 in extremely dilute solutions; generally, it is less. This is because most salts have a tendency to form "ion pairs" when dissolved in water: these result in salt dissolution without complete dissociation of all the component ions. At any instant in time, a few positive ion–negative ion pairs are interacting with each other and not moving independently. Consider the simplified drawing of an ion pair of sodium chloride in Figure 19.2. The water solvates the pair of ions simultaneously but does not completely separate the sodium ion from the chloride ion, so they continue to move together.

Because of ion pairing, freezing point depression is a poor method for determining the molar mass of unknown ionic salts. However, freezing point depression does provide a way of measuring the van't Hoff factor, and thus quantifying ion pairing, for known solids—so that is what you will do.

Discussion of the Method

In this experiment you will study the freezing point behavior of some aqueous solutions. First you will measure the freezing point of pure water, using a combination of ice and water. Then you will determine the freezing

Interaction between ions persists

Independent ions

An ion pair of Na$^+$ and Cl$^-$

Fully solvated Na$^+$ and Cl$^-$

Figure 19.2 Ion pairs exist in solution because the solvation shells around dissolved ions do not perfectly shield the charges of the solvated ions, so solvated ions of opposite charge are still loosely attracted to one another and do not always move independently.

point of an aqueous solution containing a known mass of an unknown liquid, thus determining the freezing point depression, ΔT_f. This will allow you to calculate the molality of the solution. Finally, you will separate the solution from the ice in the mixture and weigh the solution, which will contain all of the solute. This information furnishes you with the composition of the solution: it lets you calculate the mass of solute present per kilogram of water. The molar mass can then be calculated with Equations 1 and 2.

Experimental Procedure

Wear your safety glasses while performing this experiment.

Obtain a sensitive thermometer, an insulated cup, a wire gauze square or other strainer, a long stirring rod, and two unknowns: one liquid and one solid. The actual molar mass of the solid will be indicated.

A. Measuring the freezing point of pure water

You will first need to determine what your thermometer thinks the freezing point of pure water is. Be certain that your insulated cup is clean; then prepare a mixture of water and ice by first filling the cup with ice, then adding deionized water a little at a time and stirring after each addition. (This gives the ice a chance to melt as it cools the water.) Once the water level reaches the top of the ice, stir well for at least 30 seconds and then record the temperature. (Because your thermometer may not be perfectly calibrated, this temperature may not be exactly 0.0 °C; that is the point of this step: it serves to calibrate your thermometer.)

B. Finding the freezing point of a solution of liquid unknown

Pour the ice water out of your cup and shake most of the water out of it, but do not bother drying the inside. (You do want the *outside* to be dry.) Place the empty cup on a balance (top-loading, if available) and tare it out. Add 10 ± 2 g of your liquid unknown, recording the exact mass you end up using to ± 0.01 g. Bring the cup back to your workstation and add ice (for now *just* ice, no water), so that you end up with an ice layer about one centimeter thick, and stir while monitoring the temperature. Keep stirring until the temperature reading no longer systematically drops; add ice as needed to maintain about a 1-cm layer at the top of the solution. If this final temperature is warmer than -6 °C, you are ready to proceed. If not, add water a little at a time and stir, until the equilibrium temperature is warmer than -6 °C and a roughly one-centimeter layer of ice remains on top. (Do not obsess about the amount of ice—just make sure that the liquid surface is completely covered with ice, and that you do not have so much ice that it will not all fit on the wire screen or in your strainer.) Now stir well and record the lowest stable temperature that you observe. Be patient here—make sure that the temperature has stabilized, and that you are stirring when you record the depressed freezing point of the solution. However, once you have confidently determined this temperature, do not wait too long (because the ice will continue to melt, and further dilute the solution) before separating your solution from the ice by pouring it through a wire screen or other strainer into a tared beaker on a balance (top-loading, if available). Give the screen or strainer a good shake, to get as much of the liquid out of the ice as possible, then lift it above the beaker so it is not contributing to the mass measured on the balance. Record the mass of the solution, then pour the solution and the ice back into the cup.

Conduct a second trial with your unknown liquid, at a more dilute solute (unknown liquid) concentration. Use the solution and ice you just poured back into the cup. Rinse the beaker with a *small amount* of deionized water and pour the rinsing into the cup, thereby slightly lowering the concentration of the solute in your solution. Add more water as necessary to reduce the freezing point depression to *roughly one-half* of its original value, all the while maintaining a one-centimeter layer of ice in the cup. (If your initial freezing point was $-6\,°C$, you would want to aim for about $-3\,°C$. Again, do not stress about this, just aim for something near half. *Note that if you over-dilute, there is no way to undo it—you will not be able to make the freezing point of the solution colder again. Just go with what you have: do not add more ice in an effort to cool it down. The ice itself is at $0\,°C$ and is warmer than your solution!*) Accurately measure the new equilibrium freezing point for the mixture, then separate the solution from the ice and determine its mass, as before.

When you have finished this part of the experiment, dispose of the mixture of ice and solution as directed by your instructor.

C. Quantifying the dissociation of a solid

The objective of this part of the experiment will be to experimentally determine the van't Hoff factor for your solid. You will use the same general procedure as for the liquid.

On a balance (top-loading, if available), add about ten grams of your solid to a thoroughly rinsed, empty cup that is dry on the outside, recording the exact amount used. Return to your workstation with the cup and, while stirring, add a minimum amount of water to dissolve the solid. Stir until you can see that dissolution is complete, adding more water only as needed. Then proceed as you did with your unknown liquid, adding ice (and, if necessary, water) to obtain the minimum freezing point possible that is not colder than $-6\,°C$. Pour off and weigh the solution as you did with the liquid unknown, and record your results.

As with the liquid unknown, return the ice and solution to the cup and take a second measurement, aiming for a freezing point depression about half that observed in your first trial.

If the molar mass you find for your solid is appreciably less than its actual molar mass (which will be given to you by your instructor), you probably have an ionic solid. No matter what, the ratio of the true molar mass over the value you determine experimentally is the van't Hoff factor, i.

When you are finished with the experiment, pour the solution and the contents of the cup into the sink. Return the sensitive thermometer, the insulated cup, the wire gauze square or other strainer, the long stirring rod, and the containers for your unknowns.

Take it further (optional): Experimentally determine the largest possible freezing point depression for a solution of sodium chloride, NaCl. Find the composition of that solution. What phases are in equilibrium in the final mixture, at the depressed freezing point? What does this indicate about the minimum temperature at which NaCl can be used to melt ice?

Experiment 19

Data and Calculations: Determining Molar Mass by Freezing Point Depression

A. Measured freezing point of pure water   _____ °C

B. Finding the freezing point of a solution of liquid unknown

Unknown liquid #   _____

Actual mass of solute (liquid unknown) used   _____ g

Trial 1

Freezing point of solution (observed)   _____ °C

Mass of solution   _____ g

Trial 2

Freezing point of solution   _____ °C

Mass of solution   _____ g

Calculations	**Trial 1**	**Trial 2**
Freezing point depression, ΔT_f	_____ C°	_____ C°
Molality of unknown solution, b_u	_____ molal	_____ molal
Mass of solution	_____ g	_____ g
Mass of solvent (water)	_____ g	_____ g
Mass of solvent (water)	_____ kg	_____ kg
Moles of solute (liquid unknown)	_____ mol	_____ mol
Molar mass of unknown (as indicated by freezing point depression)	_____ g/mol	_____ g/mol

(continued on following page)

C. Quantifying the dissociation of a solid

Solid ID Code ======================

Actual molar mass of solid (provided by instructor) _____ g/mol

Mass of solid solute used ====================== g

Trial 1
Freezing point of solution _____ °C

Mass of solution ====================== g

Trial 2
Freezing point of solution _____ °C

Mass of solution ====================== g

Calculations	Trial 1	Trial 2
Freezing point depression, ΔT_f	_____ C°	_____ C°
(Apparent) molality of solid solution, b_s	_____ molal	_____ molal
Mass of solution	_____ g	_____ g
Mass of solid solute in solution	_____ g	_____ g
Mass of solvent (water)	_____ g	_____ g
Mass of solvent (water)	_____ kg	_____ kg
(Apparent) moles of solute	_____ mol	_____ mol
(Apparent) molar mass of solute	_____ g/mol	_____ g/mol
Value of i, the van't Hoff factor (Equation 3)	_____	_____

Experiment 19

Advance Study Assignment: Determining Molar Mass by Freezing Point Depression

1. A student determined the molar mass of an unknown non-dissociating liquid by the method described in this experiment. She found that the equilibrium temperature of a mixture of ice and pure water was indicated to be $-0.1\,°C$ on her thermometer. When she added 9.9 g of her sample to the mixture, the temperature reading, after thorough stirring, fell to $-3.9\,°C$. She then poured off the solution through a screen into a beaker. The mass of the solution was 84.2 g.

   a. What was the freezing point depression, ΔT_f?

   _____ C°

   b. What was the molality of unknown liquid, b_u, in the cold solution? (Use Equation 2.)

   _____ m

   c. What mass of unknown liquid was in the poured-off solution?

   _____ g

   d. What mass of water was in the poured-off solution?

   _____ g

   e. How many moles of unknown liquid were present in the poured-off solution? [Use Equation 1 and your answers to 1(b) and 1(d).]

   _____ moles of unknown liquid

   f. Based on these data, what value did she calculate for the molar mass of her unknown liquid, assuming she carried out the calculation correctly? [Divide the mass of the unknown liquid sample by your answer to 1(e).]

   _____ g/mol

2. A student determined the freezing point depression caused by sodium bromide in water by the method described in this experiment. She found that the equilibrium temperature of a mixture of ice and pure water was indicated to be −0.1 °C on her thermometer. When she dissolved 10.07 g of sodium bromide into the mixture, the temperature reading—after thorough stirring—fell to −5.6 °C. She then poured off the solution through a screen into a beaker. The mass of the solution was 67.77 g.

a. What was the freezing point depression, ΔT_f?

_____ C°

b. What was the (apparent) molality of independently moving particles, b_s, in the cold solution? (Use Equation 2.)

_____ m

c. What mass of water was in the poured-off solution?

_____ g

d. How many moles of independently moving particles were present in the poured-off solution? [Use Equation 1 and your answers to 1(b) and 1(c).]

_____ moles of independent particles

e. Based on these data, what value did she calculate for the (apparent) molar mass of sodium bromide, assuming she carried out the calculation correctly? [Divide the mass of NaBr by your answer to 1(d).]

_____ g/mol

f. What is the actual molar mass of sodium bromide?

_____ g/mol

g. What do these results indicate the value of the van't Hoff factor, i, for sodium bromide is in the student's (cold and quite concentrated, both of which tend to increase ion paring) solution? (Use Equation 3.)

$i =$ _____

Experiment 20

Rates of Chemical Reactions, 1. The Iodination of Acetone

The study of the rates of chemical reactions is called <u>chemical kinetics</u>. The rate at which a chemical reaction occurs depends on several factors, including the nature of the reaction, the concentrations of the reactants, the temperature, and the presence of <u>catalysts</u> (substances that increase the rate of a chemical reaction without being chemically altered themselves). Each of these factors can have a big influence on the observed rate of a chemical reaction.

Some reactions at a given temperature are very slow: the reaction of oxygen with hydrogen gas, or with wood, does not proceed to a visible extent even after 100 years at room temperature. Other reactions happen as quickly as their reactants come in contact: the precipitation of silver chloride when solutions containing silver ions and chloride ions are mixed, and the formation of water when acidic and basic solutions are mixed, are examples of extremely fast reactions. In this experiment you will study a reaction that, at room temperature, proceeds at a rate you will be able to measure: not too fast, not too slow.

For a given reaction, the rate typically increases with an increase in the concentration of any reactant. Consider the balanced chemical reaction

$$a\,A + b\,B \rightarrow c\,C$$

in which A, B, and C are chemical substances and a, b, and c are stoichiometric coefficients. The relationship between reaction rate and concentrations, called the <u>rate law</u>, can usually be expressed by the equation

$$\text{rate} = k[A]^m[B]^n \tag{1}$$

in which m and n are generally, but not always, whole numbers: specifically 0, 1, 2, or possibly 3. [A] and [B] are the concentrations of A and B (ordinarily in moles per liter), and k is a constant, called the <u>rate constant</u> of the reaction. The numbers m and n are called the <u>orders of the reaction</u> with respect to A and B. If m is 1, the reaction is said to be <u>first order</u> with respect to the reactant A. If n is 2, the reaction is <u>second order</u> with respect to reactant B. The <u>overall reaction order</u> is the sum of m and n. In this example, the reaction would be <u>third order</u> overall. Note that the orders of the reaction do not have a reliable relationship to the stoichiometric coefficients (a, b, and c) of the overall balanced reaction.

The rate of a reaction also strongly depends on the temperature at which the reaction occurs, which is reflected in a strong temperature dependence in the rate constant, k. An increase in temperature increases the rate; a general rule that a $10\,C°$ rise in temperature will double the rate works pretty well with many reactions, though it is only an approximate generalization. Nevertheless, it is clear that an increase in temperature on the order of $100\,C°$ can really change the rate of a reaction!

A more precise equation relating temperature to reaction rate is known, but it is more mathematically complicated. It is based on the idea that in order to react, reactants must collide with a certain minimum amount of energy—typically provided by the kinetic (motion) energy of the reactants, which is indicated by their temperature. This required amount of energy is called the <u>activation energy</u> of the reaction. The equation relating the rate constant k to the absolute temperature, T, and the activation energy, E_a, is called the <u>Arrhenius equation</u> and is most often written in this form:

$$\ln k = \frac{-E_a}{RT} + \text{constant} \tag{2}$$

In Equation 2, R is the gas constant [8.3145 joules/(mole·K) for E_a in joules per mole] and $\ln k$ is the natural logarithm of the rate constant, k. By measuring k at different temperatures it is possible to determine the activation energy for a reaction.

In this experiment you will study the kinetics of the reaction between iodine and acetone:

$$CH_3 - \overset{\overset{\displaystyle O}{\displaystyle \|}}{C} - CH_3(aq) + I_2(aq) \rightarrow CH_3 - \overset{\overset{\displaystyle O}{\displaystyle \|}}{C} - CH_2I(aq) + H^+(aq) + I^-(aq) \qquad (3)$$

acetone   iodine   iodoacetone   hydrogen ion   iodide ion

The rate of Reaction 3 is found to depend on the concentration of hydrogen ion in the solution; this is not because H^+ is a product of the reaction, but rather because the mechanism by which the reaction actually occurs involves hydrogen ion as a reactant in the slow, or <u>rate-determining</u>, step. Incorporating that fact into the form of Equation 1, the rate law for this reaction should be

$$\text{rate} = k[\text{acetone}]^m [H^+]^n [I_2]^p \qquad (4)$$

in which m, n, and p are the orders of the reaction with respect to acetone, hydrogen ion, and iodine, respectively, and k is the rate constant for the reaction.

Because the stoichiometric coefficient of $I_2(aq)$ in the balanced chemical reaction is 1, the rate of this reaction can be expressed as the (small) change in the concentration of I_2, $\Delta[I_2]$, that occurs, divided by the time interval, Δt, required for the change:

$$\text{rate} = \frac{-\Delta[I_2]}{\Delta t} \qquad (5)$$

The minus sign is added to make the rate positive (as $\Delta[I_2]$ is negative). Ordinarily, because rate varies with the concentrations of the reactants according to Equation 4, in a rate study it would be necessary to measure, directly or indirectly, the concentration of each reactant as a function of time; the rate would typically change a lot over time, decreasing to very low values as the concentration of at least one reactant became very low. This makes reaction rate studies difficult to carry out and mathematically complicated; however, in some cases (such as this one!) the nature of the reaction simplifies such an analysis.

Several things about the iodination of acetone make its rate easier to study than most. First of all, iodine has a color, so changes in iodine concentration can be followed visually. A second and very important characteristic of this reaction is that it turns out to be zero order in I_2 concentration. This means (see Equation 4) that the rate of the reaction does not depend on $[I_2]$ at all; $[I_2]^0 = 1$, no matter what the value of $[I_2]$ is, as long as $[I_2]$ is not zero (that is, there is still some I_2 present).

Because the rate of Reaction 3 does not depend on $[I_2]$, you can study the rate by making I_2 the <u>limiting reagent</u>, present in a large excess of acetone and H^+ ion. You then measure the time required for a known initial concentration of I_2 to be used up completely. If both acetone and H^+ are present at much higher initial concentrations than that of I_2, their concentrations will not change much during the course of the reaction, and the rate will remain, by Equation 4, effectively constant until all the iodine is gone, at which time the reaction will stop. Under such circumstances, if it takes t seconds for the color of a solution having an initial concentration of I_2 equal to $[I_2]_i$ to disappear, the rate of the reaction, by Equation 5, would be

$$\text{rate} = \frac{-\Delta[I_2]}{\Delta t} = \frac{[I_2]_i}{t} \qquad (6)$$

Although under these conditions the rate of the reaction remains constant over the entire time the reaction occurs, you can vary it by changing the initial concentrations of acetone and H^+ ion. For example, if in preparing a different mixture, Mixture 2, you double the initial concentration of acetone over that in Mixture 1, keeping $[H^+]$ and $[I_2]$ unchanged, then the rate of Mixture 2 would, according to Equation 4, differ from that in Mixture 1:

$$\text{rate}_2 = k[\text{acetone}]_2^m [H^+]_2^n [I_2]_2^0 = k(2[\text{acetone}]_1)^m [H^+]_1^n [I_2]_1^0 \qquad (7a)$$

$$\text{rate}_1 = k[\text{acetone}]_1^m [H^+]_1^n [I_2]_1^0 \qquad (7b)$$

Dividing the Equation 7a by 7b, the ks cancel, as do the terms in the hydrogen ion and iodine concentrations, because they have the same values in both reactions:

$$\frac{\text{rate}_2}{\text{rate}_1} = \frac{(2[\text{acetone}]_1)^m}{[\text{acetone}]_1^m} = \left(\frac{2[\text{acetone}]_1}{[\text{acetone}]_1}\right)^m = 2^m \qquad (7)$$

Having measured both rate$_2$ and rate$_1$ by Equation 6, their ratio, which must be equal to 2^m, can be determined. You can then solve for m, either by inspection or using logarithms, and in that way find the order of Reaction 3 with respect to acetone.

By a similar procedure you can measure the order of the reaction with respect to H^+ ion concentration and also confirm that the reaction is zero order with respect to I_2. Having found the order with respect to each reactant, you can then evaluate k, the rate constant for the reaction.

Determining the orders m and n, confirming p (the order with respect to I_2) equals zero, and evaluating the rate constant k for the reaction at room temperature are your goals in this experiment. You will be furnished with standard solutions of acetone, iodine, and hydrogen ion, and with the composition of one mixture that will give a reasonable rate. Planning and carrying out the rest of the experiment will be up to you.

An optional part of this experiment is to study the rate of Reaction 3 at different temperatures to determine its activation energy. The general procedure called for is to study the rate of reaction in a given mixture at room temperature and at two other temperatures, one above and one below room temperature. Knowing the rates, and hence the ks, at the three temperatures, you can find the activation energy, E_a, for the reaction, by plotting $\ln k$ vs $1/T$. The slope of the straight line that should result is $-E_a/R$, according to Equation 2.

Experimental Procedure

Wear your safety glasses while performing this experiment.

A. Reaction rate data

Select two medium (18×150 mm) test tubes; when filled with deionized water, they should appear to have the same color when viewed from above the opening (down the tube) against a white background.

Depending on how the stock solutions for this experiment are made available to you, you may want to set up a "local supply" of them. You can do so by pouring 50 mL of each of the following solutions into clean, dry beakers, one solution to a beaker: 4.0 M acetone, 1.0 M HCl, and 0.0050 M I_2. Then cover each beaker with a watch glass, which you should leave in place except when pouring solution out of a beaker.

With your graduated cylinder, measure out 10. mL of the 4 M acetone solution and pour it into a clean and dry 125-mL Erlenmeyer flask. Then measure out 10. mL of 1.0 M HCl and add that to the acetone in the flask. Measure out 20. mL of deionized H_2O and add it to the flask. Drain the graduated cylinder, shaking out any excess water, and then use the cylinder to measure out 10. mL of 0.0050 M I_2 solution. Be careful not to spill the iodine solution on your hands or clothes, as it will stain.

Noting the time to the nearest second, or starting a stopwatch simultaneously, pour the iodine solution into the Erlenmeyer flask and quickly swirl the flask to mix its contents well. The reaction mixture will appear yellow because of the presence of the iodine, and the color will fade slowly as the iodine reacts with the acetone. Fill one of the test tubes three-quarters full with the reaction mixture, and fill the other test tube to the same depth with deionized water. Look down the test tubes toward a well-lit piece of white paper, and note the time the color of the iodine disappears. Then measure the temperature of the reaction mixture in the test tube.

Repeat the experiment, using as a color reference the reacted solution from the first trial instead of deionized water. The amount of time required in the two runs should agree within about twenty seconds.

The rate of the reaction equals the initial concentration of I_2 *in the reaction mixture* divided by the elapsed time. Because both acetone and H^+ ion are present in great excess, the concentrations of both acetone and H^+ remain at essentially their initial values in the reaction mixture; and because Reaction 3 is also zero order in I_2, for a given mixture the reaction rate remains essentially constant throughout the time the reaction is studied.

B. Determining the reaction orders with respect to acetone, H^+ ion, and I_2

Having found the reaction rate for one composition of the system, think for a moment about what changes in composition would decrease the time (increase the rate of the reaction). In particular, how could you change the composition to allow you to determine how the rate depends upon acetone concentration? If it is not clear how to proceed, reread the discussion preceding Equation 7. In your new mixture you should keep the total volume at 50. mL, and ensure that the concentrations of H^+ and I_2 are the *same* as they were in the first mixture. Carry out the reaction twice with your new mixture, continuing to use a spent reaction mixture as your reference; the times should not differ from one another by more than about fifteen seconds. The temperature should be kept within about one degree of that in the initial trials. Calculate the rate of the reaction. Compare it with that for the first mixture, and then calculate the order of the reaction with respect to acetone, using a relationship similar to Equation 7. First, write an equation like 7a for the second reaction mixture, substituting in the values for the rate as obtained by Equation 6 and the initial concentrations of acetone, H^+, and I_2 in the reaction mixture. Then write an equation like 7b for the first reaction mixture, using the observed rate and the initial concentrations in that mixture. Obtain an equation like Equation 7 by dividing Equation 7a by Equation 7b. Solve this equation for m, the order of the reaction with respect to acetone.

Now change the composition of one of the first two reaction mixtures so that a measurement of the reaction rate will give you information about the order of the reaction with respect to H^+. Repeat the experiment with this mixture to establish the time of reaction to within 15 seconds, again making sure that the temperature is within about one degree of that observed previously. From the rate you determine for this mixture find n, the order of the reaction with respect to H^+.

Finally, change one of the first two reaction mixture compositions in a way that allows you to show that the order of the reaction with respect to I_2 is zero. Measure the rate of the reaction twice, and calculate p, the order with respect to I_2.

C. Determining the rate constant, *k*

Having found the order of the reaction for each substance on which the rate depends, evaluate k, the rate constant for the reaction, from the rate and concentration data for each of the mixtures you studied. If the temperatures at which the reactions were run are all equal to within one or two degrees, k should be nearly the same for each mixture. Calculate the average (mean) value of k and the standard deviation for the set of k values. (See Appendix H.)

Optional **D. Predicting a reaction rate**

Make up a new mixture (Mixture 5), involving a combination of reactant volumes that you did not use in any previous trials. Using Equation 4, the concentrations in the mixture, and the reaction orders and rate constant you calculated from your experimental data, predict how long it will take for the I_2 color to disappear from this mixture. Measure the time for the reaction and compare it with your prediction. ▪

Optional **E. Determining the activation energy, *E*ₐ**

If time permits, make up the first mixture again (10. mL of each stock solution plus 20. mL of water) and determine its rate of reaction at about 10 °C and at about 40 °C. From the two rates you find, plus the rate at room temperature, calculate the activation energy for the reaction, using Equation 2. ▪

Disposal of reaction products: Dispose of your solutions from this experiment as directed by your instructor.

Experiment 20

Data and Calculations: The Iodination of Acetone

A. Reaction rate data

Table 20.1

| Mixture | Volume used in mixture {mL} | | | | Time for reaction {sec} | | Temperature {°C} |
	4.0 M acetone	1.0 M HCl	0.0050 M I_2	H_2O	1st run	2nd run	
1	10.	10.	10.	20.			
2							
3							
4							

B. Determining the reaction orders with respect to acetone, H^+ ion, and I_2

$$\text{rate} = k\,[\text{acetone}]^m\,[H^+]^n\,[I_2]^p \tag{4}$$

Calculate the initial concentrations of acetone, H^+ ion, and I_2 in each of the mixtures you studied. Use Equation 6 to find the rate of each reaction.

Table 20.2

Mixture	[acetone] {M}	$[H^+]$ {M}	$[I_2]_i$ {M}	Average time {sec}	Rate = $\dfrac{[I_2]_i}{\text{avg. time}}$ {M/sec}
1	0.80	0.20	0.0010		
2					
3					
4					

Substituting in the initial concentrations and rate from this table, write Equation 4 for Reaction Mixture 2:

rate$_2$ =

Now write Equation 4 for Reaction Mixture 1, substituting in the concentrations and rate from Table 20.2:

rate$_1$ =

(continued on following page)

Divide the equation for Mixture 2 by the equation for Mixture 1; the resulting equation should have the ratio of $rate_2$ to $rate_1$ on the left side, and a ratio of acetone concentrations raised to the power m on the right. It should be similar in appearance to Equation 7. Write the resulting equation here:

$$\frac{rate_2}{rate_1} =$$

The only unknown in the equation is m. Solve for m.
(Round off to the nearest whole number.) $\qquad m = \underline{\hspace{3cm}}$

Now write Equation 4 for Reaction Mixture 3 and for Reaction Mixture 4:

$rate_3 =$

$rate_4 =$

Using the ratios of the rates of Mixtures 3 and 4 to those of Mixtures 1 or 2, find the orders of Reaction 3 with respect to H^+ ion and I_2, n and p. (Round each of these off to the nearest whole number.)

$$\frac{rate_3}{rate_{\underline{}}} =$$

$\qquad n = \underline{\hspace{3cm}}$

$$\frac{rate_4}{rate_{\underline{}}} =$$

$\qquad p = \underline{\hspace{3cm}}$

C. Determining the rate constant, k

Using the values of m, n, and p determined in Part B, calculate the rate constant k for each mixture by substituting those orders, the initial concentrations, and the observed rate from Table 20.2 into Equation 4. Then calculate the mean (average) and standard deviation of these individual rate constant measurements (see Appendix H), which should all be similar.

Mixture	1	2	3	4	Average (mean)	Standard deviation
k						

Optional **D. Predicting a reaction rate**

Composition of new reaction mixture (Mixture 5)
Volumes used {mL}:

4.0 M acetone _____   1.0 M HCl _____   0.0050 M I_2 _____ H_2O _____
Initial concentrations:

[acetone]$_5$ _____ M $[H^+]_5$ _____ M $[I_2]_{i,5}$ _____ M

Predicted rate$_5$ _____ (Equation 4, show your work)

Predicted time for reaction _____ sec (Equation 6)

Observed time for reaction _____ sec ▪

Optional **E. Determining the activation energy, E_a**

Using Reaction Mixture 1:

Time for reaction at about 10 °C _____ sec   Temperature _____ °C; _____ K

Time for reaction at about 40 °C _____ sec   Temperature _____ °C; _____ K

Time for Mixture 1 at room temp _____ sec   Temperature _____ °C; _____ K

Calculate the rate constant at each temperature from your data, following the procedure in Part C.

	Rate	k	ln k	$\dfrac{1}{T\{K\}}$
~10 °C	_____	_____	_____	_____
~40 °C	_____	_____	_____	_____
Room temp	_____	_____	_____	_____

Plot ln k vs $1/T$, using a spreadsheet or the graph paper on the following page. Find the slope of the best straight line through the points. (Consult Appendix G1, G2, or E for more help.)

Slope = _____ K

By Equation 2: E_a {J/mol} $= -8.3145 \times$ slope

E_a _____ joules/mol

(continued on following page)

Experiment 20

Data and Calculations: The Iodination of Acetone

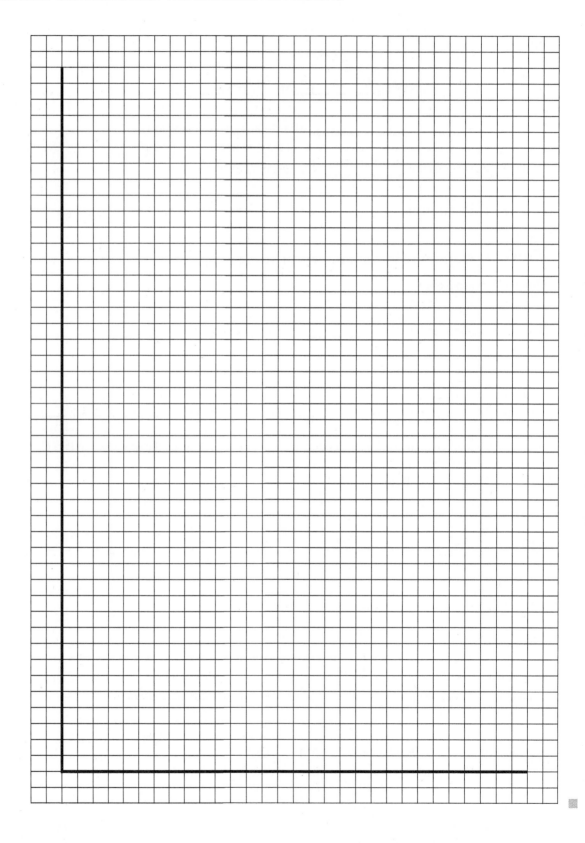

Experiment 20

Advance Study Assignment: The Iodination of Acetone

1. In a reaction involving the iodination of acetone, the reaction mixture (Mixture X) was composed of

    10. mL 4.0 M acetone + 10. mL 1.0 M HCl + 10. mL 0.0050 M I_2 + 20. mL H_2O

    a. How many moles of acetone were in the reaction mixture? Recall that, for a component A, moles of $A = [A] \times V_A$, where $[A]$ is the molarity of A in the solution that was used (in this case, in the acetone stock solution) and V_A is the volume in liters of the solution of A that was used (in this case, of the acetone stock solution added to the reaction mixture).

    _____ moles of acetone

    b. What was the molarity of acetone in the reaction mixture, $[acetone]_X$? The total volume of the reaction mixture was 50. mL, or 0.050 L, and you calculated the moles of acetone in Part (a).

    $$[A] = \frac{\text{moles of A}}{\text{volume of solution \{L\}}}$$

    _____ M acetone

    c. How could you double the molarity of acetone in the reaction mixture, keeping the total volume at 50. mL and keeping the same concentrations of H^+ ion and I_2 as in the original mixture?

2. It took 510 seconds for the color of the I_2 to disappear from Mixture X.

    a. What was the rate of the reaction? (*Hint*: First find the initial concentration of I_2 in the reaction mixture, $[I_2]_{i,X}$. Then use Equation 6.)

    $rate_X =$ _____ M/sec

    b. Given the rate from Part (a), and the initial concentrations of acetone, H^+ ion, and I_2 in the reaction mixture (you have yet to calculate $[H^+]$; do so now), write Equation 4 for Mixture X.

    $rate_X =$

    c. What unknowns remain in the equation in Part (b)? _____ _____ _____ _____

(continued on following page)

3. A second reaction mixture (Mixture Y) was made up, consisting of

    10. mL 4.0 M acetone + 20. mL 1.0 M HCl + 10. mL 0.0050 M I_2 + 10. mL H_2O

   a. What were the initial concentrations of acetone, H^+ ion, and I_2 in this reaction mixture?

   $[acetone]_{i,Y} = $ _____ M $[H^+]_{i,Y} = $ _____ M $[I_2]_{i,Y} = $ _____ M

   b. It took 127 seconds for the I_2 color to disappear from this reaction mixture when it occurred at the same temperature as the reaction in Problem 2. What was the rate of the reaction?

   $rate_Y = $ _____ M/sec

   Write Equation 4 for Mixture Y:

   $rate_Y = $

   c. Divide the $rate_Y$ equation by the $rate_X$ equation. The result should have the ratio of the two rates on the left side and a ratio of H^+ concentrations raised to the n power on the right. Write the resulting equation and solve for the value of n, the order of the reaction with respect to H^+. (Round off the value of n to the nearest whole number.)

   $n = $ _____

4. A third reaction mixture was made up (Mixture Z), consisting of

    10. mL 4.0 M acetone + 20. mL 1.0 M HCl + 5.0 mL 0.0050 M I_2 + 15. mL H_2O

   If the reaction is zero order in I_2, how long would it take for the I_2 color to disappear from Mixture Z at the same temperature at which Mixture X and Mixture Y were studied? Show and explain your work.

   _____ seconds

Experiment 21

Rates of Chemical Reactions, 2. A Clock Reaction

The study of the rates of chemical reactions is called <u>chemical kinetics</u>. The rate at which a chemical reaction occurs depends on several factors, including the nature of the reaction, the concentrations of the reactants, the temperature, and the presence of <u>catalysts</u> (substances that increase the rate of a chemical reaction without being chemically altered themselves). Each of these factors can have a big influence on the observed rate of a chemical reaction.

Some reactions at a given temperature are very slow: the reaction of oxygen with hydrogen gas, or with wood, does not proceed to a visible extent even after 100 years at room temperature. Other reactions happen as quickly as their reactants come in contact: the precipitation of silver chloride when solutions containing silver ions and chloride ions are mixed, and the formation of water when acidic and basic solutions are mixed, are examples of extremely fast reactions. In this experiment you will study a reaction that, at room temperature, proceeds at a rate you will be able to measure: not too fast, not too slow.

For a given reaction, the rate typically increases with an increase in the concentration of any reactant. Consider the balanced chemical reaction

$$a\,A + b\,B \rightarrow c\,C$$

in which A, B, and C are chemical substances and a, b, and c are stoichiometric coefficients. The relationship between reaction rate and concentrations, called the <u>rate law</u>, can usually be expressed by the equation

$$\text{rate} = k[A]^m[B]^n \tag{1}$$

in which m and n are generally, but not always, whole numbers: specifically 0, 1, 2, or possibly 3. [A] and [B] are the concentrations of A and B (ordinarily in moles per liter), and k is a constant, called the <u>rate constant</u> of the reaction. The numbers m and n are called the <u>orders of the reaction</u> with respect to A and B. If m is 1, the reaction is said to be <u>first order</u> with respect to reactant A. If n is 2, the reaction is <u>second order</u> with respect to reactant B. The <u>overall reaction order</u> is the sum of m and n. In this example, the reaction would be <u>third order</u> overall. Note that the orders of the reaction do not have a reliable relationship to the stoichiometric coefficients (a, b, and c) of the overall balanced reaction.

The rate of a reaction also strongly depends on the temperature at which the reaction occurs, which is reflected in a strong temperature dependence in the rate constant, k. An increase in temperature increases the rate; a general rule that a $10\,C°$ rise in temperature will double the rate works pretty well with many reactions, though it is only an approximate generalization. Nevertheless, it is clear that increasing the temperature by $100\,C°$ can really change the rate of a reaction!

A more precise equation relating temperature to reaction rate is known, but it is more mathematically complicated. It is based on the idea that in order to react, reactants must collide with a certain minimum amount of energy—typically provided by the kinetic (motion) energy of the reactants, which is indicated by their temperature. This required amount of energy is called the <u>activation energy</u> for the reaction. The equation relating the rate constant k to the absolute temperature, T, and the activation energy, E_a, is called the <u>Arrhenius equation</u> and is most often used in this form:

$$\ln k = \frac{-E_a}{RT} + \text{constant} \tag{2}$$

In Equation 2, $\ln k$ is the natural logarithm of the rate constant, k, and R is the gas constant [8.3145 joules/(mole · K) for E_a in joules per mole]. By measuring k at different temperatures it is possible to determine the activation energy for a reaction.

This experiment involves the study of the rate properties, or chemical kinetics, of the following reaction between iodide ion and bromate ion under acidic conditions:

$$6\,I^-(aq) \;+\; BrO_3^-(aq) \;+\; 6\,H^+(aq) \;\rightarrow\; 3\,I_2(aq) \;+\; Br^-(aq) \;+\; 3\,H_2O(\ell) \tag{3}$$
iodide ion   bromate ion   hydrogen ion   iodine   bromide ion   water

Reaction 3 proceeds reasonably slowly at room temperature, its rate depending on the concentrations of the I^-, BrO_3^-, and H^+ ions according to this rate law:

$$\text{rate} = k[I^-]^m[BrO_3^-]^n[H^+]^p \tag{4}$$

One of the main goals of this experiment will be to evaluate the rate constant k and the reaction orders m, n, and p for this reaction. You will also study how the reaction rate depends on temperature, and evaluate the activation energy of the reaction, E_a.

The method you will use for measuring the rate of Reaction 3 involves what is often called a "clock reaction". In addition to Reaction 3, the following reaction will also be made to occur, simultaneously, in the reaction flask:

$$I_2(aq) \;+\; 2\,S_2O_3^{2-}(aq) \;\rightarrow\; 2\,I^-(aq) \;+\; S_4O_6^{2-}(aq) \tag{5}$$
iodine   thiosulfate ion   iodide ion   tetrathionate ion

This reaction is much faster than Reaction 3, so the $I_2(aq)$ produced by Reaction 3 reacts completely with the thiosulfate ion, $S_2O_3^{2-}$, present in the solution as quickly as the I_2 forms, such that until all the thiosulfate ion has reacted the concentration of I_2 remains effectively zero. As soon as all the $S_2O_3^{2-}$ is consumed, however, the I_2 produced by Reaction 3 begins to accumulate in the solution. The buildup of I_2 is made easy to see by adding a starch indicator to the reaction mixture: I_2, even at low concentrations, reacts with starch in solution to produce a deep blue color.

By carrying out Reaction 1 in the presence of $S_2O_3^{2-}$ and a starch indicator, a "clock" is introduced into the system. The clock indicates when a known amount of $I_2(aq)$ has been produced by Reaction 3, and thus when a given amount of BrO_3^- ion has reacted ($\frac{1}{6}$ mole BrO_3^- per mole $S_2O_3^{2-}$). It also keeps the concentration of iodide ion, $[I^-]$, constant by regenerating it as quickly as it is consumed in Reaction 3. Because the rate of reaction can be expressed in terms of the time it takes for a particular amount of BrO_3^- to be used up, the clock allows the rate to be determined. In all of the reactions you will study in this experiment, the amount of BrO_3^- and H^+ that react over the time the clock runs will be constant and small compared to the initial amounts of those reactants, while I^- will be constantly regenerated by Reaction 5. This means that the concentrations of all reactants will be essentially constant in Equation 4, and so will the rate of Reaction 3 during each trial.

In this experiment you will carry out Reaction 3 under a variety of concentration conditions. You will combine measured amounts of each of the reactant ions (in aqueous solution) in the presence of a constant, small amount of $S_2O_3^{2-}$ and measure the time it takes for each mixture to turn blue. The time before the color change occurs in each reaction mixture will indicate the rate of Reaction 3 in that mixture. By changing the concentration of one reactant and keeping the other concentrations constant, you will investigate how the rate of Reaction 3 varies with the concentration of a particular reactant. After determining the order of Reaction 3 in each reactant you can determine the rate constant for the reaction.

In the last part of this experiment you will investigate how the rate of the reaction depends on temperature. By measuring how the rate varies with temperature you will be able to determine the activation energy, E_a, for the reaction using the Arrhenius equation, Equation 2.

$$\ln k = -\frac{E_a}{RT} + \text{constant} \tag{2}$$

By plotting $\ln k$ against $1/T$, a straight line whose slope equals $-E_a/R$ should be obtained, and from that the activation energy can be calculated.

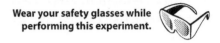

Experimental Procedure

A. How reaction rate depends on concentration

Table 21.1 summarizes the volumes to be used in carrying out the several trials whose rates are needed to determine the general rate law for Reaction 3. Depending on how the stock solutions for this experiment are made available to you, you may want to set up a "local supply" of them. You can do so by pouring about one hundred milliliters of each of the listed solutions into clean, labeled flasks or beakers; you can then use these in preparing your reaction mixtures.

Table 21.1

Reaction Mixtures at Room Temperature {liquid volumes in mL}				
	Reaction Flask A (250 mL)			**Reaction Flask B (125 mL)**	
Reaction Mixture	**0.010 M KI**	**0.0010 M $Na_2S_2O_3$**	**H_2O**	**0.040 M $KBrO_3$**	**0.10 M HCl**
1	10.0	10.0	10.0	10.0	10.0
2	20.0	10.0	0	10.0	10.0
3	10.0	10.0	0	20.0	10.0
4	10.0	10.0	0	10.0	20.0
5	15.0	10.0	5.0	5.0	15.0

The general procedure for each reaction mixture will be the same—it is described in detail here for the first reaction mixture, Reaction Mixture 1.

Because there are several liquids to mix, and you do not want the reaction to start until you are ready, you will put some of the reactants into one flask and the rest into another, selecting these so that no reaction occurs until the contents of the two flasks are mixed. Start by adding about three drops of starch indicator solution to a clean (wet with water is fine) 250-mL Erlenmeyer flask (Reaction Flask A). Then, using a 10-mL graduated cylinder or autopipet to measure volumes, measure out 10.0 mL of 0.010 M KI, 10.0 mL of 0.0010 M $Na_2S_2O_3$, and 10.0 mL of deionized water into the same flask (Reaction Flask A). When measuring out solutions, rinse the graduated cylinder or autopipet tip with deionized water after you have added the solutions to Reaction Flask A, and *before* you measure out the solutions for Reaction Flask B. (This is not necessary, however, if the last thing you measure out is H_2O.) Having rinsed out your measuring device, measure out 10.0 mL of 0.040 M $KBrO_3$ and 10.0 mL of 0.10 M HCl into a 125-mL Erlenmeyer flask (Reaction Flask B). If at any point during the measuring process Reaction Flask A turns blue (it should not do so, yet!), you probably have a cross-contamination issue that you should identify and correct before you proceed (Reaction 3 is starting before it is supposed to do so!).

Pour the contents of Reaction Flask B into Reaction Flask A and swirl the combined liquids in Flask A to mix them thoroughly. Note the time or start a stopwatch at the instant the solutions are mixed. Continue swirling the solution. It should turn blue in less than 2 minutes. Note the time or stop the stopwatch the instant the blue color appears. Record the temperature of the blue solution to the nearest 0.2 °C.

Repeat this procedure with the other mixtures in Table 21.1. *Do not forget to add the indicator* to Reaction Flask A. The reaction flasks should be rinsed with deionized water between runs. When measuring out the solutions, be certain to rinse your measuring device with deionized water after you have added the solutions to Reaction Flask A and *before* you measure out the solutions for Reaction Flask B. Try to keep the temperature just about the same in all the runs. Repeat any experiments that did not appear to proceed properly or for which you did not get a reliable time measurement.

B. How reaction rate changes with temperature: the activation energy, E_a

In this part of the experiment the reaction will be carried out at several different temperatures, using Reaction Mixture 1 in all cases. Target temperatures of 20 °C, 40 °C, 10 °C, and 0 °C.

You have already measured the time for Reaction Mixture 1 at room temperature, which is close enough to 20 °C. To determine the time at a higher temperature, make up Reaction Mixture 1 as you did in Part A, including the indicator. However, instead of mixing the solutions in the two flasks at room temperature, put the flasks into water at 40 °C (this number has one significant figure, so it indicates a temperature between 30 °C and 50 °C), from the hot-water tap, in one or more large beakers. Check to see that the water is indeed in the required temperature range, and leave the flasks in the hot water for several minutes, swirling them every few seconds, to bring them to that temperature. *Be prepared for a much shorter reaction time when you mix the two solutions*, and continue swirling the reaction flask in the warm water. When the color change occurs, record the time, and also the temperature of the solution in the flask.

Repeat the experiment at 10 ± 5 °C, cooling all the reactants in water at that temperature before starting the reaction. Record the time required for the color to change and the final temperature of the reaction mixture. Repeat once again at 0 °C (between −1 °C and 1 °C), this time using an ice-water bath to cool the reactants.

Optional   C. How reaction rate changes in the presence of a catalyst

Catalysts are substances that alter the rate of a chemical reaction without being chemically altered themselves. Certain ions can have a powerful catalytic effect on the rates of specific chemical reactions. Observe the effect of a catalyst on this reaction by once more making up Reaction Mixture 1, again starting with about three drops of starch indicator solution in Reaction Flask A. Before combining the contents of the two flasks, add 1 drop of 0.01 M $(NH_4)_6Mo_7O_{24}$, ammonium heptamolybdate (a catalyst) to Reaction Flask B. Swirl Reaction Flask B to mix the catalyst in thoroughly. Then combine the contents of the two reaction flasks at room temperature, noting the time required for the color to change. (Be prepared for that time to be *short*!) ▨

Disposal of reaction products: The reaction products in this experiment are very dilute and may be poured into the sink as you complete each part of the experiment, unless your instructor directs otherwise.

Experiment 21

Data and Calculations: Rates of Chemical Reactions, 2. A Clock Reaction

A. How reaction rate depends on concentration

$$6\,I^-(aq) + BrO_3^-(aq) + 6\,H^+(aq) \rightarrow 3\,I_2(aq) + Br^-(aq) + 3\,H_2O(\ell) \tag{3}$$

$$\text{rate} = k[I^-]^m[BrO_3^-]^n[H^+]^p = -\frac{\Delta[BrO_3^-]}{t} \tag{4a}$$

In all the reaction mixtures used in this experiment, the color change occurs once a constant, predetermined number of moles of BrO_3^- have been consumed in the reaction. The color "clock" allows you to measure the time required for this fixed quantity of BrO_3^- to react. The rate of each reaction is determined by the time t required for the color to change; because in Equation 4a the change in concentration of BrO_3^- ion, $\Delta[BrO_3^-]$, is the same for each mixture, the rate of each reaction changes by the same factor that $1/t$ does. This experiment is mainly concerned with relative rates, and relative rates equal to (1000 sec)/t will be used because they are easier to compare. Fill in the following table, first calculating the relative reaction rate for each mixture:

Table 21.2

Reaction Mixture	Time, t {sec} for color to change	Relative rate of reaction, (1000 sec)/t	Reactant concentrations in reacting mixture {M}			Temperature {°C}
			[I⁻]	[BrO₃⁻]	[H⁺]	
1	══════	_____	0.0020	_____	_____	══════
2	══════	_____	_____	_____	_____	══════
3	══════	_____	_____	_____	_____	══════
4	══════	_____	_____	_____	_____	══════
5	══════	_____	_____	_____	_____	══════

The reactant concentrations in the reaction mixtures are *not* those of the stock solutions, because the reactants are diluted by the addition of the other solutions. The final volume of the reaction mixture is 50.0 mL in all cases. Because the moles of reactant present do not change on dilution, you know that for I^- ion, for example,

$$\text{moles of } I^- = [I^-]_{stock} \times V_{stock} = [I^-]_{mixture} \times V_{mixture} \tag{6}$$

For Reaction Mixture 1,

$$[I^-]_{stock} = 0.010\ M, \quad V_{I^-\ stock,\ Mixture\ 1} = 10.0\ mL, \quad V_{mixture} = 50.0\ mL$$

Therefore,

$$[I^-]_{Mixture\ 1} = \frac{0.010\ M \times 10.0\ mL}{50.0\ mL} = 0.0020\ M$$

Calculate the rest of the concentrations in the preceding table using the same approach.

Determining the orders of the reaction

You now need to find the order of Reaction 3 in each reactant, and its rate constant. Equation 4a can be adapted for relative rates as follows:

$$\text{relative rate} = k'[I^-]^m[BrO_3^-]^n[H^+]^p = \frac{1000\ sec}{t} \tag{4b}$$

(continued on following page)

You need to determine the relative rate constant k' and the orders m, n, and p that are consistent with your data. The solution to this problem requires comparison of the reaction mixtures. Each mixture (2 to 4) differs from Reaction Mixture 1 in the concentration of only one substance (see Table 21.2). This means that for any pair of mixtures that includes Reaction Mixture 1, there is only one concentration that changes. From the ratio of the relative rates for such a pair of mixtures you can find the order with respect to the reactant whose concentration was changed.

Write Equation 4b for Reaction Mixtures 1 and 2, substituting in the relative rates and the concentrations of I^-, BrO_3^-, and H^+ ions from Table 21.2:

Relative Rate 1 = _____ = $k'[$_____$]^m[$_____$]^n[$_____$]^p$

Relative Rate 2 = _____ = $k'[$_____$]^m[$_____$]^n[$_____$]^p$

Divide the Relative Rate 1 equation by the Relative Rate 2 equation; nearly all the terms cancel out:

$$\frac{\text{Relative Rate 1}}{\text{Relative Rate 2}} =$$

You should now have an equation involving only m as an unknown. Solve this equation for m, the order of Reaction 3 with respect to I^- ion:

$$m = \underline{\qquad} \text{ (round to the nearest whole number)}$$

Apply the same approach to Reaction Mixtures 1 and 3:

Relative Rate 1 = _____ = $k'[$_____$]^m[$_____$]^n[$_____$]^p$

Relative Rate 3 = _____ = $k'[$_____$]^m[$_____$]^n[$_____$]^p$

Divide the upper equation by the lower one to determine n, the order of Reaction 3 with respect to BrO_3^- ion:

$$n = \underline{\qquad} \text{ (round to the nearest whole number)}$$

Now that you have the idea, apply the same method one last time, to Reaction Mixtures 1 and 4:

Relative Rate 1 = _____ = $k'[$_____$]^m[$_____$]^n[$_____$]^p$

Relative Rate 4 = _____ = $k'[$_____$]^m[$_____$]^n[$_____$]^p$

Again divide the upper equation by the lower one, this time to calculate p, the order with respect to H^+ ion:

$$p = \underline{\qquad} \text{ (round to the nearest whole number)}$$

Having found m, n, and p, the relative rate constant, k', can be calculated by substituting m, n, p, and the known relative rates and reactant concentrations into Equation 4b. Evaluate k' for Reaction Mixtures 1 to 4:

Mixture	**1**	**2**	**3**	**4**
k'	_____	_____	_____	_____

Average (mean) k' _____ Standard deviation in k' _____ (See Appendix H.)

Why should k' have nearly the same value for each of these reaction mixtures? (*Hint*: What was kept constant?)

Using your average k' in Equation 4b, predict the relative rate and predict the time, $t_{5,\ predicted}$, for Reaction Mixture 5. Calculate and record concentrations in Table 21.2, then use them here.

(Relative Rate 5)$_{predicted}$ _____ $t_{5,\ predicted}$ _____ sec $t_{5,\ observed}$ _____ sec

B. How reaction rate changes with temperature: the activation energy, E_a

To find the activation energy for Reaction 3, it will be helpful to complete Table 21.3.
The temperature dependence of the rate constant, k', for Reaction 3 is given by Equation 2:

$$\ln k' = -\frac{E_a}{RT} + \text{constant} \tag{2}$$

Because the reactions at the different temperatures studied all involve the same reactant concentrations, the rate constants, k', for two different mixtures will have the same ratio as the reaction rates themselves for the two mixtures. This means that in calculating E_a you can use the observed relative rates instead of rate constants. Calculate the relative rate of reaction in each of the mixtures and enter these values on line (c) in Table 21.3. Take the natural logarithm of the relative rate for each mixture and enter these values on line (d). To calculate $1/T$, fill in lines (b), (e), and (f) in Table 21.3.

Table 21.3

	Approximate temperature in °C			
	20	40	10	0
(a) Time, t, in seconds, for color to appear	_____	_____	_____	_____
(b) Temperature of the reaction mixture in °C	_____	_____	_____	_____
(c) Relative rate $= \dfrac{1000 \text{ sec}}{t}$	_____	_____	_____	_____
(d) ln of relative rate	_____	_____	_____	_____
(e) Temperature T in K	_____	_____	_____	_____
(f) $1/T\ \{K^{-1}\}$	_____	_____	_____	_____

To evaluate E_a, make a graph of (ln of relative rate) vs $1/T$ using a spreadsheet (see Appendix G1 or G2), or the graph paper provided on the following page (with the help of Appendix E).
Find the slope of the line obtained by drawing the best straight line through the experimental points.

Slope = _____ K

The slope of the line equals $-E_a/R$, where $R = (8.3145$ joules/mole·K) if E_a is to be in joules per mole. Calculate the activation energy, E_a, for the reaction.

$E_a =$ _____ joules/mole

Optional   ## C. How reaction rate changes in the presence of a catalyst

	Uncatalyzed Reaction 3	Catalyzed Reaction 3
Time for color to appear {seconds}	_____	_____

Would you expect the activation energy, E_a, for the catalyzed reaction to be greater than, less than, or equal to the activation energy for the uncatalyzed reaction? Why?

Experiment 21

Data and Calculations: Rates of Chemical Reactions,
2. A Clock Reaction

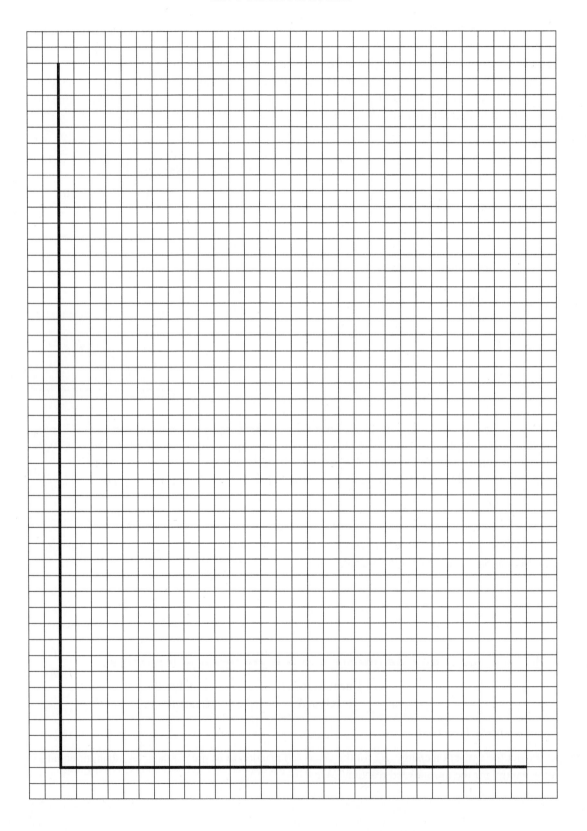

Experiment 21

Advance Study Assignment: Rates of Chemical Reactions,
2. A Clock Reaction

1. A student studied Reaction 3 using the method described in this experiment. They set up Reaction Mixture 2 by mixing 20.0 mL of 0.010 M KI, 10.0 mL of 0.0010 M $Na_2S_2O_3$, 10.0 mL of 0.040 M $KBrO_3$, and 10.0 mL of 0.10 M HCl. It took 44 sec for the color to turn blue.

   a. They found the concentration of each reactant in the reacting mixture by realizing that the moles of each reactant did not change when that reactant was mixed with the others, but its concentration did. For any reactant A,

   $$\text{moles of A} = [A]_{stock} \times V_{stock} = [A]_{mixture} \times V_{mixture} \tag{7}$$

   The volume of the mixture was 50.0 mL. Plugging this in and rearranging, they obtained

   $$[A]_{mixture} = [A]_{stock} \times \frac{V_{stock}\{mL\}}{50.0 \text{ mL}} \tag{7a}$$

   Find the concentrations of each reactant in the reaction mixture, using Equation 7a.

   $[I^-]_{mixture} = $ _____ M $[BrO_3^-]_{mixture} = $ _____ M $[H^+]_{mixture} = $ _____ M

   b. What was the relative rate of the reaction, $\dfrac{1000 \text{ sec}}{t}$? _____

   c. Knowing the relative rate of reaction for Mixture 2 and the concentrations of I^-, BrO_3^-, and H^+ in that mixture, the student was ready to use Equation 4b. The only quantities that remained unknown were k', m, n, and p. Substitute the values calculated in parts (a) and (b) into Equation 4b:

(continued on following page)

2. For Reaction Mixture 1, the student found that 92 seconds were required. On dividing Equation 4b for Reaction Mixture 1 by Equation 4b for Reaction Mixture 2, and after canceling out the common terms (k' and terms in $[BrO_3^-]$ and $[H^+]$), they got the following equation:

$$\frac{10._9}{22._7} = \left(\frac{0.020}{0.040}\right)^m = \left(\frac{1}{2}\right)^m$$

Recognizing that $10._9/22._7 = 0.48$ is about equal to one half, they obtained an approximate value for m. What was that value?

$$m = \text{_____}$$

By taking the logarithm of both sides of the equation, the student got an exact value for m. What was that value?

$$m = \text{_____}$$

Because reaction orders are usually whole numbers, the student rounded their value of m to the nearest whole number and reported that value as the order of the reaction with respect to I^-.

Experiment 22

Properties of Systems in Chemical Equilibrium—Le Châtelier's Principle

When working in the laboratory, we often make observations that at first seem surprising and hard to explain. We might add a chemical to a solution and obtain a precipitate; surprisingly, adding more of that same chemical causes the precipitate to dissolve. A violet solution might turn yellow upon adding a chemical to it, but adding another chemical produces first a green color and then the original violet. Clearly, chemical reactions are occurring, but how and why they behave as they do is not immediately obvious.

In this experiment you will examine and attempt to explain several observations of the sort just described. Central to their explanation will be recognizing that chemical systems tend to exist in a state of equilibrium. If the equilibrium is disturbed in one way or another, the reaction may shift to the left or right, producing the kinds of effects just mentioned. If the principles governing the equilibrium system can be understood, it is often possible to see how to disturb the system, such as by adding a particular chemical or form of energy, to cause it to change in a desired way.

Before proceeding to specific examples, let us examine the general situation, noting the key principle that allows us to make a system in equilibrium behave as we wish. Consider the reaction

$$A(aq) \rightleftharpoons B(aq) + C(aq) \tag{1}$$

where A, B, and C are molecules or ions in solution. If we have a mixture of these substances in equilibrium, there is a condition that their concentrations must meet, namely that

$$\frac{[\underline{B}] \times [\underline{C}]}{[\underline{A}]} = K \tag{2}$$

where K is a constant, called the <u>equilibrium constant</u> for the reaction. For a given reaction at a given temperature, K has a specific value. The concentrations, for example [$\underline{B}$], are underlined to indicate two important things: first, that for the equation to be true the concentration must be for a system *at equilibrium*, and second, that it must be measured in moles per liter but used *without units*. Therefore K has no units.

That at a given temperature K has a specific value has important consequences. For example, we might find that for a given solution in which Reaction 1 can occur the equilibrium values for the molarities of A, B, and C have certain values. If these are substituted into Equation 2, we get a value of 10 for K. Now, suppose that more of substance A is added to that solution. What will happen? Remember, K can not change if the temperature stays the same. If we substitute the new, higher molarity of A into Equation 2 we get a value that is smaller than K. This means that the system is not in equilibrium, and it *must* change in some way to get back to equilibrium. How can it do this? By shifting Reaction 1 to the right: producing more B and C, and using up some A. It *must* do this, and *will* (though it may be quickly or slowly), until the molarities of C, B, and A reach values that, on substitution into Equation 2, equal 10. At that point the system is once again in equilibrium. In the new equilibrium state, [B] and [C] are greater than they were initially, and [A] is larger than its initial value but smaller than if Reaction 1 had not shifted to the right, partly offsetting the increase.

The conclusion you should reach on reading the previous paragraph is that *we can always cause an equilibrium to shift to the right by increasing the concentration of a reactant. An increase* in the concentration of a *product* will force a *shift* to the *left*. By a similar argument a *decrease* in a *reactant* concentration causes a *shift* to the *left*; a *decrease* in a *product* concentration produces a *shift* to the *right*. This is all true because K does not change (unless the temperature changes). The changes in concentration that we can produce by adding particular chemicals may be huge, so the shifts in the equilibrium may also be huge. A lot of otherwise mysterious chemical behavior can be understood using this concept.

Another way an equilibrium system can be disturbed is by changing its temperature. When this happens, the value of K changes. The change in K depends upon the enthalpy change, ΔH, for the reaction. If ΔH is positive, greater than zero (an "underline{endothermic} reaction"), K increases with increasing T. If ΔH is negative (an "exothermic reaction"), K decreases with an increase in T. Let us return to our original equilibrium between A, B, and C, for which $K = 10$. Let us assume that ΔH for Reaction 1 is -40 kJ. If we raise the temperature, K will go down ($\Delta H < 0$), say to a value of 1. This means the system will no longer be in equilibrium. Substituting the initial values of [A], [B], and [C] into Equation 2 produces a value that is too big: 10 instead of 1. How can the system change itself to regain equilibrium? It must shift Reaction 1 to the left, lowering [B] and [C] and raising [A]. This will make the concentration ratio in Equation 2 smaller. The shift will continue until the concentrations of A, B, and C, on substitution into Equation 2, give the fraction a value of 1. Note that warming an exothermic reaction causes it to shift toward the reactants: meaning it goes in the reverse direction, which is endothermic and cools the system, partly offsetting the change made. Warming an endothermic reaction causes the reaction to shift in the forward direction, creating more products and absorbing thermal energy, also cooling the system and partly offsetting the imposed change.

The discussion in the previous paragraph indicates that an equilibrium system will shift to the left on being warmed if the reaction is exothermic ($\Delta H < 0$, K decreases with temperature). It will shift to the right if the reaction is endothermic ($\Delta H > 0$, K increases with temperature). Again, because we can change temperatures over a wide range, we can shift equilibria (for reactions with ΔH far from zero) a lot, too. An endothermic reaction that at 25 °C has an equilibrium state that consists mainly of reactants might, at 1000 °C, shift to consist of almost entirely products.

The effects of concentration and temperature on systems in chemical equilibrium are often summarized by Le Châtelier's principle, which states:

If you change a system in chemical equilibrium, it will react in such a way as to partially counteract the change you made.

Le Châtelier's principle predicts the same behavior as our analysis based on the properties of K. Increasing the concentration of a reactant will cause a change that decreases that concentration; that change must be a shift to the right. Increasing the temperature of a reaction mixture will cause a change that tends to absorb thermal energy: that change must be a shift in the endothermic direction. In some cases the principle requires more careful reasoning than the more direct approach we employed. For the most part we will find it more useful to base our arguments on the properties of K.

In working with aqueous (water-as-the-solvent) systems, the most important equilibrium is often the dissociation (chemical breaking apart) of water into H^+ and OH^- ions:

$$H_2O(\ell) \rightleftharpoons H^+(aq) + OH^-(aq) \qquad K = [\underline{H^+}]\,[\underline{OH^-}] = 1 \times 10^{-14} = K_w \qquad (3)$$

In this reaction the concentration of water is essentially constant at 55 M, so it is incorporated into K_w. The value of K_w is very small, which means that in *any* aqueous system at equilibrium the product of [$\underline{H^+}$] and [$\underline{OH^-}$] must be very small. In pure water at 25 °C, [H^+] and [OH^-] both equal 1×10^{-7} M.

Although the product $[\underline{H^+}] \times [\underline{OH^-}]$ is small, this does not mean that both concentrations are necessarily small. If, for example, we dissolve HCl in water, the HCl in the solution will dissociate completely to H^+ and Cl^- ions; in 1.0 M HCl, [H^+] will become 1.0 M, and there is nothing that Reaction 3 can do about changing that concentration appreciably. Rather, Reaction 3 must shift to maintain equilibrium. It does this by lowering [OH^-] by reaction to the left; this uses up a very small amount of H^+ ion and drives [OH^-] to the value it must have when [H^+] is 1.0 M: namely, 1×10^{-14} M. In 1.0 M HCl, [OH^-] is a factor of 10 million times *smaller* than it is in water. This makes the properties of 1.0 M HCl very different from those of pure water, particularly where H^+ and OH^- ions are involved.

If we take 1.0 M HCl and add an aqueous solution of NaOH to it, an interesting situation develops. Like HCl, NaOH dissociates completely in water: so in 1.0 M NaOH, [OH^-] is equal to 1.0 M. If we add 1.0 M NaOH to 1.0 M HCl, we will initially raise [OH^-] way above 1×10^{-14} M. However, Reaction 3 cannot be in equilibrium when both [H^+] and [OH^-] are high; reaction must occur to re-establish equilibrium. The added OH^- ions react with H^+ ions (by the reverse of Reaction 3) to form H_2O, decreasing both concentrations until equilibrium is re-established. If only a small amount of OH^- ion is added, it will essentially all be used up; [H^+] will remain high, and [OH^-] will still be very small, but somewhat larger than 10^{-14} M. If we add OH⁻ ion until the amount added (in moles) exactly equals the amount of H⁺ contributed by the HCl, then Reaction 3 will go to the left until $[H^+] = [OH^-] = 1 \times 10^{-7}$ M, and both concentrations will be very small. Adding more OH⁻ ion will raise [OH⁻] to much higher values, easily as high as 1 M. At that point, [H⁺] would be very low: 1×10^{-14} M. So, in aqueous solution, depending on the <u>solutes</u> (dissolved substances) present, we can have [H⁺] and [OH⁻] range from 1 M to 10^{-14} M—spanning 14 orders of magnitude! This will have a tremendous effect on any *other* equilibrium system in which [H⁺] or [OH⁻] ions are reactants or products. Similar situations arise in other equilibrium systems in which the concentration of a reactant or product can be changed significantly by adding a particular chemical.

In many equilibrium systems, several equilibria are present simultaneously. For example, in aqueous solution, Reaction 3 must *always* be in equilibrium. There may, in addition, be equilibria between solutes in the aqueous solution. Some examples are those in Reactions 4, 5, 7, 8, 9, and 10 in the Experimental Procedure section. In some of those reactions, H⁺ and OH⁻ ions appear; in others, they do not. In Reaction 4, for example, H⁺ ion is a product. The molarity of H⁺ in Reaction 4 is *not* determined by the indicator (in this case methyl violet, or HMV), because it is only present in a tiny amount. Reaction 4 will have an equilibrium state that is fixed by the state of Reaction 3, which as we have seen depends strongly on the presence of acidic or basic solutes such as HCl or NaOH. Reactions 8 and 9 can also be controlled by Reaction 3. Reactions 5 and 7, which do not involve H⁺ or OH⁻ ions, are not dependent on Reaction 3 for their equilibrium state. Reaction 10 is sensitive to NH₃ concentration and can be driven far to the right by adding a source of it, such as 6 M NH₃.

Experimental Procedure

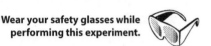

Wear your safety glasses while performing this experiment.

In this experiment you will work with several equilibrium systems, each of which is similar to the A-B-C system of Reaction 1. You will alter these systems in various ways, forcing shifts to the right and left by changing concentrations and temperature. You will be asked to interpret your observations in terms of the principles presented in the introduction of this experiment.

A. Acid-base indicators

There is a large group of chemical substances, called acid-base indicators, that change color in solution in response to the concentration of hydrogen ion, [H⁺]. A typical substance of this sort is called methyl violet, to which we will give the abbreviated chemical formula HMV. In aqueous solution, HMV dissociates as follows:

$$HMV(aq) \rightleftharpoons H^+(aq) + MV^-(aq) \qquad (4)$$
$$\text{yellow} \qquad\qquad\qquad \text{violet}$$

HMV has an intense yellow color, while the negatively charged ion (technical term: <u>anion</u>) MV⁻ is violet. Because Reaction 4 is an equilibrium, the color of the indicator in solution depends very strongly on [H⁺].

Step 1. Add about five milliliters of deionized water to a medium (18×150 mm) test tube. Add a few drops of methyl violet (HMV) indicator. Report the color of the solution on the report page.

Step 2. How could you force the equilibrium system to go to the other form (color)? Select a solution that should do this and add it to the solution, drop by drop, until the color change is complete. If your chosen solution works, record it on the report page. If it does not, start over and try another, until you find one that does. Work with 6 M solutions if they are available.

Step 3. Equilibrium systems are reversible. That is, the reaction can be driven to the left and right at will, over and over, by changing the conditions in the system. How can you force the system in Step 2 to return to its original color? Select a solution that should do this and add it, drop by drop, until the color has become the original one. Again, if your first choice does not work, try another solution. On the report page record the solution that was effective. Answer all the questions for Part A before going on to Part B.

B. Solubility equilibria; finding a value for K_{sp}

Many ionic salts have limited solubility in water. A typical example is lead(II) chloride, $PbCl_2$, which dissolves to a limited extent in water according to the reaction

$$PbCl_2(s) \rightleftharpoons Pb^{2+}(aq) + 2\,Cl^-(aq) \tag{5}$$

The equilibrium expression for Reaction 5 is

$$K = [\underline{Pb^{2+}}] \times [\underline{Cl^-}]^2 = K_{sp} \tag{6}$$

The $PbCl_{2(s)}$ does not enter into the equilibrium expression because it is an undissolved solid, and so has a constant effect on the system, independent of how much there is—provided there is *some* solid present. The equilibrium constant for a solubility equilibrium is called the <u>solubility product</u>, and is given the special symbol K_{sp}.

For Reaction 5 to be in equilibrium, and for Equation 6 to apply, there *must* be some solid $PbCl_2$ present in the system. If there is no solid, there is no equilibrium: Equation 6 is *not* obeyed, and $[Pb^{2+}] \times [Cl^-]^2$ must be *less* than the value of K_{sp}. If solid $PbCl_2$ *is* present, even in a tiny amount, then the values of $[Pb^{2+}]$ and $[Cl^-]$ *are* subject to the requirements of Equation 6.

Step 1. Set up a hot-water bath, using a 400-mL beaker half full of water. Start heating the water while you continue to Step 2.

Step 2. To a medium (18×150 mm) test tube add about five milliliters of 0.3 M $Pb(NO_3)_2$; this is a depth of about three centimeters in these tubes. In this solution $[Pb^{2+}]$ equals 0.3 M. Add 5.0 mL of 0.3 M HCl to a 10-mL graduated cylinder. In this solution $[Cl^-]$ equals 0.3 M. Add 1.0 mL of the HCl solution to the $Pb(NO_3)_2$ solution. What happens? Swirl the test tube, then wait for about fifteen seconds. What happens? Record your observations.

Step 3. Add the HCl solution to the test tube in 1.0-mL increments until a noticeable amount of white solid $PbCl_2$ remains after swirling the tube. Record the total volume of HCl solution added at that point.

Step 4. Put the test tube with the $PbCl_2$ precipitate in it into the hot-water bath. Swirl the test tube for a few moments. What happens? Record your observations. Cool the test tube under the cold-water tap. What happens? Record this observation as well.

Step 5. Rinse out your graduated cylinder, then add 5.0 mL of deionized water to it. Add a bit of this water at a time, in 1.0-mL increments, to the test tube, swirling well after each addition. When the precipitate completely dissolves, record the total amount of water you added to it in this step. Answer the questions and do the calculations in Part B before proceeding.

C. Complex ion equilibria

Many metallic ions in solution exist not as simple ions but as <u>complex ions</u>, surrounding themselves with other ions or molecules, called <u>ligands</u>. For example, the Co^{2+} ion in water exists as the pink $Co(H_2O)_6^{2+}$ complex ion, and Cu^{2+} as the blue $Cu(H_2O)_6^{2+}$ complex ion. In both of these complex ions the ligands are H_2O molecules. Complex ions are reasonably stable but may be converted to other complex ions upon adding ligands that form more stable complexes than the original ones. Among the common ligands that may form complexes, OH^-, NH_3, and Cl^- are important.

An interesting Co(II) complex is the $CoCl_4^{2-}$ ion, which is blue. This ion is stable in concentrated Cl^- solutions. Depending upon conditions, Co(II) in solution may exist as either $Co(H_2O)_6^{2+}$ or as $CoCl_4^{2-}$. The principles of chemical equilibrium can be used to predict which ion will be present:

$$Co(H_2O)_6^{2+}(aq) + 4\,Cl^-(aq) \rightleftharpoons CoCl_4^{2-}(aq) + 6\,H_2O(\ell) \tag{7}$$

Step 1. Put a few small crystals (about one tenth of a gram) of $CoCl_2 \cdot 6H_2O$ in a medium (18×150 mm) test tube. Add 2 mL of 12 M HCl. **CAUTION:** **12 M HCl is a concentrated solution of an acidic gas; avoid contact with it and its fumes. HCl is a strong acid with a choking odor.** Swirl the test tube to dissolve the crystals. Record the color of the solution.

Step 2. Add 2-mL portions of deionized water, swirling after each dilution, until no further color change occurs. Record the new color.

Step 3. Place the test tube into a hot-water bath and note any change in color. Cool the tube under cold tap water and report your observations. Complete the questions in Part C before continuing.

D. Dissolving insoluble solids

You saw in Part B that you can dissolve more $PbCl_2$ into a saturated solution by either heating it or adding water. These procedures will not work very well in general, however, because most solids are much less soluble than $PbCl_2$.

There are, however, some very powerful methods for dissolving solids; their effectiveness depends upon the principles of equilibrium. As an example of a nearly insoluble substance, consider $Zn(OH)_2$:

$$Zn(OH)_2(s) \rightleftharpoons Zn^{2+}(aq) + 2\,OH^-(aq) \qquad K_{sp} = 5 \times 10^{-17} = [Zn^{2+}] \times [OH^-]^2 \qquad \textbf{(8)}$$

The equilibrium constant K_{sp} for Reaction 8 is very small, which tells us that the reaction does not go very far to the right or, equivalently, that $Zn(OH)_2$ is almost completely insoluble in water. Adding a single drop of a solution containing OH^- ion to one containing Zn^{2+} ion will cause $Zn(OH)_2$ to precipitate.

At first you might wonder how you could possibly dissolve, say, 1 mole of $Zn(OH)_2$ in an aqueous solution. However, examining Equation 8 you can see, from the equation for K_{sp}, that in the saturated solution $[Zn^{2+}] \times [OH^-]^2$ must equal 5×10^{-17}. If that product can somehow be lowered to a value *below* 5×10^{-17}, then $Zn(OH)_2$ will dissolve until the product again becomes equal to K_{sp}—at which point equilibrium will again exist. To lower the product, the concentration of either Zn^{2+} or OH^- must be lowered drastically. One way to lower $[OH^-]$ is to add H^+ ions from an acid. That drives Reaction 3 to the left, making $[OH^-]$ very small—small enough to dissolve substantial amounts of $Zn(OH)_2$.

Alternately, $[Zn^{2+}]$ can be lowered by taking advantage of the fact that zinc(II) forms stable complex ions with both OH^- and NH_3:

$$Zn^{2+}(aq) + 4\,OH^-(aq) \rightleftharpoons Zn(OH)_4^{2-}(aq) \qquad K_f = K_9 = 3 \times 10^{15} \qquad \textbf{(9)}$$

$$Zn^{2+}(aq) + 4\,NH_3(aq) \rightleftharpoons Zn(NH_3)_4^{2+}(aq) \qquad K_f = K_{10} = 1 \times 10^9 \qquad \textbf{(10)}$$

In high concentrations of OH^- ion, Reaction 9 is driven strongly to the right, making $[Zn^{2+}]$ very low. The same thing would happen with Reaction 10 in solutions containing high concentrations of NH_3. In both situations we would therefore expect that $Zn(OH)_2$ might dissolve, because if $[Zn^{2+}]$ is very low, Reaction 8 must go to the right.

Step 1. To each of three small (13×100 mm) test tubes add about two milliliters (a depth of about two centimeters in these tubes) of 0.10 M $Zn(NO_3)_2$. In this solution $[Zn^{2+}]$ equals 0.10 M. To each test tube add 1 drop of 6 M NaOH and swirl the tube to mix its contents. Record your observations.

Step 2. To the first tube add 6 M HCl, drop by drop, swirling to mix the contents after each addition. To the second tube add 6 M NaOH, again drop by drop. To the third tube add 6 M NH_3, a drop at a time, swirling after each drop. Note what happens in each case.

Step 3. Repeat Steps 1 and 2, this time starting with three test tubes containing 0.10 M $Mg(NO_3)_2$. Record your observations. Answer the questions in Part D.

Disposal of reaction products: The waste materials from Parts B and C should be poured into a waste container. Those from Parts A and D may be poured down the sink unless your instructor directs otherwise.

Experiment 22

Observations and Analysis: Properties of Systems in Chemical Equilibrium—
Le Châtelier's Principle

A. Acid-base indicators

1. Color of methyl violet in water _____

2. Solution causing color change _____

3. Solution causing shift back _____.

   Explain, by considering how changes in $[H^+]$ will cause Reaction 4 to shift to the right and left, why the solutions in Steps 2 and 3 caused methyl violet to change color. Note that Reactions 3 and 4 must both come to equilibrium after a solution is added.

B. Solubility equilibria; finding a value for K_{sp}

2. 5 mL of 0.3 M $Pb(NO_3)_2$:   moles of Pb^{2+} = $[Pb^{2+}]_{stock} \times V_{stock}$ = 0.3 M $\times$ 0.005 L = 1.5×10^{-3} moles

   1.0 mL of 0.3 M HCl: moles of Cl^- = $[Cl^-]_{stock} \times V_{stock}$ = 0.3 M $\times$ 0.0010 L = 0.3×10^{-3} moles

   Observations:

3. Total volume of 0.3 M HCl used: _____ mL; total moles of Cl^- added: _____ moles

4. Observations in hot water: _____

   in cold water: _____

5. Volume of H_2O added to dissolve $PbCl_2$: _____ mL

   Total volume of solution: _____ mL

   a. Explain why $PbCl_2$ did not precipitate immediately upon adding HCl. (What condition must be met by $[Pb^{2+}]$ and $[Cl^-]$ if solid $PbCl_2$ is to be present at equilibrium?)

(continued on following page)

b. Explain your observations in Step 4. (In which direction did Reaction 5 shift when heated? What must have happened to the value of K_{sp} in the hot solution? What does this tell you about the sign of ΔH in Reaction 5?)

c. Explain why the $PbCl_2$ dissolved when water was added in Step 5. (What was the effect of the added water on $[Pb^{2+}]$ and $[Cl^-]$? In what direction would such a change drive Reaction 5?)

d. Given the moles of Pb^{2+} and Cl^- present in the final solution in Step 5, and the volume of that solution, calculate $[Pb^{2+}]$ and $[Cl^-]$ in that solution.

$[Pb^{2+}] = $ _____ M; $[Cl^-] = $ _____ M

Noting that the molarities just calculated are essentially those in equilibrium with solid $PbCl_2$, calculate $[Pb^{2+}] \times [Cl^-]^2$. This is equal to K_{sp} for $PbCl_2$.

$K_{sp} = $ _____

C. Complex ion equilibria

1. Color of solid $CoCl_2 \cdot 6H_2O$ ＿＿＿＿＿＿

Color in solution in 12 M HCl ＿＿＿＿＿＿

2. Color in diluted solution ＿＿＿＿＿＿

3. Color of hot solution ＿＿＿＿＿＿

Color of cooled solution ＿＿＿＿＿＿

Formula of Co(II) complex ion present in solution in

a. 12 M HCl ＿＿＿＿＿＿＿＿＿

b. diluted solution ＿＿＿＿＿＿＿＿＿

c. hot solution _____

d. cooled solution _____

Explain the color change that occurred when

a. water was added in Step 2. (Consider how a change in [Cl$^-$] and [H$_2$O] will shift Reaction 7.)

b. the diluted solution was heated. (How would Reaction 7 shift if K_f went up? How did increasing the temperature affect the value of K_f? What does this indicate the sign of ΔH is for Reaction 7?)

D. Dissolving insoluble solids

1. Observations upon adding 1 drop of 6 M NaOH to 0.10 M Zn(NO$_3$)$_2$ solution:

2. Effect on solubility of Zn(OH)$_2$:

   a. of added HCl solution

   b. of added NaOH solution

   c. of added NH$_3$ solution

3. Observations upon adding 1 drop of 6 M NaOH to 0.10 M Mg(NO$_3$)$_2$ solution:

   Effect on solubility of Mg(OH)$_2$:

   a. of added HCl solution

   b. of added NaOH solution

   c. of added NH$_3$ solution

(continued on following page)

Explain your observations in Step 1. (Consider Reaction 8; how is it affected by adding OH^- ion?)

In Step 2(a), how does an increase in $[H^+]$ affect Reaction 3? (What does that do to Reaction 8?) Explain your observations in Step 2(a).

In Step 2(b), how does an increase in $[OH^-]$ affect Reaction 9? What does that do to Reaction 8? Explain your observations in Step 2(b).

In Step 2(c), how does an increase in $[NH_3]$ affect Reaction 10? What does that do to Reaction 8? Explain your observations in Step 2(c).

In Step 3, you probably found that $Mg(OH)_2$ was similar in some ways in its behavior to that of $Zn(OH)_2$, but different in others.

   a. How was it similar? Explain that similarity. (In particular, why would any insoluble hydroxide tend to dissolve in acidic solution?)

   b. How was it different? Explain that difference. (In particular, does Mg^{2+} appear to form complex ions with OH^- and/or NH_3? What would you observe if it did? If it did not?)

Experiment 22

Advance Study Assignment: Properties of Systems in Chemical
Equilibrium—Le Châtelier's Principle

1. Methyl red, HMR, is a common acid-base indicator. In solution it ionizes according to the equation:

$$HMR(aq) \rightleftharpoons H^+(aq) + MR^-(aq) \tag{11}$$
$$\text{red} \qquad\qquad\qquad \text{yellow}$$

   If methyl red is added to deionized water, the solution turns yellow. If one or two drops of 6 M HCl are added to the yellow solution, it turns red. If a few drops of 6 M NaOH are then added to the solution, the color reverts to yellow.

   a. Why does adding 6 M HCl to the yellow solution of methyl red cause the color to change to red? (Note that in solution HCl exists as H^+ and Cl^- ions.)

   b. Why does adding 6 M NaOH to the red solution tend to make it turn back to yellow? Note that in solution NaOH exists as Na^+ and OH^- ions. (*Hint*: How does increasing $[OH^-]$ shift Reaction 3? How would the resulting change in $[H^+]$ affect Reaction 11? Explain these as part of your answer.)

(continued on following page)

2. Magnesium hydroxide is only slightly soluble in water. The reaction by which it goes into solution is:

$$Mg(OH)_2(s) \rightleftharpoons Mg^{2+}(aq) + 2OH^-(aq) \qquad (12)$$

a. Write the equation for the equilibrium constant, K_{sp}, for Reaction 12.

b. It is possible to dissolve significant amounts of $Mg(OH)_2$ in solutions in which the concentration of either Mg^{2+} or OH^- is kept very small. Explain, using K_{sp}, why this is the case.

c. Explain why $Mg(OH)_2$ might have very appreciable solubility in 1.0 M HCl. (*Hint*: Consider the effect of Reaction 3 on Reaction 12.)

Experiment 23

Determining the Equilibrium Constant for a Chemical Reaction

When chemical substances react, the reaction typically does not go to completion. Rather, the system goes to some intermediate state in which both the reactants and products have concentrations that do not change with time. Such a system is said to be in <u>chemical equilibrium</u>. When in equilibrium at a particular temperature, a reaction mixture must satisfy a mathematical condition on the concentrations of reactants and products requiring that some ratio of them be equal to the equilibrium constant K for the reaction.

In this experiment you will study the equilibrium properties of the reaction between the iron(III) ion, Fe^{3+}, and the thiocyanate ion, SCN^-:

$$Fe^{3+}(aq) + SCN^-(aq) \rightleftharpoons FeSCN^{2+}(aq) \qquad (1)$$
$$\text{colorless} \qquad \text{colorless} \qquad \text{red-orange}$$

When solutions containing Fe^{3+} and SCN^- are mixed, Reaction 1 occurs to some extent, forming the iron (III) thiocyanate complex ion, $FeSCN^{2+}$, which has a deep red-orange color. As a result of the reaction, the equilibrium amounts of Fe^{3+} and SCN^- will be less than they would have been if no reaction had occurred; for every mole of $FeSCN^{2+}$ that is formed, 1 mole of Fe^{3+} and 1 mole of SCN^- will be consumed in Reaction 1. The equilibrium constant for Reaction 1 is called a <u>formation constant</u>, K_f, because it describes the formation of a complex ion:

$$\frac{[FeSCN^{2+}]}{[Fe^{3+}][SCN^-]} = K_f \qquad (2)$$

The value of K_f in Equation 2 is constant at a given temperature. This means that mixtures containing Fe^{3+} and SCN^- will react until Equation 2 is satisfied, so that the same value of K_f will be obtained no matter what initial amounts of Fe^{3+} and SCN^- were used. Your goal in this experiment will be to find K_f for this reaction for several different mixtures, and to show that K_f does indeed have the same value in each of the mixtures. This reaction is a good one to study because the equilibrium does not strongly favor either products or reactants, and the color of the $FeSCN^{2+}$ ion makes analysis of the equilibrium mixture relatively easy.

You will be preparing the mixtures by combining stock solutions containing known concentrations of iron(III) nitrate, $Fe(NO_3)_3$, and potassium thiocyanate, KSCN. The color of the $FeSCN^{2+}$ ion formed will allow you to determine its equilibrium concentration. Knowing the initial composition of a mixture and the equilibrium concentration of $FeSCN^{2+}$, you will be able to calculate the equilibrium concentrations of the reactants and then determine K_f.

Because the calculations required in this experiment are complicated, we will go through a step-by-step procedure by which they can be carried out. As a specific example, suppose you prepare a mixture by combining 10.0 mL of 2.00×10^{-3} M $Fe(NO_3)_3$ with 10.0 mL of 2.00×10^{-3} M KSCN. As a result of Reaction 1, some $FeSCN^{2+}$ is formed. Based on the intensity of its color, you find the concentration of $FeSCN^{2+}$ at equilibrium to be 1.50×10^{-4} M. Your goal is to find K_f for the reaction from this information. To do this you first need to find the initial moles of each reactant in the mixture. Second, you determine how many moles of product were present at equilibrium. Because the product was formed from the reactants, you can calculate the amount of each reactant that was used up. In the third step, you calculate the moles of each reactant remaining in the equilibrium mixture. Fourth, you determine the equilibrium concentration of each reactant. Finally, in the fifth step, you evaluate K_f for Reaction 1, using Equation 2.

Step 1. Finding the initial moles of each reactant From the volumes and concentrations of the stock solutions that were mixed, the moles of each reactant initially present can be calculated. The molarity, [A], of a substance A is defined as

$$[A] := \frac{\text{moles of A}}{\text{liters of solution, } V} \quad \text{so} \quad \text{moles of A} = [A] \times V \tag{3}$$

Using Equation 3, we find the initial moles of Fe^{3+} and SCN^-. For each stock solution the volume used was 10.0 mL, or 0.0100 L. The molarity of each of the stock solutions was 2.00×10^{-3} M, so $[Fe^{3+}]_{\text{stock}} = 2.00 \times 10^{-3}$ M and $[SCN^-]_{\text{stock}} = 2.00 \times 10^{-3}$ M. Therefore,

initial moles $Fe^{3+} = [Fe^{3+}]_{\text{stock}} \times V_{\text{stock}} = 2.00 \times 10^{-3}$ M $\times 0.0100$ L $= 20.0 \times 10^{-6}$ moles

initial moles $SCN^- = [SCN^-]_{\text{stock}} \times V_{\text{stock}} = 2.00 \times 10^{-3}$ M $\times 0.0100$ L $= 20.0 \times 10^{-6}$ moles

Step 2. Finding the moles of product formed The concentration of $FeSCN^{2+}$ was found to be 1.50×10^{-4} M at equilibrium. The volume of the mixture at equilibrium is the *sum* of the two stock solution volumes: 20.0 mL, or 0.0200 L. So, by Equation 3,

moles $FeSCN^{2+} = [FeSCN^{2+}] \times V_{\text{mixture}} = 1.50 \times 10^{-4}$ M $\times 0.0200$ L $= 3.00 \times 10^{-6}$ moles

The moles of Fe^{3+} and SCN^- *used up* in producing the $FeSCN^{2+}$ also must both be equal to 3.00×10^{-6} moles because, by Reaction 1, it takes *one mole* of Fe^{3+} and *one mole* of SCN^- to make each mole of $FeSCN^{2+}$.

Step 3. Finding the moles of each reactant present at equilibrium In Step 1 we determined that 20.0×10^{-6} moles Fe^{3+} and 20.0×10^{-6} moles SCN^- were initially present. In Step 2 we found that in the reaction 3.00×10^{-6} moles of Fe^{3+} and 3.00×10^{-6} moles of SCN^- were used up. The moles present at equilibrium must equal the number we started with minus the number that reacted. Therefore, *at equilibrium*,

$$\text{moles at equilibrium} = \text{initial moles} - \text{moles used up} \tag{4}$$

equilibrium moles of $Fe^{3+} = 20.0 \times 10^{-6} - 3.00 \times 10^{-6} = 17.0 \times 10^{-6}$ moles

equilibrium moles of $SCN^- = 20.0 \times 10^{-6} - 3.00 \times 10^{-6} = 17.0 \times 10^{-6}$ moles

Step 4. Find the concentrations of all substances at equilibrium Experimentally, we obtained the equilibrium concentration of $FeSCN^{2+}$ directly. $[FeSCN^{2+}] = 1.50 \times 10^{-4}$ M. The concentrations of Fe^{3+} and SCN^- can be calculated using Equation 3. How many moles of each of these substances were present at equilibrium was found in Step 3. The volume of the mixture being studied was 20.0 mL, or 0.0200 L. So, *at equilibrium*,

$$[Fe^{3+}]_{\text{equil}} = \frac{\text{equilibrium moles of } Fe^{3+}}{\text{volume of mixture}} = \frac{17.0 \times 10^{-6} \text{ moles}}{0.0200 \text{ L}} = 8.50 \times 10^{-4} \text{ M}$$

$$[SCN^-]_{\text{equil}} = \frac{\text{equilibrium moles of } SCN^-}{\text{volume of mixture}} = \frac{17.0 \times 10^{-6} \text{ moles}}{0.0200 \text{ L}} = 8.50 \times 10^{-4} \text{ M}$$

Step 5. Finding the value of K_f for the reaction Once the equilibrium concentrations of the reactants and products are all known, substituting their values into Equation 2 yields K_f:

$$K_f = \frac{[FeSCN^{2+}]}{[Fe^{3+}][SCN^-]} = \frac{1.50 \times 10^{-4}}{(8.50 \times 10^{-4}) \times (8.50 \times 10^{-4})} = 208$$

In this experiment you will obtain data similar to that shown in this example and process it in the same way. However, your results are likely to differ because the data in this example was probably obtained at a different temperature, and so reflects a different value of K_f.

In carrying out this analysis we made the assumption that the reaction that occurred was Reaction 1. There is no inherent reason why the reaction might not have been

$$Fe^{3+}(aq) + 2\ SCN^-(aq) \rightleftharpoons Fe(SCN)_2^+(aq) \tag{5}$$

You might ask how we know whether Reaction 1 or Reaction 5 is actually taking place. The line of reasoning is that if Reaction 1 is occurring, K_f for that reaction as we calculate it should remain constant with different mixtures. If, however, Reaction 5 is happening, K_f as calculated for *that* reaction should remain constant. The optional section of the report page assumes Reaction 5 occurs and analyzes K_f on that basis. The results of the two sets of calculations should indicate that it is Reaction 1 that yields a consistent value for K_f.

Two different analytical methods can be used to determine $[FeSCN^{2+}]$ in the equilibrium mixtures. The more precise method uses a spectrophotometer, which measures the amount of light absorbed by the red-orange $FeSCN^{2+}$ ion at 447 nm (the wavelength at which it most strongly absorbs). The absorbance, A, of the complex is proportional to its concentration, $[FeSCN^{2+}]$, and can be measured directly on the spectrophotometer:

$$A = k[FeSCN^{2+}] \tag{6}$$

Your instructor will show you how to operate the spectrophotometer, if one is available in your laboratory, and may provide you with a calibration curve or equation from which you can find $[FeSCN^{2+}]$ once you have determined the absorbance of your solutions. (Optionally, they may instead have you prepare your own calibration curve, which will provide more accurate results.) See Appendix D for more information about spectrophotometers.

In the other analytical method, a solution with a known concentration of $FeSCN^{2+}$ is prepared. The $FeSCN^{2+}$ concentrations in the solutions being studied are found by comparing the color intensities of these solutions with that of the known. The method involves matching the color intensity of a fixed depth of unknown solution with that for a measured depth of known solution. The actual procedure and method of calculation are discussed in the Experimental Procedure section.

In preparing the mixtures for this experiment you will maintain the concentration of H^+ ion at 0.5 M. The hydrogen ion does not participate directly in Reaction 1, but its concentration does influence the concentrations of both Fe^{3+} and SCN^- through other (acid–base) reactions; thus, maintaining a constant H^+ concentration is important.

Experimental Procedure

Wear your safety glasses while performing this experiment.

Label five medium (18 × 150 mm) test tubes 1 through 5, or choose to keep track of which is which based on their positions in your test tube rack. Pour about thirty milliliters of 2.00×10^{-3} M $Fe(NO_3)_3$ in 1.0 M HNO_3 into a clean and dry 100-mL beaker. Pipet 5.00 mL of that solution into each test tube. Then add about twenty milliliters of 2.00×10^{-3} M KSCN to another clean and dry 100-mL beaker. Pipet 1.00, 2.00, 3.00, 4.00, and 5.00 mL from the KSCN beaker into the corresponding test tube, labeled 1 through 5. Then pipet the proper number of milliliters of water into each test tube to bring the total volume in each tube to 10.00 mL. The volumes to be added to each tube are summarized in Table 23.1, which you should complete by filling in the required volumes of water. See Appendix D for information about pipets and their proper use.

Mix each solution thoroughly by swirling its tube several times, or with the help of a vortex mixer.

Table 23.1

Stock Solution and Dilution Volumes To Be Used in Preparing Samples					
	\multicolumn{5}{c}{**Test tube number**}				
	1	**2**	**3**	**4**	**5**
Volume of $Fe(NO_3)_3$ solution {mL}	5.00	5.00	5.00	5.00	5.00
Volume of KSCN solution {mL}	1.00	2.00	3.00	4.00	5.00
Volume of H_2O {mL}	_____	_____	_____	_____	_____

Method 1. Analysis by spectrophotometric measurement

Transfer a portion of the mixture in Tube 1 into a spectrophotometer cell, as demonstrated by your instructor, and measure the absorbance of the solution at 447 nm. Determine the concentration of $FeSCN^{2+}$ from a calibration curve or equation furnished to you, or from one that you prepare yourself by following the optional instructions in the next paragraph. Record the value on the report page. Repeat the measurement using the mixtures in each of the other test tubes. For a discussion of how absorbance and concentration are related, see Appendix D.

Optional   Prepare a calibration curve as follows: In a 100-mL volumetric flask, combine 2.00 mL of 2.00×10^{-3} M KSCN with 50. mL of a special calibration solution, 0.200 M $Fe(NO_3)_3$ in 1.0 M HNO_3, then dilute to the mark on the volumetric flask with deionized water. Repeat this procedure using 4.00 mL, and then with 8.00 mL, of 2.00×10^{-3} M KSCN. This will give three solutions, which may be assumed to have equilibrium $FeSCN^{2+}$ concentrations equal to 4.00×10^{-5}, 8.00×10^{-5}, and 1.60×10^{-4} M because $[Fe^{3+}] \gg [SCN^-]$, which drives Reaction 1 strongly to the right. If your instructor directs you to make a 5-point calibration curve, also use 1.00 mL and 6.00 mL of 2.00×10^{-3} M KSCN, which will make 2.00×10^{-5} M and 1.20×10^{-4} M $FeSCN^{2+}$. Measure the absorbances of these calibration solutions at 447 mm, using deionized water as the reference (blank). Plot absorbance vs $[FeSCN^{2+}]$ and draw the best straight line possible that passes through the origin. Use the plot itself, or the equation of the best-fit line, to calculate the concentration of $FeSCN^{2+}$ in each of your five test tubes. ▨

Method 2. Analysis by visual comparison with a standard

Prepare a solution of known $FeSCN^{2+}$ concentration by pipetting 10.00 mL of a special calibration solution, 0.200 M $Fe(NO_3)_3$ in 1.0 M HNO_3, into a test tube and adding 2.00 mL of 2.00×10^{-3} M KSCN and 8.00 mL of water. Mix the solution thoroughly by swirling the test tube multiple times or using a vortex mixer.

Because in this solution $[Fe^{3+}] \gg [SCN^-]$, Reaction 1 is driven strongly to the right. You can assume without serious error that essentially all the SCN^- added is converted to $FeSCN^{2+}$. Assuming this to be the case, calculate $[FeSCN^{2+}]$, in the standard solution and record this value on the report page.

The concentration of $FeSCN^{2+}$, $[FeSCN^{2+}]$, in the unknown mixture in test tubes 1 to 5 can be found by comparing the intensity of the red-orange color in those mixtures with that of the standard solution. This can be done by placing the test tube containing Mixture 1 next to a test tube containing the standard. Look down both test tubes toward a well-illuminated piece of white paper on the laboratory bench. Transfer standard solution from its test tube into a clean, dry beaker until the color intensity you see when looking down the tube containing the standard matches the intensity you see when looking down the tube containing the unknown. When the colors match, the following relation is valid:

$$[FeSCN^{2+}]_{unknown} \times \text{depth of unknown solution} = [FeSCN^{2+}]_{standard} \times \text{depth of standard solution} \qquad (7)$$

Measure the depths of the matching solutions with a ruler and record them. Repeat the measurement for Mixtures 2 through 5, recording the depth of each unknown and of the standard solution that matches it in intensity.

Disposal of reaction products: Dispose of your solutions from this experiment as directed by your instructor.

Experiment 23

Data and Calculations: Determining the Equilibrium Constant for a Chemical Reaction

Mixture	Volume of 2.00×10^{-3} M Fe(NO$_3$)$_3$ {mL}	Volume of 2.00×10^{-3} M KSCN {mL}	Volume of water {mL}	Method 1 (if not using Method 2) Absorbance	Method 2 (if not using Method 1) Depth in mm		[FeSCN]$^{2+}$ {M}
					Standard	Unknown	
1	5.00	1.00					
2	5.00	2.00					
3	5.00	3.00					
4	5.00	4.00					
5	5.00	5.00					

If Method 2 was used: [FeSCN^{2+}]$_{standard}$ = _____ $\times 10^{-4}$ M; Calculate [FeSCN^{2+}] in Mixtures 1 to 5 using Equation 7.

Calculations

A. Calculating K_f assuming that Reaction 1 describes what is happening:

This calculation is most easily done by following the steps listed here. If you are using a spreadsheet, set it up so it resembles the tables on these pages and follow these steps (with the help of Appendix G1 or G2):

$$Fe^{3+}(aq) + SCN^-(aq) \rightleftharpoons FeSCN^{2+}(aq) \qquad (1)$$

Step 1. Find the initial moles of Fe^{3+} and SCN$^-$ in the mixtures in test tubes 1 through 5. Use Equation 3 and enter the values in the first two columns of the table on the following page. There is room to show your calculation methods on the last (blank) report page.

Step 2. Calculate the experimentally determined value of [FeSCN^{2+}] at equilibrium for each of the mixtures and record it in the rightmost column of the table on this page, and in the next-to-last column of the table on the following page. There, if not using a spreadsheet, scale all the molarities to 10—4 M, which will simplify the math. Use Equation 3 to find the moles of FeSCN^{2+} in each of the mixtures, and enter those values in the fifth column of the table on the following page. Note that this is also how many moles of Fe^{3+} and SCN$^-$ are used up in the process of Reaction 1 reaching equilibrium.

(continued on following page)

Step 3. From the moles of Fe^{3+} and SCN^- initially present in each mixture, and the moles of Fe^{3+} and SCN^- used up in forming $FeSCN^{2+}$, calculate the moles of Fe^{3+} and SCN^- that remain in each mixture at equilibrium. Use Equation 4. Enter the results in the third and fourth columns of the table on this page.

Step 4. Use Equation 3 and the results of Step 3 to find the concentrations of all of the substances at equilibrium. The volume of the mixture is 10.00 mL, or 0.0100 liter, in all cases. Enter the equilibrium concentrations in the sixth and seventh columns of the table on this page.

Step 5. Calculate K_f for the reaction for each of the mixtures by substituting values for the equilibrium concentrations of Fe^{3+}, SCN^-, and $FeSCN^{2+}$ into Equation 2. Record these results in the last column of the table on this page.

Step 6. Calculate and record the average (mean) value and the standard deviation of K_f. (See Appendix H.)

Mixture	Initial moles {moles}		Equilibrium moles {moles}			Equilibrium concentrations {M}			K_f
	Fe^{3+}	SCN^-	Fe^{3+}	SCN^-	$FeSCN^{2+}$	$[Fe^{3+}]$	$[SCN^-]$	$[FeSCN^{2+}]$	
1	___ $\times 10^{-6}$	___ $\times 10^{-6}$	___ $\times 10^{-6}$	___ $\times 10^{-6}$	___ $\times 10^{-6}$	___ $\times 10^{-4}$	___ $\times 10^{-4}$	___ $\times 10^{-4}$	___
2	___ $\times 10^{-6}$	___ $\times 10^{-6}$	___ $\times 10^{-6}$	___ $\times 10^{-6}$	___ $\times 10^{-6}$	___ $\times 10^{-4}$	___ $\times 10^{-4}$	___ $\times 10^{-4}$	___
3	___ $\times 10^{-6}$	___ $\times 10^{-6}$	___ $\times 10^{-6}$	___ $\times 10^{-6}$	___ $\times 10^{-6}$	___ $\times 10^{-4}$	___ $\times 10^{-4}$	___ $\times 10^{-4}$	___
4	___ $\times 10^{-6}$	___ $\times 10^{-6}$	___ $\times 10^{-6}$	___ $\times 10^{-6}$	___ $\times 10^{-6}$	___ $\times 10^{-4}$	___ $\times 10^{-4}$	___ $\times 10^{-4}$	___
5	___ $\times 10^{-6}$	___ $\times 10^{-6}$	___ $\times 10^{-6}$	___ $\times 10^{-6}$	___ $\times 10^{-6}$	___ $\times 10^{-4}$	___ $\times 10^{-4}$	___ $\times 10^{-4}$	___

Average (mean) value of K_f _____   Standard deviation _____

On the basis of the results of Part A, does it appear that Reaction 1 is reaching equilibrium, as mathematically described by Equation 2, in all five mixtures?

B. Optional

In Part A it is (correctly) assumed that the formula of the complex ion is $FeSCN^{2+}$. This is not necessarily the case, however. Another possibility is that $Fe(SCN)_2^{+}$ is the product formed. The reaction would then be

$$Fe^{3+}(aq) + 2\ SCN^-(aq) \rightleftharpoons Fe(SCN)_2^{+}(aq) \qquad (5)$$

If the equilibrium analysis is carried out assuming that Reaction 5 occurs rather than Reaction 1, obtaining nonconstant values of K_f would be a strong indication that Reaction 5 does not describe the equilibrium being studied. Using the same kind of procedure as in Part A, calculate K_f for the mixtures based on the assumption that $Fe(SCN)_2^{+}$ is the formula of the complex ion formed by the reaction between Fe^{3+} and SCN^-. As a result of the procedure used for calibrating the system by Method 1 or Method 2, $[Fe(SCN)_2^{+}]$ will equal *half* of the $[FeSCN^{2+}]$ obtained for each solution in Part A. Note that *two* moles of SCN^- are needed to form *one* mole of $Fe(SCN)_2^{+}$. This changes the expression for K_f. Also, in calculating the equilibrium moles of SCN^- you will need to subtract ($2 \times$ moles $Fe(SCN)_2^{+}$) from the initial moles of SCN^-.

Mixture	Initial moles {moles}		Equilibrium moles {moles}			Equilibrium concentrations {M}			K_f
	Fe^{3+}	SCN^-	Fe^{3+}	SCN^-	$Fe(SCN)_2^{+}$	$[Fe^{3+}]$	$[SCN^-]$	$[Fe(SCN)_2^{+}]$	
1	___ $\times 10^{-6}$	___ $\times 10^{-6}$	___ $\times 10^{-6}$	___ $\times 10^{-6}$	___ $\times 10^{-6}$	___ $\times 10^{-4}$	___ $\times 10^{-4}$	___ $\times 10^{-4}$	___
2	___ $\times 10^{-6}$	___ $\times 10^{-6}$	___ $\times 10^{-6}$	___ $\times 10^{-6}$	___ $\times 10^{-6}$	___ $\times 10^{-4}$	___ $\times 10^{-4}$	___ $\times 10^{-4}$	___
3	___ $\times 10^{-6}$	___ $\times 10^{-6}$	___ $\times 10^{-6}$	___ $\times 10^{-6}$	___ $\times 10^{-6}$	___ $\times 10^{-4}$	___ $\times 10^{-4}$	___ $\times 10^{-4}$	___
4	___ $\times 10^{-6}$	___ $\times 10^{-6}$	___ $\times 10^{-6}$	___ $\times 10^{-6}$	___ $\times 10^{-6}$	___ $\times 10^{-4}$	___ $\times 10^{-4}$	___ $\times 10^{-4}$	___
5	___ $\times 10^{-6}$	___ $\times 10^{-6}$	___ $\times 10^{-6}$	___ $\times 10^{-6}$	___ $\times 10^{-6}$	___ $\times 10^{-4}$	___ $\times 10^{-4}$	___ $\times 10^{-4}$	___

What do you conclude about the formula of the iron(III) thiocyanate complex ion, based on what you have calculated in Part B?

Experiment 23

Advance Study Assignment: Determining the Equilibrium Constant
for a Chemical Reaction

1. A student mixes 5.00 mL of 2.00×10^{-3} M $Fe(NO_3)_3$ in 1.0 M HNO_3 with 4.00 mL of 2.00×10^{-3} M KSCN and 1.00 mL of water. He finds that in the equilibrium mixture the concentration of $FeSCN^{2+}$ is 9.2×10^{-5} M. Find K_f for Reaction 1, $Fe^{3+}(aq) + SCN^{-}(aq) \rightleftharpoons FeSCN^{2+}(aq)$.

    Step 1. Calculate the moles of Fe^{3+} and SCN^{-} initially present. (Use Equation 3.)

    _____ moles of Fe^{3+}; _____ moles of SCN^{-}

    Step 2. What is the volume of the equilibrium mixture? How many moles of $FeSCN^{2+}$ are in the mixture at equilibrium? (Use Equation 3.)

    _____ mL; _____ moles of $FeSCN^{2+}$

    How many moles of Fe^{3+} and SCN^{-} are used up in making the $FeSCN^{2+}$?

    _____ moles of Fe^{3+}; _____ moles of SCN^{-}

    Step 3. How many moles of Fe^{3+} and SCN^{-} remain in the solution at equilibrium? (Use Equation 4 and the results of Steps 1 and 2.)

    _____ moles of Fe^{3+}; _____ moles of SCN^{-}

    Step 4. What are the concentrations of Fe^{3+}, SCN^{-}, and $FeSCN^{2+}$ at equilibrium? (Use Equation 3 and the results of Steps 2 and 3.)

    $[Fe^{3+}]$ = _____ M; $[SCN^{-}]$ = _____ M; $[FeSCN^{2+}]$ = _____ M

    Step 5. What is the value of K_f for Reaction 1? (Use Equation 2 and the results of Step 4.)

    K_f = _____

(continued on following page)

2.   Optional   Assume that the reaction studied in Problem 1 is Reaction 5,

     $Fe^{3+}(aq) + 2\ SCN^-(aq) \rightleftharpoons Fe(SCN)_2^+(aq)$. Find K_f for this reaction, given the data in Problem 1.

   a.   Write out the K_f equilibrium expression for Reaction 5.

   b.   Find K_f as you did in Problem 1; as you carry out the calculations on the remainder of this page, remember that *two* moles of SCN^- are used up for each mole of $Fe(SCN)_2^+$ formed, and that $[Fe(SCN)_2^+]$ will be half of what $[Fe(SCN)^{2+}]$ was in Problem 1!

     Step 1. Results are as in Problem 1.

     Step 2. How many moles of $Fe(SCN)_2^+$ are in the mixture at equilibrium? (This will *not* be the same as your answer for Step 2 in Problem 1! Read the paragraph following Reaction 5 on the optional report page.)

                    _____ moles of $Fe(SCN)_2^+$

     How many moles of Fe^{3+} and SCN^- are used up in making the $Fe(SCN)_2^+$?

        _____ moles of Fe^{3+}; _____ moles of SCN^-

     Step 3. How many moles of Fe^{3+} and SCN^- remain in solution at equilibrium? Use the results of Steps 1 and 2.

        _____ moles of Fe^{3+}; _____ moles of SCN^-

     Step 4. What are the concentrations of Fe^{3+}, SCN^-, and $Fe(SCN)_2^+$ at equilibrium? (Use Equation 3 and the results of Step 3.)

   $[Fe^{3+}] =$ _____ M; $[SCN^-] =$ _____ M; $[Fe(SCN)_2^+] =$ _____ M

     Step 5. Calculate K_f based on the assumption that Reaction 5 occurs. [Use the answer to 2(a).]

                            $K_f =$ _____

Experiment 24

Standardizing a Basic Solution and Determining the Molar Mass of an Acid

When a solution of a strong acid is mixed with a solution of a strong base, a chemical reaction occurs that can be represented by the following net ionic equation:

$$H^+(aq) + OH^-(aq) \rightarrow H_2O(\ell) \tag{1}$$

This is called a <u>neutralization reaction</u>, and chemists use it often, especially to change the acidic or basic properties of solutions. The equilibrium constant for Reaction 1 is on the order of 10^{14} at room temperature, so it can be considered to proceed completely to the right, using up whichever of the reactant ions is present in the lesser amount and leaving the solution either acidic or basic, depending on whether H^+ or OH^- ion was in excess.

Because Reaction 1 goes essentially to completion, it can be used to determine the concentrations of acidic or basic solutions. A frequently used procedure involves <u>titrating</u> an acid with a base. In such a <u>titration</u>, a basic solution is added from a volume dispensing device called a <u>buret</u> (see Appendix D) to a measured volume of acid solution until the moles of OH^- ion added is just equal to the moles of H^+ ion present in the acid. At that point the volume of basic solution that has been added is read off the buret.

The <u>molarity</u>, [A], of a substance A is defined as

$$[A] := \frac{\text{moles of A}}{\text{liters of solution, } V} \quad \text{or} \quad \text{moles of A} = [A] \times V \tag{2}$$

At the <u>equivalence point</u> of an acid–base titration in which an acid is titrated with a base,

$$\text{moles of } H^+ \text{ originally present} = \text{moles of } OH^- \text{ added} \tag{3}$$

So, by Equation 2, $\qquad [H^+]_{\text{in acid}} \times V_{\text{acid}} = [OH^-]_{\text{in base}} \times V_{\text{base}} \tag{4}$

Therefore, if the molarity of either the H^+ ion in the acid solution or the molarity of the OH^- ion in the basic solution is known, the molarity of the other solution can be found from the titration volumes.

In many acid–base titrations, a chemical called an acid-base <u>indicator</u> is used: it changes color in response to the acidity or basicity of the solution; when that change is first detected the titration is stopped, and is said to have reached its <u>endpoint</u>. If the indicator is chosen carefully, the equivalence point and the titration's endpoint are effectively identical. One of the most common indicators is phenolphthalein, which is colorless in acidic solutions but becomes red when the solution it is in becomes basic enough that $[OH^-]$ rises above 10^{-5} M.

When a solution of a <u>strong</u> acid (one that dissociates completely in water, such as hydrochloric acid, HCl) is titrated with a solution of a strong base (such as sodium hydroxide, NaOH), at the equivalence point both $[H^+]$ and $[OH^-]$ are 10^{-7} M. At the equivalence point a single drop of acid or base added to the solution will change both of these concentrations by a factor of 10000 (in opposite directions), changing the color of phenolphthalein; this makes it a very effective indicator in such titrations. If a <u>weak</u> acid (one that does not dissociate completely in water, such as acetic acid, CH_3COOH) is titrated with a strong base, $[OH^-]$ at the equivalence point is somewhat larger than 10^{-7} M, perhaps 10^{-6} or 10^{-5} M, and phenolphthalein is still a very satisfactory indicator, changing color within a drop of the equivalence point. If, however, a solution of a weak base such as ammonia, NH_3, is titrated with a strong acid, $[OH^-]$ will be around 10^{-9} M at the equivalence point and phenolphthalein will change color more than a drop or two away from the equivalence point. It will not be as good an indicator for that titration as, for example, methyl red, whose color changes from red to yellow as $[OH^-]$ changes from about 10^{-8} M to 10^{-9} M. A good indicator for a given acid–base titration changes color at about the pH of its equivalence point.

In this experiment you will determine the molarity of OH^- ion in a titrant NaOH solution by titrating that solution against a <u>standardized</u> solution of HCl, meaning one whose molarity is both accurately and precisely (see Appendix H) known. Because in these solutions one mole of acid in solution furnishes one mole of H^+ ion, and one mole of base produces one mole of OH^- ion, $[HCl]_{standard} = [H^+]_{standard}$ in the acidic solution and $[NaOH]_{titrant} = [OH^-]_{titrant}$ in the basic solution. Therefore, the titration will allow you to accurately and precisely determine $[NaOH]_{titrant}$ as well as $[OH^-]_{titrant}$: in other words, to standardize your titrant.

In the second part of this experiment you will use your NaOH titrant solution, whose concentration you now know, to titrate a sample of a pure solid acid. By titrating a weighed sample of the unknown acid with your standardized NaOH titrant you can, by Equation 3, find the moles of H^+ ion that the acid sample can provide.

If your acid has one acidic hydrogen atom, then the moles of acid will equal the moles of H^+ that react during the titration. The molar mass of the acid, MM, will equal the number of grams of acid that contain one mole of H^+ ion:

$$MM = \frac{\text{grams of acid}}{\text{moles of } H^+ \text{ ion furnished}} \tag{5}$$

Many acids release one mole of H^+ ion per mole of acid on titration with a strong base. Such acids are called <u>monoprotic</u>. Acetic acid, CH_3COOH, is a classic example of a monoprotic acid (only the last H atom in the formula is acidic). Like all organic acids, acetic acid is weak: it only ionizes to a small extent in water.

Some acids contain more than one acidic hydrogen atom per molecule. Sulfuric acid, H_2SO_4, is an example of an inorganic <u>diprotic acid</u>. It is the acid used in lead-acid car batteries. Maleic acid, $HOOCCH{=}CHCOOH$ or $C_4H_4O_4$, is a diprotic organic acid commonly used to make foods taste sour. (Only the H atoms on each end of the formula are acidic.) If samples of these acids are titrated with a solution of a strong base such as NaOH, it takes $\underline{2}$ moles of OH^- ion, or $\underline{2}$ moles of NaOH, to neutralize $\underline{1}$ mole of acid because each mole of acid releases two moles of H^+ ion. If you do not know the formula of an acid, you cannot be sure it is monoprotic, so you can only calculate the mass of acid that will react with one mole of OH^- ion. That is the mass capable of releasing $\underline{1}$ mole of H^+ ion and is called the <u>equivalent mass</u> of the acid. The molar mass and the equivalent mass are related by counting number multiples. Because sulfuric acid gives up $\underline{2}$ moles of H^+ ion when titrated with NaOH, its molar mass is *twice* the equivalent mass—the amount that gives up $\underline{1}$ mole of H^+ ion.

To simplify matters, in this experiment only monoprotic acids are used: so $\underline{1}$ mole of your acid will react with $\underline{1}$ mole of NaOH, and you can find the molar mass of your acid using Equation 5.

Experimental Procedure

Wear your safety glasses while performing this experiment.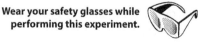

Obtain two burets and a sample of solid unknown acid.

A. Standardizing a solution of NaOH

Pour about four hundred milliliters of deionized water into a 500-mL flask (any type). Next, transfer 7 mL of the stock 6 M NaOH solution provided in the laboratory into a small graduated cylinder, then add this to the 500-mL flask. Stopper the 500-mL flask tightly and mix the solution thoroughly at several-minute intervals over a period of at least 15 minutes before using it: this will be your NaOH titrant solution. Fill the graduated cylinder with deionized water and let it sit, full of water, until you finish mixing the 500-mL flask. Pour it out into the 500-mL flask before mixing for the last time. Then rinse the graduated cylinder a few more times before putting it away, pouring the rinsings down the sink. (This is necessary because 6 M NaOH is difficult to rinse out of a container, and it will eat away at the container over time if it is not rinsed out completely.)

Transfer about seventy-five milliliters of standardized HCl solution (provided for you) into a clean, *dry* 125-mL Erlenmeyer flask. This amount should provide all the standardized acid you will need; do not waste it. Record the precise molarity of the standardized HCl, which will be given to you.

Prepare your burets for the titration by the following procedure, which is described in greater detail in Appendix D. (The purpose of this procedure is to make sure that the solution in each buret has the same concentration as it has in the container from which it was poured.) Clean the two burets and rinse them with deionized water. Then rinse one buret three times with a few milliliters of the HCl solution, in each case thoroughly wetting the walls of the buret with the solution and then letting it out through the stopcock. Fill the same buret completely with standardized HCl; open the stopcock momentarily to fill the tip and bring the liquid level below the zero mark. Next, clean and fill the other buret with your NaOH titrant solution, using the same rinsing procedure. Put the acid buret, A, on the left side of your buret clamp, and the base buret, B, on the right side. Check to see that your burets do not leak and that there are no air bubbles in either buret tip. Read and record the levels in the two burets to ± 0.01 mL.

Allow about twenty-five milliliters of the HCl solution from buret A to flow into a clean 250-mL Erlenmeyer flask or white titration cup that has been rinsed several times with fresh deionized water. Add about twenty-five milliliters of fresh deionized water and two or three drops of phenolphthalein indicator solution. If using a transparent glass flask, titrate over a white sheet of paper or other white surface to help you see the color change. Allow about ten milliliters of the NaOH titrant solution to flow from buret B into the flask or cup, noting the pink phenolphthalein color that appears and disappears as the titrant drops hit the liquid and are mixed into it. Begin swirling the liquid in the flask or cup gently and continuously as you add more NaOH titrant from buret B. When the pink color begins to last for a little while, slow down the rate of addition. In the final stages of the titration add NaOH drop by drop, until the entire solution turns a pale pink color and stays pink for about thirty seconds while being swirled. If you go past the endpoint and obtain a red solution, add a few drops of the HCl solution from buret A to completely remove the color, then add NaOH 1 drop at a time again until the solution stays pink. Carefully record the *final* readings on the HCl and NaOH burets.

Now, add about ten more milliliters of the standardized HCl solution from buret A to the 250-mL Erlenmeyer flask or titration cup containing the titrated solution. Titrate this as before with the NaOH titrant from buret B to a pink endpoint, and carefully record both buret readings once again. To this solution add about ten more milliliters of HCl and titrate a third time with NaOH, again recording the final buret readings.

You have now completed three titrations, with *total* HCl volumes of about twenty-five, thirty-five, and forty-five milliliters. Using Equation 4, calculate the molarity of your titrant, $[OH^-]_{titrant}$, for each of the three titrations. In each case, use the *total volumes* of acid and base that were added to reach that endpoint. At least two of these concentrations should agree to within 1%. If they do, proceed to the next part of the experiment. If they do not, perform more titrations until two calculated concentrations do agree. Calculate the average (mean) value of $[OH^-]_{titrant}$ and its standard deviation. (See Appendix H.)

B. Determining the molar mass of an unknown acid

Weigh the vial containing your unknown solid acid to ± 0.0001 g. Carefully pour out about half the contents of the vial into a clean but not necessarily dry 250-mL Erlenmeyer flask or titration cup. (You can use the same one from before, provided you first rinse it a few times with deionized water.) Weigh the vial again, accurately. Add about fifty milliliters of fresh deionized water and two or three drops of phenolphthalein to the flask or cup. The acid may be somewhat insoluble, so do not worry if it does not all dissolve.

Refill buret B (on the right side) with NaOH titrant solution from the 500-mL flask. Add more of the standardized HCl to buret A, on the left side, until it is about half full. Read both levels carefully and record them.

Titrate the solution of the solid acid with NaOH from buret B. As the acid is neutralized it will tend to dissolve in the solution. If the acid appears to be relatively insoluble, add NaOH until the pink color persists, and then swirl to dissolve the solid. If the solid still will not dissolve completely, and the solution remains pink, add about twenty-five milliliters of ethanol to increase the solubility of the acid. If you go past the endpoint, add HCl as necessary, from buret A, to get back to a colorless solution, then titrate again with NaOH titrant from buret B. The final pink endpoint should appear upon adding a single drop of NaOH titrant. Record the final levels in the NaOH and HCl burets.

Refill buret B again, as described previously, and record the initial readings on both burets in the Trial 2 column on the report sheet. Pour the rest of your unknown acid sample from its vial into a clean but not necessarily dry 250-mL Erlenmeyer flask or titration cup (you can use the same one again if you rinse it) and weigh the empty vial accurately. Titrate this second sample of acid as before, using the NaOH and HCl solutions in your burets. *If doing the optional Part C of this experiment, do not dispose of this final titrated solution yet!*

If you use HCl in these titrations (which can be hard to avoid), the calculations needed are a bit more complicated than in the standardization of the NaOH solution. To find the moles of H^+ ion in the solid acid, you must subtract the moles of HCl used from the moles of NaOH. For a back-titration, which is what adding some HCl is called in this case:

$$\text{moles of } H^+ \text{ in solid acid} = \text{moles of } OH^- \text{ in NaOH titrant} - \text{moles of } H^+ \text{ in standardized HCl} \quad \textbf{(6)}$$

In terms of the titrant concentrations and volumes (be mindful of units!), Equation 6 becomes:

$$\text{moles of } H^+ \text{ in solid acid} = ([\text{NaOH}]_\text{titrant} \times V_\text{titrant}) - ([\text{HCl}]_\text{standard} \times V_\text{standard}) \quad \textbf{(7)}$$

Optional ## C. Determining K_a for an unknown acid

Your instructor will tell you in advance if you are to do this part of the experiment. If you do this part, you will need to *save* one of the titrated solutions from Part B. You will add HCl to that titrated solution until half of the NaOH titrant you added to it is neutralized. In the resulting solution, $[H^+]$ is equal to the acid dissociation constant, K_a, for your acid. You will then measure $[H^+]$, and in doing so determine your acid's K_a.

A weak monoprotic acid, HB, dissociates in water by Reaction 8:

$$HB(aq) \rightleftharpoons H^+(aq) + B^-(aq) \quad \textbf{(8)}$$

The equilibrium constant for this reaction is called the acid dissociation constant of HB and is given the symbol K_a. A solution of HB and/or B^- will obey the equilibrium condition given by Equation 9:

$$K_a = \frac{[H^+][B^-]}{[HB]} \quad \textbf{(9)}$$

Your acid is a weak monoprotic acid, so it will react according to Reaction 8 and obey Equation 9.

In the titration in Part B you converted a solution of HB into one containing B^- by adding sufficient NaOH to reach the endpoint. If, to your titrated solution, you now add some HCl, it will convert some of the B^- ions back into HB. If you add *half* as many moles of HCl as there were moles of HB in your original sample, then *half* of the B^- ions will be converted to HB. In the resulting solution, $[HB]$ will equal $[B^-]$, and so, by Equation 9,

$$K_a = [H^+] \quad \text{in the half-neutralized solution} \quad \textbf{(10)}$$

Using this approach, calculate how much HCl is needed and add it to a titrated unknown acid solution such that $[HB]$ equals $[B^-]$. Measure the pH of that solution as indicated by your instructor. If you use a pH meter, follow the procedure described in Appendix D. From the pH, you can calculate K_a for your unknown acid because pH is a measure of $[H^+]$:

$$pH := -\log[H^+] \quad \text{or} \quad [H^+] = 10^{(-pH)} \text{ M} \quad \textbf{(11)}$$

Take it further (optional): Using the procedure in this experiment, find the molarity of acetic acid in household vinegar. What is the mass percent of acetic acid in this vinegar? Compare your result to the concentration indicated on the container.

Disposal of reaction products: The neutralized solutions from this experiment may be poured down the drain, unless your instructor directs otherwise.

Experiment 24

Data and Calculations: Standardizing a Basic Solution and Determining the Molar Mass of an Acid

A. Standardizing a solution of NaOH

	Trial 1	Trial 2	Trial 3
Initial reading, HCl buret (A, left side)	_____ mL		
Initial reading, NaOH buret (B, right side)	_____ mL		
Final reading, HCl buret (A, left side)	_____ mL	_____ mL	_____ mL
Final reading, NaOH buret (B, right side)	_____ mL	_____ mL	_____ mL

B. Determining the molar mass of an unknown acid

Mass of vial plus contents	_____ g
Mass of vial plus contents less Sample 1	_____ g
Mass less Sample 2 (empty sample vial)	_____ g

	Trial 1	Trial 2
Initial reading, HCl buret (A, left side)	_____ mL	_____ mL
Initial reading, NaOH buret (B, right side)	_____ mL	_____ mL
Final reading, HCl buret (A, left side)	_____ mL	_____ mL
Final reading, NaOH buret (B, right side)	_____ mL	_____ mL

Calculations
A. Standardizing a solution of NaOH

	Trial 1	Trial 2	Trial 3
Total volume of HCl	_____ mL	_____ mL	_____ mL
Total volume of NaOH	_____ mL	_____ mL	_____ mL
Molarity of standardized HCl, $[HCl]_{standard}$			_____ M
Molarity of H^+ in standardized HCl, $[H^+]_{standard}$			_____ M

(continued on following page)

By Equation 4,

$$[H^+]_{standard} \times V_{HCl\ standard} = [OH^-]_{titrant} \times V_{NaOH\ titrant} \quad \text{or} \quad [OH^-]_{titrant} = [H^+]_{standard} \times \frac{V_{HCl\ standard}}{V_{NaOH\ titrant}} \quad \textbf{(4a)}$$

Use Equation 4a to find the molarity of your NaOH titrant solution, $[OH^-]_{titrant}$. Note that the volumes do not need to be converted to liters: as long as the same units are used for both volumes, the units will cancel out.

	Trial 1	**Trial 2**	**Trial 3**

$[OH^-]_{titrant}$ _____ M _____ M _____ M   (at least two of these should agree within 1%)

The molarity of the NaOH, $[NaOH]_{titrant}$, will equal $[OH^-]_{titrant}$ because 1 mol NaOH → 1 mol OH⁻.

Based only on the trials that agree within 1%:
   Average (mean) molarity of NaOH solution, $[NaOH]_{titrant}$ _____ M

   Standard deviation in $[NaOH]_{titrant}$ _____ M

B. Determining the molar mass of an unknown acid

	Trial 1	**Trial 2**

Mass of acid sample used in trial _____ g _____ g

Volume of $NaOH_{titrant}$ used in trial _____ mL = _____ L _____ mL = _____ L

Moles of $OH^- = V_{NaOH\ titrant} \times [OH^-]_{titrant}$ _____ moles _____ moles

Volume of HCl used in trial _____ mL = _____ L _____ mL = _____ L

Moles of $HCl = V_{HCl\ standard} \times [H^+]_{standard}$ _____ moles _____ moles

Moles of H^+ in sample (use Equation 6) _____ moles _____ moles

$MM = \dfrac{\text{grams of acid}}{\text{moles of } H^+}$ _____ g/mol _____ g/mol

Unknown acid ID code ======== → _____

Optional   C. Determining K_a for an unknown acid

Moles of H^+ in acid sample (from Part B) _____ moles   Moles of HB in acid sample _____ moles

Moles of HCl to be added _____ moles   Volume of HCl added _____ mL

pH of half-neutralized solution _____ K_a of acid _____

Experiment 24

Advance Study Assignment: Standardizing a Basic Solution and Determining
the Molar Mass of an Acid

1. 6.9 mL of 6.0 M NaOH are diluted with water to a total volume of 40_0 mL. You are asked to find the molarity of the resulting solution.

   a. First find out how many moles of NaOH are present in 6.9 mL of 6.0 M NaOH. Use Equation 2. Note that the volume must be in liters.

                                          _____ moles

   b. Because the total moles of NaOH present do not change on dilution, the molarity after dilution can also be found by Equation 2—using the final volume of the solution (which is 400 mL with two significant figures, written 40_0 mL—see Appendix E). Calculate that molarity.

                                          _____ M

2. In an acid–base titration, 22.44 mL of an NaOH titrant solution were needed to neutralize 22.13 mL of a 0.0997 M standardized HCl solution. To find the molarity of the titrant, use the following procedure:

   a. First note the value of $[H^+]_{standard}$ in the standardized HCl solution.

                                          _____ M

   b. Find $[OH^-]_{titrant}$ in the NaOH titrant solution. (Use Equation 4.)

                                          _____ M

   c. Obtain $[NaOH]_{titrant}$ from $[OH^-]_{titrant}$.

                                          _____ M

(continued on following page)

3. A 0.5891-g sample of an unknown monoprotic acid required a total of 38.50 mL of 0.1011 M NaOH titrant for neutralization to a pink phenolphthalein endpoint. The endpoint was initially overshot, so 0.39 mL of 0.0997 M standardized HCl was used for back-titration.

a. How many moles of OH^- were added? How many moles of H^+ from HCl?

_____ moles of OH^- _____ moles of H^+

b. How many moles of H^+ were in the solid acid? (Use Equation 6.)

_____ moles of H^+ in solid

c. What is the molar mass of the unknown acid? (Use Equation 5.)

_____ g/mol

Experiment 25

pH Measurements—
Buffers and Their Properties

An important property of an aqueous (water-as-the-solvent) solution is its concentration of hydrogen ion. The H^+ ion (more precisely, the hydronium ion, H_3O^+, but in this manual we will use H^+) has a great impact on the solubility of many inorganic and organic substances, on the formation of complex metallic ions found in solution, and on the rates of many chemical reactions. It is therefore important to know how to measure the concentration of hydrogen ion and understand its effect on solution properties.

The concentration of H^+ ion is frequently expressed as the pH of a solution rather than as a molarity. The pH of a solution is defined by the equation

$$pH := -\log[H^+] \tag{1}$$

in which the logarithm is base 10. If $[H^+]$ is 1×10^{-4} moles per liter, the pH of the solution is 4.0. If $[H^+]$ is 5×10^{-2} M, the pH is 1.3. pH is convenient because H^+ concentrations are rarely more than 1 M, and often orders of magnitude less. pH = 10.3 is more easily communicated than is 5×10^{-11} M. Given a pH value, the concentration of H^+ ion can be calculated from the inverse of Equation 1:

$$[H^+] := 10^{-pH} \text{ M} \tag{1a}$$

Equation 1a indicates that in order to get $[H^+]$ from a pH value, you should raise 10 to the power $-pH$. The result will have units of molarity, M (moles per liter).

Basic solutions can also be described in terms of pH. In aqueous solutions the following equilibrium expression will always be obeyed:

$$[H^+] \times [OH^-] = K_w = 1 \times 10^{-14} \text{ at } 25\,°C \tag{2}$$

In pure (deionized) water $[H^+]$ equals $[OH^-]$, so, by Equation 2, $[H^+]$ must be 1×10^{-7} M. Therefore, the pH of deionized water is 7. Solutions in which $[H^+] > [OH^-]$ are said to be acidic, and have a pH of less than 7; if $[H^+] < [OH^-]$, the solution is described as basic, and its pH is greater than 7. A solution with a pH of 10 will have $[H^+] = 1 \times 10^{-10}$ M and $[OH^-] = 1 \times 10^{-4}$ M at $25\,°C$.

The pH of a solution can be measured experimentally in two ways. The first of these involves a chemical called an indicator, which is sensitive to pH. These substances have colors that change over a relatively narrow range of pH values (about two pH units) and can, when properly chosen, be used to roughly determine the pH of a solution. Two common indicators are litmus, usually supplied on indicator paper, and phenolphthalein, the most common indicator in acid–base titrations. Litmus changes from red to blue as the pH of a solution goes from about six to about eight. Phenolphthalein changes from colorless to red as the pH goes from 8 to 10. A given indicator is useful for determining numerical pH values (rather than a pH range) only over the pH region in which it changes color. Indicators are available for measurement of pH in all the important ranges of acidity and basicity. By matching the color of a suitable indicator in a solution of known pH with that in an unknown solution, the pH of the unknown can be determined to within about 0.3 pH units.

The other common method for measuring pH is with a device called a pH meter. This device has two electrodes, one of which is sensitive to $[H^+]$, which are immersed in a solution. The voltage between the two electrodes is related to the pH. A pH meter is designed to directly display the pH of the solution. It can provide a more precise measurement of pH than a typical indicator, and works over a much wider range of pH values.

Some acids and bases ionize very effectively in water, and are called strong because of their essentially complete ionization in reasonably dilute solutions. Other acids and bases, because they ionize only partially (often far less than even 1% in 0.10 M solution), are called <u>weak</u>. Hydrochloric acid, HCl, is a common example of a strong acid, while sodium hydroxide, NaOH, is a common example of a strong base. Acetic acid, CH_3COOH, is a commonly used weak acid, while ammonia, NH_3, is a common weak base.

A weak acid, HB, will <u>ionize</u> (break apart into ions in solution) in a reaction that reaches equilibrium:

$$HB(aq) \rightleftharpoons H^+(aq) + B^-(aq) \tag{3}$$

The B^- ion is called the <u>conjugate base</u> of the weak acid HB. It is formed when HB loses a proton, H^+. At equilibrium,

$$\frac{[H^+][B^-]}{[HB]} = K_a \tag{4}$$

K_a is called the <u>acid dissociation constant</u> of the acid HB; in solutions containing HB, the ratio of concentrations in Equation 4 will remain constant at equilibrium, independent of how the solution was made. A similar relation can be written for solutions of a weak base.

The value of the acid dissociation constant K_a for a weak acid can be found experimentally, in several ways. In this experiment you will determine K_a for a weak acid by studying the properties of a partly neutralized solution of the acid, called a <u>buffer</u>.

Salts formed by reacting *strong* acids with *strong* bases—such as NaCl, KBr, and $NaNO_3$—ionize completely when dissolved in water, but release ions that do not react with water. They form neutral solutions with a pH of about seven. When dissolved in water, salts of *weak* acids or *weak* bases release conjugate base or conjugate acid ions that tend to react to some extent with water, producing molecules of the weak acid or base and changing the H^+ concentration in the solution.

If HB is a weak acid, the conjugate base B^- ion produced when NaB is dissolved in water will react with water to some extent, by Reaction 5:

$$B^-(aq) + H_2O(\ell) \rightleftharpoons HB(aq) + OH^-(aq) \tag{5}$$

Solutions of sodium acetate, $NaCH_3COO$, the salt formed by reacting the strong base sodium hydroxide, NaOH, with the weak acid acetic acid, CH_3COOH, are slightly *basic* because the acetate ion, CH_3COO^- (B^-), reacts with water to produce acetic acid (HB) and the hydroxide ion, OH^-. Similarly, solutions of ammonium chloride, NH_4Cl, the salt formed by reacting the strong acid hydrochloric acid, HCl, with the weak base ammonia, NH_3, are slightly *acidic* because of Reaction 6 (which is very similar to Reaction 3):

$$NH_4^+(aq) \rightleftharpoons NH_3(aq) + H^+(aq) \tag{6}$$

Buffers

Some solutions, called buffers, resist pH changes. Water is not a buffer, because its pH is very sensitive to the addition of any acidic or basic substance. Even bubbling your breath through a straw into deionized water can lower its pH by at least one pH unit, due to the small amount of the acidic gas carbon dioxide, CO_2, present in exhaled air. However, you could exhale through a straw into a buffer solution for half an hour and not change the pH measurably. All living systems contain buffer solutions, because a stable pH is essential for many of the biochemical reactions that maintain living organisms.

All that is required to make a buffer is a solution containing a weak acid and its conjugate base. An example of such a solution is one containing the weak acid HB, and its conjugate base, B^- ion—which can come from mixing the salt NaB into the solution or adding a strong base to the solution of weak acid HB.

The pH of a buffer is established by the relative concentrations of HB and B^- in the solution. Equation 4 can be rearranged to solve for the concentration of H^+ ion:

$$[H^+] = K_a \times \frac{[HB]}{[B^-]} \tag{4a}$$

This equation is often further modified when working with buffers, by taking the negative base 10 logarithm of both sides:

$$pH = pK_a + \log\frac{[B^-]}{[HB]} \tag{4b}$$

where pK_a equals $-\log K_a$ and $pH := -\log [H^+]$. Equation 4b is called the <u>Henderson-Hasselbach equation</u>, and it has many applications. Equation 4b indicates that the pH of an HB/B^- buffer is controlled by the ratio of weak acid to conjugate base, $[HB]/[B^-]$, and allows quick, direct calculation of the pH.

A weak acid, HB, whose K_a equals 1×10^{-5}, has a pK_a of 5.0. If equal volumes of 0.10 M HB and 0.10 M NaB are mixed, the pH of the resulting solution will equal the pK_a of the acid, in this case 5.0, because the concentrations of HB and B^- are equal in the final solution, and the logarithm of 1 is 0. $[H^+]$ in this buffer will be 1×10^{-5} M. The H^+ ion in the solution has to come from ionizing HB, but the amount is tiny (in this case it is on the order of 10^{-5} M) relative to the concentration of HB (0.10 M)—so this ionization does not measurably change the HB/B^- ratio.

It can generally be safely assumed that *in any buffer the weak acid and its conjugate base do not react appreciably with one another when their solutions are mixed.* Therefore their relative concentrations can be calculated from the way the buffer was put together.

Using Equation 4b, you can answer many questions about buffers. For example, how does the pH of a given buffer solution change if you dilute it with water? How does it change if you add an HB solution? What if you add a solution of NaB, containing the conjugate base? What is the pH range over which such a buffer would be useful, if you assume that the ratio of $[B^-]$ to $[HB]$ must lie between 10:1 and 1:10? How sensitive is the pH of water to the addition of a strong acid, such as HCl, or the addition of a strong base, such as NaOH? What happens when you add a strong acid or a strong base to a buffer? Why and how does the pH of a buffer resist changing when small amounts of a strong acid or strong base are added?

Some of these questions you can answer by just looking at Equation 4b. For others you need to realize that although a weak acid and its conjugate base do not react with each other, a weak acid, such as HB, will react completely (technical term: <u>quantitatively</u>) with a strong base, such as NaOH:

$$HB(aq) + OH^-(aq) \rightarrow B^-(aq) + H_2O(\ell) \tag{7}$$

and a weak base, such as NaB, will react completely (quantitatively) with a strong acid, such as HCl:

$$B^-(aq) + H^+(aq) \rightarrow HB(aq) \tag{8}$$

In Reaction 7, a small amount of NaOH will not raise the pH very much because the OH^- ion is essentially soaked up by the acid HB, producing some B^- ion and increasing the value of $[B^-]/[HB]$. But the buffer will not be destroyed, or the $[HB]/[B^-]$ ratio changed significantly, provided the (effective—that is, after-dilution) concentration of NaOH added is significantly smaller than the concentrations of HB and B^- in the buffer.

Reactions 7 and 8 can also be used to *prepare* a buffer from a solution of HB by adding NaOH (or another strong base), or from a solution of NaB by adding HCl (or another strong acid).

In this experiment you will determine the approximate pH of several solutions using acid-base indicators. Then you will find the pH of some other solutions with a pH meter. In the rest of the experiment you will carry out some reactions that will allow you to answer all the questions raised previously, about buffers. Finally, you will prepare one or two buffers having specific pH values.

Wear your safety glasses while performing this experiment.

Experimental Procedure

A. Determining pH using acid-base indicators

To each of five small (13×100 mm) test tubes add about one milliliter of 0.10 M HCl (in these tubes, a depth of about one centimeter). To each tube add one or two drops of one of the indicators listed in Table 25.1, one indicator to a tube. Note the color of the solution you obtain in each case. By comparing the colors you observe with the information in Table 25.1, estimate the pH of the solution. In making your estimate, note that the color of an indicator is most indicative of pH in the pH region over which the indicator changes color.

Repeat this procedure with each of the following solutions:

0.10 M NaH$_2$PO$_4$	0.10 M CH$_3$COOH	0.10 M ZnSO$_4$
sodium dihydrogenphosphate	acetic acid	zinc sulfate

Record the colors you observe and your pH estimate for each solution.

Table 25.1

Useful pH Ranges for Some Common Acid-Base Indicators								

	Useful pH range (approximate)							
Indicator	0	1	2	3	4	5	6	7
Methyl violet	yellow [_____] violet							
Thymol blue	magenta [_____] yellow-orange							
Methyl yellow	red-orange [_____] yellow							
Congo red	violet [_____] red-orange							
Bromocresol green	yellow [_____] blue							

Note: Beyond the left and right ends of the boxes drawn in Table 25.1, the color of the indicator remains that indicated on the nearest end of the box. So, for example, methyl violet remains violet at pH values above 2, and methyl yellow remains red-orange at pH values below 2.

B. Measuring the pH of some common solutions

In the rest of this experiment you will use a pH meter to measure pH. Your instructor will show you how to operate it. The electrodes may be fragile, so use caution when handling the meter's probe. See Appendix D for a discussion of pH meters.

Using a 25-mL sample in a small beaker, measure and record the pH of a 0.10 M solution of each of the following substances:

NaCl	Na$_2$CO$_3$	NaCH$_3$COO	NaHSO$_4$
sodium chloride	sodium carbonate	sodium acetate	sodium hydrogensulfate

Rinse the pH probe in deionized water between measurements. After you have completed a measurement, add a drop or two of bromocresol green or universal indicator to the solution and record the color you observe. Is the color consistent with the measured pH?

Some of these solutions are nearly neutral; others are acidic or basic. For the two solutions with pH furthest from 7, write a net ionic equation that shifts the pH of water in the direction needed to get that result.

C. Some properties of buffers

The lab contains 0.10 M stock solutions that can be used to make three different common buffer systems. These are

CH$_3$COOH/CH$_3$COO$^-$	NH$_4^+$/NH$_3$	HCO$_3^-$/CO$_3^{2-}$
acetic acid/acetate ion	ammonium ion/ammonia	hydrogencarbonate ion/carbonate ion

The sources of the ions will be sodium or chloride salts containing those ions. Select *one* of these buffer systems for this part of the experiment.

1. Using a graduated cylinder, measure 15 mL of a solution of the acid component of your buffer into a 100-mL beaker. The acid will be one of the following solutions: 0.10 M CH$_3$COOH, 0.10 M NH$_4$Cl, or 0.10 M NaHCO$_3$. Rinse out the graduated cylinder with deionized water and then use it to add 15 mL of a solution of the conjugate base of your buffer. Mix well, then measure the pH of the mixture and record it. Calculate the buffer's hydrogen ion concentration, [H$^+$], as well as pK_a for the acid.

2. Add 30 mL of water to your buffer mixture, mix, and pour half of the resulting solution into another small beaker. Measure the pH of the diluted buffer. Calculate $[H^+]$ and pK_a once again. Add 5 drops of 0.10 M NaOH to one portion of the diluted buffer and measure the pH again. To the other portion of the diluted buffer add 5 drops of 0.10 M HCl, and again measure the pH. Record your results.

3. Make a new buffer mixture containing 2.0 mL of the 0.10 M acid component solution and 20. mL of the solution containing the conjugate base. Mix, and measure the pH. Calculate a third value for pK_a. To this solution add 3 mL of 0.10 M NaOH. Measure the pH. Explain your results.

4. Put 25 mL of fresh deionized water into a well-rinsed small beaker. Measure the pH. Add 5 drops of 0.10 M HCl and measure the pH again. To that solution add 10 drops of 0.10 M NaOH, mix, and measure the pH.

5. Select a pH different from any of those you observed in your experiments, but within half a pH unit of the pK_a you calculated for your acid. Design a buffer that should have this target pH by selecting appropriate volumes of your acidic and basic components. Make up the buffer and measure its pH.

D. Preparing a buffer from a solution of a weak acid

So far in this experiment the buffers used have been made using a solution of a weak acid and a solution of its conjugate base. When making a buffer, most experienced chemists use a different approach. They start with a solution of a weak acid with a pK_a within half a pH unit of the pH of the buffer that is needed. To the acid they slowly add a strong base solution, usually NaOH, from a buret, stirring well, while at the same time measuring the pH. When the desired pH is reached, they stop adding NaOH. The buffer is then ready to use.

In this part of the experiment you will be given a 0.50 M solution of a weak acid with a known pK_a, and a target pH for a buffer you will be asked to prepare with it. First, dilute the acid solution to 0.10 M by adding 10. mL of the acid to 40. mL of water in a 100-mL or larger beaker and stirring well.

Using Equation 4b, calculate the ratio of $[B^-]$ to $[HB]$ you should need in the buffer to get the target pH. Once you know this ratio, use it to calculate how much 0.10 M NaOH you will have to add to 20. mL of your 0.10 M acid solution to produce your buffer. The NaOH reacts with HB, converting it to B^-—so if you add x mL of the NaOH, the value of $[B^-]/[HB]$ will become equal to $x/(20. - x)$ because each mL of 0.10 M NaOH produces the equivalent of 1 mL of 0.10 M B^- and consumes the equivalent of 1 mL of 0.10 HB. On the report page, record the volume of NaOH that you calculate should be needed.

Now actually prepare the buffer, to check your prediction. Use the buret containing 0.10 M NaOH solution that has been set up by the pH meter. Record the initial volume reading on the buret, before starting to add the base, as well as the initial pH of 20.0 mL of your acid solution, measured out into a clean but not necessarily dry container. Then slowly add NaOH to the acid, stirring well and watching the pH as it slowly goes up. When you obtain a solution with the target pH, stop adding NaOH. Record the final volume reading on the buret. Report the actual volume of NaOH solution required to produce your buffer, and compare it with the amount you calculated would be necessary.

Take it further (optional): Vitamin C ($MM = 176$ g/mol) is a weak acid, called ascorbic acid. Study the buffering properties of vitamin C using a solution made from a 500-mg tablet. Determine the acid dissociation constant, K_a, for vitamin C and the pH range over which ascorbic acid might be useful as a buffer.

> **Disposal of reaction products:** When you are finished with this experiment, you may pour the solutions down the drain, unless your instructor directs otherwise.

Experiment 25

Data and Calculations: pH Measurements: Buffers and Their Properties

A. Determining pH using acid-base indicators

Acid-base indicator	Indicator color in a 0.10 M solution of			
	hydrochloric acid	sodium dihydrogenphosphate	acetic acid	zinc sulfate
	HCl	NaH$_2$PO$_4$	CH$_3$COOH	ZnSO$_4$
methyl violet	═══════	═══════	═══════	═══════
thymol blue	───────	═══════	═══════	═══════
methyl yellow	═══════	═══════	═══════	═══════
Congo red	═══════	═══════	═══════	═══════
bromocresol green	═══════	═══════	═══════	═══════
Estimated pH of solution	═══════	═══════	═══════	═══════

Circle the observation(s) for each solution that was (were) most useful in estimating its pH.

(continued on following page)

B. Measuring the pH of some common solutions

Record the pH and the color observed with bromocresol green or universal indicator for each of the 0.10 M solutions that was tested. Is the color consistent with the pH indicated by the pH meter?

Indicator used: _____

	NaCl	Na_2CO_3	$NaCH_3COO$	$NaHSO_4$
pH	═══════	═══════	═══════	═══════
Color	═══════	═══════	═══════	═══════
Consistent? (yes/no)	_____	_____	_____	_____

For the two solutions with pH values furthest from 7, write a net ionic equation that shifts the pH of water in the direction needed to get that result.

Solution _____ Reaction _____

Solution _____ Reaction _____

C. Some properties of buffers

Buffer system selected _____   HB is _____ (name the acid)

1. pH of buffer _____ $[H^+]$ _____ M pK_a _____
                                                    (by Equation 4b)

2. pH of diluted buffer _____ $[H^+]$ _____ M pK_a _____

   pH after adding 5 drops of NaOH   ══════════

   pH after adding 5 drops of HCl   ══════════
   Comment on your observations in Parts 1 and 2:

3. pH of buffer in which $[B^-]/[HB] = 10$.   ══════════

                                                              pK_a _____
   pH after adding excess NaOH   ══════════
   Explain these observations:

4. Observed pH of deionized water (or "unstable")   ══════════

   pH after adding 5 drops of HCl   ══════════

   pH after adding 10 drops of NaOH   ══════════
   Explain these observations:

5. pH of buffer solution to be prepared   ══════════

   Average (mean) value of pK_a (as found in Parts 1, 2, and 3)   ══════════

   $\dfrac{[B^-]}{[HB]}$ needed in buffer   ══════════

   $\dfrac{\text{Volume of 0.10 M NaB}}{\text{Volume of 0.10 M HB}}$ needed in buffer   ══════════

   Volume of 0.10 M NaB actually used in preparing buffer   ══════════ mL

   Volume of 0.10 M HB actually used in preparing buffer   ══════════ mL

   Actual pH of the buffer you prepared   ══════════

D. Preparing a buffer from a solution of a weak acid

pH of buffer to be prepared _____ pK_a of acid _____

Ratio of $\dfrac{[B^-]}{[HB]}$ required in buffer   _____

Volume of 0.10 M NaOH calculated   _____ mL

pH of acid solution before titration   _____

Initial volume reading of NaOH   _____ mL

Final volume reading of NaOH   _____ mL

Volume of NaOH actually required   _____ mL

How does the volume you actually used compare to the calculated value?

Experiment 25

Advance Study Assignment: pH Measurements—Buffers and Their Properties

1. A solution of a weak acid was tested with the indicators used in this experiment. The colors observed were as follows:

methyl violet	violet	Congo red	red-orange
thymol blue	yellow-orange	bromocresol green	blue-green
methyl yellow	yellow		

Based on these colors, what is the approximate pH of this solution? Explain your reasoning.

approximate pH = _____

2. The ammonia molecule, NH_3, is the conjugate base of the NH_4^+ (called "ammonium") ion, a weak acid. An aqueous solution of NH_3 has a pH of 11.3. Write the net ionic equation for the reaction that makes an aqueous solution of NH_3 basic. (The answer looks very similar to Equation 5; replace B^- with the base and HB with the acid.)

3. The pH of a 0.010 M solution of (just) HOCl in water (that has reached equilibrium) is 4.8.
   a. What is $[H^+]$ in that solution? (Use Equation 1a.)

$[H^+] =$ _____ M

   b. What is $[OCl^-]$? What is [HOCl]? (*Hint*: Where do the H^+ and OCl^- ions come from?)

$[OCl^-] =$ _____ M; $[HOCl] =$ _____ M

   c. What is the value of K_a for HOCl? What is the value of pK_a for HOCl?

$K_a =$ _____ $pK_a =$ _____

(*continued on following page*)

4. Hydrofluoric acid, HF, has a K_a value of 7.2×10^{-4}. You are asked to prepare a buffer having a pH of 3.85 from a solution of hydrofluoric acid and a solution of sodium fluoride, NaF, having the same molarity. How many milliliters of the sodium fluoride solution should you add to 20.0 mL of the hydrofluoric acid solution to make the buffer? (Any version of Equation 4 can get you there!) [*Note*: Perhaps surprisingly, this would be *extremely* dangerous to do! While safe and helpful at the very low concentrations found in fluoridated water, fluoride ions at much higher concentrations—such as those used to make a buffer—can easily kill you! Never attempt to actually make the buffer described here unless you are specially trained (buffers like these are used in the silicon semiconductor industry).]

_____ mL

5. How many mL of 0.10 M NaOH should you add to 20.0 mL of 0.10 M HF if you wish to prepare a buffer with a pH of 3.85, the same as in Problem 4? (*Hint*: Read Part D in the experimental procedure for help!)

_____ mL

Experiment 26

Determining the Solubility Product of $Ba(IO_3)_2$

When an ionic solid is placed in water, it will (eventually) reach a solubility equilibrium if the solid does not all dissolve. The equilibrium constant for the dissolution reaction is called the <u>solubility product</u>, K_{sp}. Barium iodate, $Ba(IO_3)_2$, dissolves according to this dissolution reaction:

$$Ba(IO_3)_2(s) \rightleftharpoons Ba^{2+}(aq) + 2\ IO_3^-(aq) \tag{1}$$

Recognizing that the pure solid is left out of the equilibrium expression but must be present for the equilibrium to hold, the equilibrium expression for Reaction 1 is

$$K_{sp} = [\underline{Ba^{2+}}]\ [\underline{IO_3^-}]^2 \tag{2}$$

in which the concentrations of the barium and iodate ions are underlined to indicate two important things: first, that the concentration must be for a system *at equilibrium*, and second, that it must be measured in moles per liter but used *without units*. Therefore, K_{sp} has no units. Equation 2 essentially states that at a given temperature, in a mixture of barium and iodate ions in equilibrium with solid barium iodate the product of the molarity of the Ba^{2+} ion times the molarity of the IO_3^- ion squared cannot vary, no matter how the ion concentrations in the system change. If one ion concentration goes up, the other must go down, keeping the ion concentration product in Equation 2 equal to K_{sp}. That is rather amazing! For barium iodate, K_{sp} has a small value, less than 10^{-7}. So, if one ion has a typical concentration, say 0.1 M, the other must have a very low concentration. If you add 0.10 M $BaCl_2$ to a small volume of a solution of 0.05 M KIO_3, you can be sure that a reaction will occur in which $Ba(IO_3)_2$ precipitates, producing a final solution containing almost no iodate ion and with the product of the ion concentrations in Equation 2 equal to K_{sp}.

Solubility product calculations are critical in determining whether a given material will be stable as a solid or will dissolve in a given solution—whether that is desired or not. In pharmacology they often underlie why certain drugs can not be taken together. At a water treatment plant, such calculations are critical in treating both incoming water and water to be released back into the environment.

In this experiment you will be examining the solubility behavior of $Ba(IO_3)_2$ in various amounts of $Ba(NO_3)_2$ and KIO_3 solutions with water (the <u>reagents</u> in this experiment), thereby measuring the solubility product (K_{sp}) of barium iodate and observing whether it really does remain constant.

Experimental Procedure

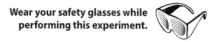
Wear your safety glasses while performing this experiment.

This experiment is quite challenging, but can give excellent results if done with sufficient care. You will be working with small amounts of the chemicals involved, so small errors will be important.

A. Preparing the reaction mixtures

In this experiment you will precipitate solid $Ba(IO_3)_2$ under several different conditions, varying the proportions of the reagents used. You will analyze the solutions remaining after the precipitation to determine their iodate ion concentrations at equilibrium and—from those values and the way the mixtures were prepared—find the equilibrium concentrations of the barium ions and the value of K_{sp} for barium iodate.

Differentiate five dry medium (18×150 mm) test tubes, either by labeling them or by noting their locations in your test tube rack.

Transfer about twenty milliliters of the 0.0350 M KIO₃ stock solution into a small, dry beaker. Carefully pipet 1.00 mL of this solution into Test Tube 1, then transfer 2.00, 3.00, 4.00, and 5.00 mL of it into Test Tubes 2, 3, 4, and 5, respectively, as shown in Table 26.1.* Visually confirm the liquid levels in the tubes make sense before continuing. Next, transfer about thirty-five milliliters of the 0.0200 M Ba(NO₃)₂ stock solution into a small, dry beaker. Add 5.00 mL of this solution to each of the test tubes, using your 5-mL pipet. Again, look at the liquid levels in the tubes to ensure you did not miss one. Finally, pipet 6.00, 5.00, 4.00, 3.00, and 2.00 mL of deionized water into Test Tubes 1 through 5, respectively. The final volume in each tube should be 12.00 mL, with the compositions of the mixtures as shown in Table 26.1.

Table 26.1

Volumes of Reagents Used in Precipitating Ba(IO₃)₂ {mL}		
Test Tube	**0.0200 M Ba(NO₃)₂**	**0.0350 M KIO₃**	**H₂O**
1	5.00	1.00	6.00
2	5.00	2.00	5.00
3	5.00	3.00	4.00
4	5.00	4.00	3.00
5	5.00	5.00	2.00

With a stirring rod, stir each solution for 30 seconds, up and down as well as in circles, wiping the rod clean on a paper towel before proceeding to the next solution. Touch the wall of the tube occasionally as you stir. Alternatively, a vortex mixer can be used to mix the solutions.

Look at each tube. In most of them you should see a crystalline white precipitate, which slowly settles out. It will be most noticeable in the higher number tubes, but it should be present as a slight cloudiness in Test Tubes 1 and 2 as well. There will only be a tiny amount of precipitate: only a few milligrams, so it can be hard to see. Try looking through the tube toward the light: you may observe tiny particles, very slowly settling out. The crystallization occurs slowly, so give it time.

Put a stopper in each of the tubes in which there is a precipitate, and slowly rotate the tubes so that the liquid goes from end to end twenty times or so, to ensure thorough mixing and help bring the system to equilibrium.

(*Note*: For any tubes in which you observe only a clear liquid, and no solid, repeat the stirring operation using a glass stirring rod and scratching the walls of the tube as you stir, which may get things started. If that still does not do the job, touch your stirring rod to the sample of pure solid barium iodate in the lab and swirl the rod in the reluctant liquid. Then remove the rod, insert a stopper in the tube, and shake it with vigor. Eventually this should get the cloudiness to form.)

Repeat the rotation step, turning each tube end to end to do what you can to make the liquid the same throughout (technical term: homogeneous) and bring it to equilibrium with the precipitate. You can not over-stir, so be patient. Give yourself five minutes with this. It will be time well spent. Finally, let the test tubes stand for 15 minutes to allow the solid precipitate particles to settle out completely, leaving a clear solution above the solids.

While you are waiting for the precipitates to settle out, you can begin the calculations you will need later. Fill in as many blanks as you can on the report page or in a spreadsheet. From the way you made your mixtures, you can determine the moles of Ba^{2+} and IO_3^- that were initially present in each tube. For example, Test Tube 1 contains 1.00 mL of 0.0350 M KIO₃. Using the definition of molarity, that means

$$\text{initial moles of } IO_3^-(aq) = [IO_3^-]_{stock} \times V_{stock} = 0.0350 \text{ M} \times 1.00 \text{ mL} \times \left(\frac{1 \text{ L}}{1000 \text{ ml}}\right) \times \left(\frac{\text{mol/L}}{\text{M}}\right) = 3.50 \times 10^{-5} \text{ moles}$$

are in that tube before any Ba(IO₃)₂ precipitate forms. After you complete the analysis of the solutions at equilibrium (in Parts B and C), you will be able to calculate the equilibrium concentrations that lead to K_{sp}.

---

* If not using an autopipet, you can use a 10-mL graduated pipet for these additions. (The easiest way to use a graduated pipet for this is to fill it to the 0.00 level, then let the level fall to 1.00 with the pipet in Test Tube 1. In Tube 2, let the level fall to 3.00, thereby adding 2.00 mL. In Tube 3, let the level fall to 6.00, and in Tube 4 let it go to 10.00. Refill the pipet to the 5.00 level and in Tube 5 let it fall to 10.00.) Use a similar procedure to measure out the water.

B. Preparing the equilibrium mixtures for analysis

At this point your test tubes should contain essentially clear solutions above the precipitates that formed. You now need to remove some of the liquid from each tube without disturbing the solid at the bottom. This is the most difficult part of this experiment, and must be done very carefully. Use a 5-mL pipet. If using a glass pipet, first rinse it with deionized water. Then drain it and blow out the water remaining in the tip. (You would not want to do this when *measuring* with a glass pipet, but the goal here is to dry the pipet so your samples are not diluted by water retained in it. In this situation, you do not have enough sample to easily rinse the pipet with it, so drying the pipet—by blowing it out—is the next best option.)

Clamp Test Tube 1 on a ring stand, so that you can work on it without it moving. Immerse the 5-mL pipet about half-way down into the clear solution and use it to transfer 5.00 mL of the liquid into a clean, dry test tube. Pull the liquid up slowly to make sure no solid is sucked up with it. (See Appendix D for more details on how to use pipets.) Repeat this process for each of the five test tubes, using five labeled, clean, and dry test tubes to receive the 5.00 mL liquid samples. If using a glass pipet, be sure to rinse, drain, and blow it out between each test tube.

This operation is easier to describe than do. If something goes wrong while you are pulling the liquid into the pipet, let the liquid drain back into the test tube. After the precipitate re-settles to the bottom, you can try again. To increase your chances of accomplishing this transfer properly, you may wish to practice doing it with water in a clamped test tube.

C. Analyzing the equilibrium solutions

Having extracted 5.00 mL of each of the five equilibrium solutions, it is now possible to analyze for the concentration of the IO$_3^-$ ion in those solutions. You will do this by reducing iodate ion to iodine, I$_2$, which is colored and can be analyzed spectroscopically (based on how much light it absorbs). The reaction you will use is:

$$IO_3^-(aq) + 5\,I^-(aq) + 6\,H^+(aq) \rightarrow 3\,H_2O(\ell) + 3\,I_2(aq)^* \tag{3}$$

To each of the test tubes add, by pipet, 1.00 mL of 1.0 M KI, 1.00 mL of 1.0 M HCl, and 3.00 mL of deionized water, bringing the total volume in each to 10.00 mL. During this step the solutions will become orange as Reaction 3 takes place. Stir each mixture well with a stirring rod, cleaning the rod on a paper towel before you go on to the next tube, or use a vortex mixer.

Use a spectrophotometer to determine the absorbance of the orange solution in each test tube at 500 nm. If you have not used a spectrophotometer before, read Appendix D to learn how to use it to measure light absorption. Unless your spectrophotometer has an immersion probe, pick up a spectrophotometer cell (technical term: cuvet), rinse and then fill it with deionized water, and use it to blank the spectrophotometer. Pour out the water from the cell and shake it to remove what liquid you can. Then add about a milliliter of the orange liquid from Tube 1, tilt the cell to rinse the walls with this solution, and then empty the cell into a waste beaker. Rinse the cell again with another roughly one-milliliter portion of the solution from Tube 1, then drain it again. Finally, fill the cell up to the measuring level with the solution from Tube 1. Put the cell into the spectrophotometer and read the absorbance of the solution. The concentration of the IO$_3^-$ ion in the *equilibrium 12.00-mL mixture in Tube 1* will equal the value on the graph or calculated with the calibration equation provided. (The graph or equation takes into account the dilution associated with carrying out Reaction 3, and will give the actual concentration in Tube 1.) Record the absorbance and concentration for Tube 1. Repeat this measurement process, *including the rinsing steps* but not the setting of the blank, for the other four tubes you worked with, and record your values for those solutions.

Disposal of reaction products: Pour all of the solutions and solids in the 10 test tubes you used in a waste container, unless your instructor directs otherwise.

---

* Given the excess of I$^-$, the dominant form of iodine in solution is actually I$_3^-$; however, Reaction 3 does essentially describe what happens.

D. Processing the data

From the volume and concentration of the KIO_3 stock solutions used in making the five reagent mixtures, you know the moles of iodate ion present in each mixture. Having now measured the concentration of *free* (not in a precipitate) IO_3^- ion in each of the 12.00-mL *equilibrium* solutions you can now calculate the moles of free IO_3^- in those solutions. The difference between the total moles of IO_3^- in each tube and the equilibrium moles of free IO_3^- in that same tube must be the moles of iodate ion in the precipitate at the bottom of the tube. The moles of barium in the precipitate must equal *half* of that amount—because the formula of barium iodate, $Ba(IO_3)_2$, indicates there are two iodate ions for every barium ion in the precipitate.

The moles of free barium ion that remain in each solution at equilibrium will equal the difference between the moles added and the moles present in the precipitate. Knowing the volume of the equilibrium solution, 12.00 mL, you can calculate the equilibrium concentration of free Ba^{2+} ion in each tube at equilibrium. Record the results of these calculations on the report page or in your spreadsheet. Finally, calculate the values of K_{sp} indicated by the equilibrium concentrations in each tube. Calculate the average (mean) value of K_{sp} and its standard deviation. (See Appendix H.)

Experiment 26

Data and Calculations: The Solubility Product of $Ba(IO_3)_2$

A. Preparing the reaction mixtures

While waiting for your precipitates to settle out, fill in this table and calculate the initial moles of each reactant at the start of the Part D section of this report page. If you are using a spreadsheet, set it up to resemble the tables here.

Test Tube Number	1	2	3	4	5
mL of 0.0350 M KIO_3	_____	_____	_____	_____	_____
mL of 0.0200 M $Ba(NO_3)_2$	_____	_____	_____	_____	_____

C. Analyzing the equilibrium solutions

	1	2	3	4	5
Absorbance of solution	=====	=====	=====	=====	=====
$[IO_3{}^-]$ in solution {M}	=====	=====	=====	=====	=====

D. Processing the data

	1	2	3	4	5
Initial moles of Ba^{2+} {moles}	_____	_____	_____	_____	_____
Initial moles of $IO_3{}^-$ {moles}	_____	_____	_____	_____	_____
Moles of free $IO_3{}^-$ in equilibrium solution (12.00 mL) {moles}	_____	_____	_____	_____	_____
Moles of $IO_3{}^-$ precipitated {moles}	_____	_____	_____	_____	_____

(continued on following page)

Test Tube Number	1	2	3	4	5
Moles of Ba^{2+} precipitated {moles}	_____	_____	_____	_____	_____
Moles of free Ba^{2+} in equilibrium solution (12.00 mL) {moles}	_____	_____	_____	_____	_____
[Ba^{2+}] in equilibrium solution {M}	_____	_____	_____	_____	_____
K_{sp} for Ba(IO$_3$)$_2$	_____	_____	_____	_____	_____

Average (mean) value of K_{sp} _____   Standard deviation in K_{sp} _____

What was the mass of solid barium iodate in the precipitate in Test Tube 1? $MM_{Ba(IO_3)_2} = 487.14$ g/mol

_____ mg

The same amount of Ba^{2+} was added to each tube. The amount of IO$_3^-$ added increased with the tube number. Why did the *equilibrium* concentrations of free Ba^{2+} decrease with tube number?

Based on your experimental K_{sp} value, what is the solubility of Ba(IO$_3$)$_2$ in water, in grams per liter? [How much solid barium iodate, Ba(IO$_3$)$_2$(s), would dissolve if you put a lot of it in pure water?]

_____ g/L

Experiment 26

Advance Study Assignment: Determining the Solubility Product of $Ba(IO_3)_2$

1. State in words what is meant by the solubility product equation for $Ba(IO_3)_2$: $K_{sp} = [\underline{Ba^{2+}}] [\underline{IO_3^-}]^2$ (2)

2. If 5.0 mL of 0.10 M $Ba(NO_3)_2$ is mixed with 5.0 mL of 0.10 M KIO_3, a precipitate forms. Which ion will still be present at appreciable concentration in the equilibrium mixture if K_{sp} for barium iodate is very small? Indicate your reasoning. What would that concentration be?

_____ _____ moles/L

3. Lead bromide, $PbBr_2$, is moderately soluble, with K_{sp} equal to 1.86×10^{-5} at 20.°C.

   a. Predict the solubility of lead bromide in pure water at 20.°C, based on this K_{sp} value. (How many moles of solid $PbBr_2$ could be completely dissolved in one liter of solution?) *Ignore the effect of any other equilibria that may impact the solubility (for example, interactions between Pb^{2+} and OH^-). Your result will not match literature values for the actual solubility of $PbBr_2$!*

_____ moles/L

(continued on following page)

b. What would the predicted solubility of $PbBr_2$ be in 1.0 M NaBr? (How many moles of solid $PbBr_2$ could be completely dissolved in 1 L of solution?) *Again, ignore all other equilibria; assume that the K_{sp} expression for $PbBr_2$ completely describes the situation.*

_____ moles/L

c. The difference in the results obtained in 3(a) and 3(b) is caused by what is known as the common ion effect. State in words what the common ion effect predicts.

Experiment 27

Relative Stabilities of Copper(II) Complex Ions and Precipitates

In aqueous (water-as-the-solvent) solutions, most positively charged ions (technical term: cations), particularly those of the transition metals, exist as the (metal) ion surrounded in an organized way by some water molecules. Such structures are called complex ions. The water molecules, usually two, four, or six in number, are bound chemically to the (metal) ion—but often rather loosely, with both electrons in the chemical bonds being provided by one of the unshared electron pairs on the oxygen atoms in the H_2O molecules. Such bonds are called coordinate covalent bonds, and the molecules or ions contributing the electrons for them are called ligands. Copper ion in aqueous solution typically exists as $Cu(H_2O)_6^{2+}$, with the six water molecules arranged in an octahedron around (one above, one below, one to the right of, one to the left of, one behind, and one in front of) the copper ion at the center.

If a hydrated metal ion (one having water as its ligands) such as $Cu(H_2O)_6^{2+}$ is mixed with other substances that can, like water, form coordinate covalent bonds with Cu^{2+}, those new ligands may displace one or more H_2O molecules and form new complex ions containing them. For example, ammonia, NH_3, is a reasonably good coordinating ligand and may displace H_2O from the hydrated copper ion, $Cu(H_2O)_6^{2+}$, to form $Cu(H_2O)_5(NH_3)^{2+}$, $Cu(H_2O)_4(NH_3)_2^{2+}$, $Cu(H_2O)_3(NH_3)_3^{2+}$, or $Cu(H_2O)_2(NH_3)_4^{2+}$. In aqueous ammonia, $NH_3(aq)$, the substitution stops there, with ammonia not displacing the last two water molecules; it is only in pure, liquefied ammonia, $NH_3(\ell)$, that the last two water molecules can be replaced, forming $Cu(H_2O)(NH_3)_5^{2+}$ and $Cu(NH_3)_6^{2+}$. Because water is almost always present as a ligand in aqueous solutions, it is common practice to omit the water ligands when writing the formula of an aqueous metal ion: "$Cu^{2+}(aq)$" is actually shorthand for $Cu(H_2O)_6^{2+}(aq)$, and "$Cu(NH_3)_3^{2+}(aq)$" is used as shorthand for $Cu(NH_3)_3(H_2O)_3^{2+}(aq)$.

Coordinating ligands differ in their tendencies to form bonds with positive ions, so in a solution containing a given metal ion and several possible ligands, an equilibrium will develop in which more of the metal ions are coordinated with those ligands with which they form the most stable (strongest) bonds. There are many kinds of ligands, but they all possess an unshared pair of electrons that they can donate to form a coordinate covalent bond with a positively charged metal ion. In addition to H_2O and NH_3, other uncharged coordinating ligands include carbon monoxide, CO, and ethylenediamine, $C_2H_4(NH_2)_2$; some common negatively charged ions (technical term: anions) that can form complexes include hydroxide, OH^-; chloride, Cl^-; cyanide, CN^-; thiocyanate, SCN^-; and sulfite, $S_2O_3^{2-}$.

When solutions containing metal ions are mixed with other solutions containing ions, precipitates are sometimes formed. When a solution of 0.10 M copper nitrate, $Cu(NO_3)_2(aq)$, is mixed with a little 1.0 M aqueous ammonia solution, $NH_3(aq)$, a precipitate forms that then dissolves after more ammonia is added. The formation of the precipitate helps us understand what is occurring as NH_3 is added. The precipitate is copper(II) hydroxide, $Cu(OH)_2(s)$, formed by reaction of the hydrated copper ion, $Cu(H_2O)_6^{2+}(aq)$, with the small amount of hydroxide ion, OH^-, present in the basic NH_3 solution. The fact that this reaction occurs means that—even at very low OH^- ion concentrations—$Cu(OH)_2(s)$ is more stable than the $Cu^{2+}(aq)$ ion.

Adding more NH_3 causes the solid to redissolve. At that point the copper substance in solution cannot be the hydrated copper ion, because it is less stable than $Cu(OH)_2(s)$ in the presence of ammonia. It must be some other complex ion, and it turns out to be the $Cu(NH_3)_4^{2+}$ ion. This observation implies that the $Cu(NH_3)_4^{2+}$ ion is also more stable in NH_3 solution than the hydrated copper ion, Cu^{2+}. To also conclude that the copper ammonia complex ion, $Cu(NH_3)_4^{2+}(aq)$, is generally more stable than copper(II) hydroxide, $Cu(OH)_2(s)$, is not warranted, however. This is because under the conditions in the solution the concentration of ammonia, $[NH_3]$, is much larger than the concentration of hydroxide ion, $[OH^-]$; and given a higher concentration of hydroxide ion, $[OH^-]$, solid copper(II) hydroxide, $Cu(OH)_2(s)$, might precipitate even in the presence of a similar concentration of NH_3.

To investigate this, you might add a little 1.0 M NaOH solution to the solution containing the $Cu(NH_3)_4^{2+}$ ion. If you do this you will find that solid copper(II) hydroxide, $Cu(OH)_2(s)$, does indeed precipitate. We can conclude from these observations that $Cu(OH)_2(s)$ is more stable than $Cu(NH_3)_4^{2+}(aq)$ in solutions in which the ligand concentrations (OH^- and NH_3) are roughly equal.

The copper-containing substances and their amounts that will be present in a system depend, as we have just seen, on the conditions in the system. We cannot say in general that one substance will be more stable than another; the stability of a given substance depends on the kinds and concentrations of other substances that are also present with it.

Another way of looking at the matter of stability is through equilibrium theory. Each copper substance we have mentioned can be formed in a reaction between the hydrated copper ion, $Cu^{2+}(aq)$, and a complexing or precipitating ligand; each reaction will have an associated equilibrium constant, either a formation constant, K_f, or a solubility product, K_{sp}, for that substance. The reactions and their equilibrium constants for the copper substances we have been considering here are

$$Cu^{2+}(aq) + 4\,NH_3(aq) \rightleftharpoons Cu(NH_3)_4^{2+}(aq) \qquad K_f = 5 \times 10^{12} \qquad (1)$$

$$Cu^{2+}(aq) + 2\,OH^-(aq) \rightleftharpoons Cu(OH)_2(s) \qquad K_{sp} = 2 \times 10^{19} \qquad (2)$$

The large size of the equilibrium constants indicates that the tendency for the hydrated copper ion to react with the ligands is very high.

In terms of these data, let us compare the stability of the $Cu(NH_3)_4^{2+}$ complex ion with that of solid $Cu(OH)_2(s)$. This can be done by considering Reaction 3:

$$Cu(NH_3)_4^{2+}(aq) + 2\,OH^-(aq) \rightleftharpoons Cu(OH)_2(s) + 4\,NH_3(aq) \qquad (3)$$

We can find the value of the equilibrium constant for this reaction by noting that it is the sum of Reaction 2 and the reverse of Reaction 1. Therefore the equilibrium constant for Reaction 3, K_3, is given by the equation

$$K_3 = \frac{K_{sp}}{K_f} = \frac{2 \times 10^{19}}{5 \times 10^{12}} = 4 \times 10^6 = \frac{[NH_3]^4}{[Cu(NH_3)_4^{2+}][OH^-]^2} \qquad (4)$$

From the expression in Equation 4, we can calculate that in a solution in which the NH_3 and OH^- ion concentrations are both about one molar,

$$[Cu(NH_3)_4^{2+}] = \frac{1}{4 \times 10^6} = 2.5 \times 10^{-7}\,M \qquad (5)$$

Because the concentration of the copper ammonia complex ion is very low, the vast majority of copper(II) in the system will exist as the solid hydroxide. In other words, the solid hydroxide is more stable under such conditions than the ammonia complex ion. That matches up well with what we observed when we treated the hydrated copper ion with ammonia and then with an equivalent amount of hydroxide ion.

This experimentally observed behavior of the copper ion allows us to conclude that because the solid hydroxide is the substance that exists when copper ion is exposed to equal concentrations of ammonia and hydroxide ion, the hydroxide is more stable under those conditions, *and* the equilibrium constant for the formation of the hydroxide is larger than the constant for the formation of the ammonia complex. By determining which substance is present when a positive ion is in the presence of equal ligand concentrations, we can assess relative stability under such conditions and can rank the equilibrium constants for the possible complex ions and precipitates in order of increasing size.

In this experiment you will carry out reactions involving a group of complex ions and precipitates of the Cu^{2+} ion. You can make these substances by mixing a solution of copper(II) nitrate, $Cu(NO_3)_2(aq)$, with solutions containing ligands: ammonia, NH_3, or negatively charged ions—which may form either precipitates or complex ions with the $Cu^{2+}(aq)$ ion present in aqueous solutions of copper(II) nitrate. By examining whether the precipitates or complex ions formed by reacting $Cu^{2+}(aq)$ with a given ligand can, upon adding a second ligand, be dissolved or transformed to another substance, you will be able to rank the relative stabilities of the precipitates and complex ions made from Cu^{2+} with respect to one another, and thus rank the equilibrium constants for each ligand in order of increasing magnitude. The ligands you will react with Cu^{2+} ion in aqueous solution are NH_3, Cl^-, OH^-, $C_2O_4^{2-}$, S^{2-}, NO_2^-, and PO_4^{3-}. In each case the test for relative stability will be made in the presence of essentially equal concentrations of two ligands. Once you have completed your ranking of the known substances, you will test an unknown substance and determine where it belongs in your ordered stability list.

Experimental Procedure

Wear your safety glasses while performing this experiment.

Obtain an unknown and seven small (13×100 mm) test tubes.

Add about one milliliter of 0.10 M copper(II) nitrate, $Cu(NO_3)_2(aq)$, solution to each of the test tubes (a depth of about one centimeter in these tubes).

To the first of the test tubes add about one milliliter (about one additional centimeter of liquid) of 1.0 M aqueous ammonia, $NH_3(aq)$, drop by drop. Note whether a precipitate forms initially, and if it dissolves in excess (relative to the moles of Cu^{2+} ion present) NH_3. Shake the tube side to side to mix the contents. In the NH_3-NH_3 space in the Table of Observations on the report page, write a "P" in the upper right-hand corner if a precipitate is formed initially. In the rest of the space, describe what you observed after all of the NH_3 was added—particularly the color of the resulting solution and whether it was clear or cloudy, or that there was no change. Below that, write the chemical formula of the substance present in the *final* solution. If the final solution is clear, rather than cloudy, with no precipitate at the bottom, the copper ion will be in a complex. For all the complexes in this experiment, Cu(II) will attach to four non-water ligands, so the formula with NH_3 would be $Cu(NH_3)_4^{2+}$, which is shorthand for $Cu(NH_3)_4(H_2O)_2^{2+}$. If a precipitate is present, it will be <u>neutral</u> (it will have no net charge), and in the case of NH_3 it would be a hydroxide with the formula $Cu(OH)_2$. Add about one milliliter of 1.0 M NH_3 to the rest of the test tubes and shake them side to side to mix them well.

Now you will test the stability of the substance present in excess NH_3 relative to those that might form with other ligands. Add, drop by drop, about one milliliter of a solution of each of the ligands in the top (heading) row of the Table of Observations to the test tubes you have just prepared, one ligand to a test tube. Note any changes that occur in the appropriate spaces in the Table of Observations. (Compare against the test tube to which only NH_3 was added, and that contains no second ligand.) A change in color or the formation of a precipitate implies that a reaction has occurred between the added ligand and the copper-containing substance originally present. As before, put a "P" in the upper right-hand corner if a precipitate initially forms upon adding the new ligand. In the rest of the space, record what you observe and write the chemical formula of the copper-containing substance that is stable in the presence of equal concentrations of NH_3 and the second ligand. Again, assume that in complex ions Cu(II) will attach to four non-water ligands, and that copper(II) precipitates will be neutral (have zero net charge). If a new copper-containing substance forms upon adding the second ligand, you should write down its chemical formula in the box. If no change occurs, the original copper-containing substance is more stable, so put its chemical formula there.

Repeat the preceding series of experiments, using 1.0 M Cl^- as the first ligand added to the $Cu(NO_3)_2$ solution. You will only need six test tubes this time, and will start by filling in the Cl^--Cl^- box of the Table of Observations with the observations you make upon adding 1.0 M Cl^- to the first test tube. Then you will add 1.0 M Cl^- to the remaining test tubes, mix well, and then proceed to test these tubes with the second ligands from the top (heading) row of the Table of Observations. In each case, record whether a precipitate forms upon adding the first drop, the color of any precipitates or solutions formed upon adding the rest of the ligand solution, and the formula of the copper-containing substance that is most stable when an excess of both Cl^- ion and any added second ligand are present in the solution. Because these reactions are reversible, it is not necessary to retest Cl^- with NH_3: the same information would be obtained (which ligand forms the more stable substance) as when the NH_3 solution was tested with Cl^- solution.

Repeat this same series of experiments with each of the ligands in the leftmost column in the Table of Observations, omitting those tests where decisions as to relative stabilities are already clear. In cases where you are unsure about what happened, particularly when both ligands produce copper-containing precipitates, it may be helpful to add the ligands in reverse order (that is, to "repeat" a test in reverse, filling in a box in the lower left half of the Table of Observations)—doing so may make it clearer what is going on. When complete, your Table of Observations should have at least 28 entries.

Examine your Table of Observations and decide on the relative stabilities of the copper-containing substances you observed, of which there should be seven. Rank them as best you can in order of increasing stability. The equilibrium constants of the formation or precipitation reactions will increase in the same order as the relative stabilities of the substances they produce.

When you are satisfied that your stability ranking order for the seven copper-containing substances is correct, carry out the necessary tests on your unknown to determine its proper position in your stability ranking order list. Your unknown may be one of the Cu(II) substances you have already observed, or it may be a new one, present in a solution of its ligand. If your unknown contains a precipitate, shake it well before using a portion of it to make a test.

Disposal of reaction products: Dispose of all reaction products in the waste container provided, or as otherwise directed by your instructor.

Optional **Alternate Procedure Using a Microscale Approach**

Obtain an unknown and two plastic well plates (4 wells × 6 wells).

Align the two plates into an 8 × 6 well configuration, with eight wells across the top. To each of the first seven wells in the top row of wells, add 6 drops of 0.10 M $Cu(NO_3)_2$. To the first (leftmost) of these wells add 1 drop of 1.0 M NH_3. Write a "P" in the upper right-hand corner of the corresponding (upper left, NH_3-NH_3) box in the Table of Observations on the report page if a precipitate is formed. Then add 5 more drops of 1.0 M NH_3 to the same well, and mix the contents of the well by moving the well plate it is in back and forth on the lab bench. In the NH_3-NH_3 box of the Table of Observations, describe what you observe—particularly the color of the resulting solution and whether it is clear or cloudy, or that there was no change. Below that in the same box, write the chemical formula of the substance present after adding all 6 drops of 1.0 M NH_3 and mixing. If the final solution is clear, rather than cloudy, with no precipitate at the bottom, the copper ion will be in a complex. For all the complexes in this experiment, Cu(II) will attach to four non-water ligands, so the formula with NH_3 would be $Cu(NH_3)_4^{2+}$, which is shorthand for $Cu(NH_3)_4(H_2O)_2^{2+}$. If a precipitate is present, it will be neutral (it will have no net charge), and in the case of NH_3 it would be a hydroxide with the formula $Cu(OH)_2$. Add 6 drops of 1.0 M NH_3 to the remaining six wells on the top row and mix their contents by moving the well plates back and forth on the lab bench.

Now you will test the stability of the substance present in excess NH_3 relative to those that might form with other ligands. First add one drop of 1.0 M NaCl to the second well in the top row. Write a "P" in the upper right-hand corner of the corresponding (Cl^--NH_3) box in the Table of Observations on the report page if a precipitate is formed. Then add 5 more drops of 1.0 M NaCl to the same well, and mix the contents by moving the well plate back and forth on the lab bench. In the Cl^--NH_3 box of the Table of Observations, describe any changes that occurred (compare the well against the first one, to which no Cl^- was added), or document that no change was evident. Below that, write the chemical formula of the copper-containing substance that is stable in the presence of equal concentrations of NH_3 and Cl^-. Again, assume that in complex ions Cu(II) will attach to four non-water ligands, while copper(II) precipitates will be neutral (have zero net charge). If a new copper-containing substance forms upon adding the second ligand, you should write down its chemical formula in the box. If no change occurs, the original copper-containing substance (formed in the presence of excess NH_3) is more stable, so put its chemical formula there. Repeat this same procedure with the next well and 1.0 M NaOH, and then the with the rest of the ligands in the top (heading) row of the table.

Now move to the second row of wells, starting with the well in the *second* column, corresponding to the Cl^--Cl^- box in the Table of Observations. Add six drops of 0.10 M $Cu(NO_3)_2$ to the wells in the second through seventh columns. Add a single drop of 1.0 M NaCl to the well in the second column; if a precipitate forms, document that by putting a "P" in the upper right-hand corner of the Cl^--Cl^- box in the Table of Observations. Then add 5 more drops of 1.0 M NaCl to the well in the second column (on the second row) and mix the contents by moving the well plate back and forth on the lab bench. Document what you observe in the Cl^--Cl^- box in the Table of Observations and write the chemical formula of the copper-containing substance present. Add 6 drops of 1.0 M NaCl to the copper-containing wells on the remainder of that (second) row and mix well. Then add first one drop and then 5 more drops of the ligands in the top (header) row of the Table of Observations, documenting whether the first drop forms a precipitate, what you observe when excess ligand is added, and finally the chemical formula of the stable copper-containing substance in the presence of equal amounts of both ligands. You will not be testing with NH_3, in the first column of the second row, because the ligand-exchange reactions are reversible and the same information would be obtained (which ligand forms the more stable substance) as when the NH_3 solution was tested with Cl^- solution, which you have already done.

You should repeat this process going down the rows of the Table of Observations, each time starting at the well one to the right of the first one used in the row above. You will not have a row for S^{2-}, though your work up to this point has already made it clear how S^{2-} acts as a ligand. All that remains is to test how it acts on 0.10 M $Cu(NO_3)_2$ itself, which you can do in the eighth (and so far unused) column of wells.

If you remain uncertain about the outcome of any ligand-ligand competition, particularly when both ligands produce precipitates, try reversing its order using the dry well corresponding to the appropriate box in the (empty) lower left portion of the Table of Observations. Document what you observe in the box; hopefully that will help clarify what is going on.

Examine your Table of Observations and decide on the relative stabilities of the copper-containing substances you observed, of which there should be seven. Rank them as best you can in order of increasing stability. The equilibrium constants of the formation or precipitation reactions will increase in the same order as the relative stabilities of the substances they produce.

When you are satisfied that your stability ranking order for the seven copper-containing substances is correct, use empty wells to carry out tests on your unknown to determine its proper position in your stability ranking order list. Your unknown may be one of the Cu(II) substances you have already observed, or it may be a new one, present in a solution of its ligand. If your unknown contains a precipitate, shake it well before using a portion of it to make a test.

Disposal of reaction products: Pour the contents of all of the wells into a waste container, unless your instructor directs otherwise. ▨

Experiment 27

Observations and Analysis: Relative Stabilities of Copper(II) Complex Ions and Precipitates

Table of Observations

second ligand → ↓ first ligand	NH_3 (ammonia)	Cl^- (chloride)	OH^- (hydroxide)	$C_2O_4^{2-}$ (oxalate)	PO_4^{3-} (phosphate)	NO_2^- (nitrite)	S^{2-} (sulfide)
NH_3 (ammonia)							
Cl^- (chloride)							
OH^- (hydroxide)							
$C_2O_4^{2-}$ (oxalate)							
PO_4^{3-} (phosphate)							
NO_2^- (nitrite)							
S^{2-} (sulfide)							
Unknown							

(continued on following page)

Determining relative stabilities

In each row of the Table of Observations you can compare the stabilities of copper-containing substances involving the ligand in the column headings of the top row with those of the copper-containing substances containing the ligand from the row heading in the leftmost column. In the first row of the Table of Observations, the copper(II)-NH_3 reaction product can be seen to be more stable than some of the substances obtained by adding the other ligands, and less stable than others. Examining each row, make a list of all the complex ions and precipitates you have in the Table of Observations, in order of increasing stability and equilibrium constant.

Reason(s)

Least
stable _____   _____

_____   _____

_____   _____

_____   _____

_____   _____

_____   _____

Most
stable _____   _____

Stability of unknown

Indicate the position your unknown would occupy in your relative stability list, and explain your reasoning:

Unknown ID code ____

Experiment 27

Advance Study Assignment: Relative Stabilities of Copper(II) Complex Ions and Precipitates

1. In testing the relative stabilities of Cu(II) substances using a well plate, a student adds 6 drops of 1.0 M NH_3 to 6 drops of 0.10 M $Cu(NO_3)_2$. He observes that a blue precipitate initially forms (after the first drop of 1.0 M NH_3 is added), but in excess NH_3 (five more drops) the precipitate dissolves and the solution turns clear again, and a deep blue color. Adding 6 drops of 1.0 M NaOH to the deep blue solution forms a blue precipitate of the same color as that seen after adding a single drop of NH_3 solution, and causes the liquid to become colorless.

   a. What is the formula of the Cu(II)-containing substance formed after adding a single drop of 1.0 M NH_3 to the 1.0 M $Cu(NO_3)_2$ solution? (Review the introduction if you are uncertain, as it is not obvious!)

   b. What is the formula of the Cu(II) substance in the deep blue solution obtained with excess NH_3?

   c. What is the formula of the blue precipitate present after adding 1.0 M NaOH?

   d. Which Cu(II)-containing substance is more stable in equal concentrations of NH_3 and OH^-, the one in Part 1(b) or the one in Part 1(c)? Explain the basis for your answer.

2. Given the following two reactions and their equilibrium constants,

$$Cu(H_2O)_4^{2+}(aq) + 4\,NH_3(aq) \rightleftharpoons Cu(NH_3)_4^{2+}(aq) + 4\,H_2O(\ell) \qquad K_1 = 5 \times 10^{12} \qquad \textbf{(1)}$$

$$Cu(H_2O)_4^{2+}(aq) + CO_3^{2-} \rightleftharpoons CuCO_3(s) + 4\,H_2O(\ell) \qquad K_6 = 7 \times 10^9 \qquad \textbf{(6)}$$

a. Evaluate the equilibrium constant for this reaction (see the discussion leading up to Equation 4 in the introduction of this experiment for help):

$$Cu(NH_3)_4^{2+}(aq) + CO_3^{2-}(aq) \rightleftharpoons CuCO_3(s) + 4\,NH_3(aq) \qquad \textbf{(7)}$$

$$K_7 = \underline{\hspace{3cm}}$$

b. When 1.0 M concentrations of all dissolved substances are present, which is more stable, $CuCO_3$ or $Cu(NH_3)_4^{2+}$? Explain your reasoning.

Experiment 28

Determining the Hardness of Water

One important factor in water quality is called <u>hardness</u>, defined as the sum of the calcium and magnesium ion concentrations in the water. In the past, when soap was commonly used for washing clothes and people usually bathed in tubs instead of showers, water hardness was more often directly observed than it is now. This is because Ca^{2+} and Mg^{2+} form insoluble salts with soap, precipitating as a sticky white solid (called "soap scum") that sticks to clothes or a bathtub. Synthetic detergents have the distinct advantage of not precipitating in hard water, and this is what allowed them to displace soaps for laundry purposes. When water contains a low concentration of Ca^{2+} and Mg^{2+} ions, it is said to be <u>soft</u>. Water hardness analyses do not distinguish between Ca^{2+} and Mg^{2+}, and because most hardness is caused by carbonate deposits in the earth, hardness is usually reported as total parts per million of calcium carbonate, by weight. A water supply with a hardness of 100 parts per million would contain the equivalent of 100 g of $CaCO_3$ in one million grams of water, or 0.1 g in 1 L of water.

Water hardness can be determined by <u>titration</u> (measuring what volume of a liquid, the <u>titrant</u>, is needed to react with a sample) using an actual soap solution as the titrant, but in most chemistry labs the multiple-coordinate-bond-forming (technical term: polydentate or multidentate) metal-binding (technical term: <u>chelating</u>) agent EDTA (ethylenediaminetetraacetic acid) is more often used. EDTA is a weak acid that loses four protons on complete neutralization; its structural formula is

$$HOOC-CH_2 \diagdown \\ N-CH_2-CH_2-N \diagup ^{\textstyle CH_2-COOH} \\ HOOC-CH_2 \diagup \qquad \diagdown CH_2-COOH$$

but we will abbreviate it chemically as H_4Y. The four acid sites and the two nitrogen atoms all contain unshared electron pairs, so that a single EDTA ion can react with up to six coordination sites on a given metal ion. The resulting complex is typically quite stable, and the conditions of its formation can ordinarily be controlled so that it contains EDTA and the metal ion in a 1:1 mole ratio. In a titration to measure the concentration of a metal ion, EDTA binds to the metal ion quickly and aggressively, in a 1:1 ratio, to form a complex. The endpoint occurs when essentially all of the metal ions have been complexed by EDTA.

In this experiment you will <u>standardize</u> a solution of EDTA (determine its concentration accurately and precisely) by using it to titrate a <u>standard solution</u> (one with an accurately and precisely known concentration) made from calcium carbonate, $CaCO_3$. You will then use this EDTA solution (the titrant) to titrate an unknown water sample, which will allow you to determine the water sample's hardness.

Because both EDTA and Ca^{2+} are colorless, it is necessary to use a rather special indicator to detect the endpoint of this titration. The indicator you will use is called Eriochrome Black T, chemically abbreviated In^{3-}, and it forms a wine-red complex, $MgIn^-$, with magnesium ion. A tiny amount of this complex will be present in the solution during the titration. As EDTA is added, EDTA will <u>complex</u> (chemically bind to) free Ca^{2+} and Mg^{2+} ions, leaving the $MgIn^-$ complex alone until essentially all of the free calcium and magnesium ions have been complexed by EDTA. At that point the EDTA concentration will have increased enough to displace Mg^{2+} from the indicator complex; the indicator then reverts to an acid form, HIn^{2-}, which is sky blue, and this establishes the endpoint of the titration.

The titration is carried out at a pH of 10, in an NH_3/NH_4^+ buffer that keeps the EDTA (H_4Y) mainly in the triply deprotonated form, HY^{3-}, in which it complexes the Group 2 (alkaline earth metal) ions very well but does not tend to react as easily with other metal ions, such as Fe^{3+}, that might be present as impurities in the water. The equations for the reactions that occur during the titration are:

$$HY^{3-}(aq) + Ca^{2+}(aq) \rightarrow CaY^{2-}(aq) + H^+(aq) \qquad \text{(main reaction)}$$

(The Mg^{2+} ion undergoes a similar reaction; however, it is less
favorable than that for Ca^{2+}, so Ca^{2+} ions are consumed first.)

$$HY^{3-}(aq) + MgIn^-(aq) \rightarrow MgY^{2-}(aq) + HIn^{2-}(aq) \qquad \text{(at endpoint)}$$
$$\underset{\text{wine red}}{\phantom{HY^{3-}(aq) + MgIn^-(aq)}} \qquad \underset{\text{sky blue}}{\phantom{MgY^{2-}(aq) + HIn^{2-}(aq)}}$$

Because the indicator requires a trace of Mg^{2+} to operate properly, you will add the same small, measured amount of magnesium ion to each solution, as well as to a <u>blank</u> (containing zero hardness) that you will titrate to determine what volume of titrant is needed to react with the added Mg^{2+}; you will then subtract that "blank volume" from all of the other titrant volumes.

Experimental Procedure

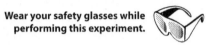

Wear your safety glasses while performing this experiment.

Obtain a 50-mL buret, a 250-mL volumetric flask, and 25- and 50-mL pipets.

A. Preparing the calcium carbonate standard

Using a spatula, transfer 0.4 g (between 0.3 g and 0.5 g) of solid calcium carbonate, $CaCO_3$(s), into a 250-mL or larger beaker, determining the mass of the $CaCO_3$ sample in the beaker to ± 0.001 g or better.

Add 25 mL of deionized water to the large beaker and then, *slowly*, about forty drops of 6 M hydrochloric acid, HCl. Cover the beaker with a watch glass (with its center lower than the edges) and allow the reaction to proceed until all of the solid carbonate has dissolved. Rinse the walls of the beaker down with deionized water from your wash bottle and heat the solution, with the watch glass over it again (in the same orientation), until the solution just begins to boil. (*Note*: Dissolved gases, particularly CO_2, will bubble out of the solution as it is heated to boiling. Make sure you actually boil the water: do not stop just because you see these escaping gas bubbles!) Add 50 mL of deionized water to the beaker and carefully transfer the solution, using a stirring rod as a pathway, into the volumetric flask. Use small amounts of deionized water to rinse the face of the watch glass that was facing the solution down into the beaker, as well as to rinse down the beaker walls; then empty the beaker into the volumetric flask. Do this several times, transferring each rinsing into the volumetric flask. It is important that all of the Ca^{2+} that was originally in the beaker end up in the volumetric flask. Fill the volumetric flask with deionized water up to the horizontal mark on its neck, approaching the mark slowly and carefully so as not to overshoot it. (See Appendix D for more details.) Stopper the flask and mix the solution in it thoroughly by inverting the flask at least thirteen times, each time waiting for the air bubble trapped in the flask to move all the way from one end of the flask to the other. This is your standard solution.

B. Preparing the EDTA titrant

Prepare your buret for the titration by the following procedure, which is described in greater detail in Appendix D. (The purpose of this procedure is to make sure that the solution in the buret has the same concentration as the stock solution.) Clean the buret and rinse it with deionized water. Then rinse it three times with a few milliliters of the stock EDTA titrant solution, each time thoroughly wetting the walls of the buret with the titrant and then letting it out through the stopcock. Fill the buret with EDTA stock solution titrant; open the stopcock momentarily to fill the tip and bring the liquid level to or below the zero mark on the buret. Confirm that the buret does not leak and that there are no air bubbles in the buret tip.

C. Determining the blank volume

Determine how much titrant is needed to cause the indicator to change color in a blank by adding 25 mL of deionized water and 5 mL of the pH 10 buffer to a clean 250-mL Erlenmeyer flask. (It is not a problem if this flask starts out wet with deionozed water.) Add a small amount of solid Eriochrome Black T indicator mixture from the stock bottle: you need only a small portion, about twenty-five milligrams, which is just enough to cover the end of a small spatula. The solution should turn blue; if the color is weak, add a bit more indicator. Add 15 drops of 0.03 M magnesium chloride, $MgCl_2$, which should contain enough Mg^{2+} to turn the solution wine red. Read the buret to ± 0.01 mL and record the reading on the report page, then add EDTA titrant from the buret to the blank in the Erlenmeyer flask until the last tinge of purple disappears. There can be some delay in the color change, so titrate slowly near the endpoint. Only a few milliliters will be needed to titrate the blank. Read the buret again to determine the volume of titrant required for the blank, recording it on the report page.

The blank is the volume of EDTA necessary to cause the indicator to change color in the presence of 15 drops of 0.03 M $MgCl_2$. When calcium is also present in a solution, the EDTA reacts first with the calcium, and then, when all the calcium is gone, it reacts with any free magnesium ions; once those are all complexed, the EDTA begins to react with the magnesium in the magnesium-indicator complex, $MgIn^-$, changing the color of the solution. The total volume of EDTA titrant required is the sum of the amount needed to react with the hardness in the water and that required to react with the 15 drops of magnesium solution intentionally added to make the indicator work. Therefore, the volume of titrant needed to titrate the blank (the "blank volume") must be subtracted from the total EDTA titrant volume used in each titration that determines hardness, and the report page will help you remember to do this. Save the blank solution and use it as a color reference for the endpoint in all your titrations.

D. Standardizing the EDTA titrant against the calcium carbonate standard

Pipet a 25.0-mL portion of the standard Ca^{2+} solution in the volumetric flask into each of three clean 250-mL Erlenmeyer flasks. (It is not a problem if these are wet with deionozed water.) To each flask add 5 mL of the pH 10 buffer, a small amount of indicator, and 15 drops of 0.03 M $MgCl_2$. Titrate the solution in one of the flasks until its color matches that of your reference solution; the endpoint is a reasonably good one, and you should be able to hit it within a few drops if you are careful. Read the buret and record the actual volumes you read on the report page; you will correct for the blank volume later. Refill the buret, read it, and record the volume reading on the report page; then proceed to titrate the second sample, and finally, repeat for the third.

E. Determining total hardness in a water sample

Your instructor will furnish you with a sample of water for hardness analysis. Because the concentration of Ca^{2+} in this sample is likely to be lower than that in the standard calcium solution you prepared, pipet a 50.0 mL water sample for the first titration: more reliable results are obtained if the volume of titrant needed is not small. As before, add some indicator, 5 mL of pH 10 buffer, and 15 drops of 0.03 M $MgCl_2$ before titrating. Carry out as many titrations as necessary to obtain two volumes of EDTA that agree within 3%. If the volume of EDTA required in the first titration is less than 20 mL (because the water sample is relatively soft), further increase the volume of the water sample used so that in the other titrations it takes at least 20 mL of EDTA titrant to reach the endpoint.

Take it further (optional): Find the mass percent calcium carbonate in an antacid tablet or a sample of limestone.

Take it further (optional): Carry out the analysis described in this experiment using soap solution as your titrant, instead of an EDTA solution. You should use the foaming of the soap as your indicator, rather than a color change in Eriochrome Black T. (When enough soap has been added to precipitate all of the Ca^{2+} and Mg^{2+} as soap scum, the titrated solution will begin to foam when agitated.) You will not need to add or use either pH 10 buffer or 0.03 M $MgCl_2$. Assume that soap and hardness react in a 2:1 molar ratio to form soap scum. The uncertainty associated with this method will be much higher, because the endpoint is far less distinct. You are very unlikely to get results that agree with one another to within 3%.

Take it further (optional): Determine the (average, effective) molar mass of a soap based on how much hard water (or calcium carbonate standard) it takes to make it stop foaming (at which point it will have all been precipitated as soap scum). Note that for this analysis, your titrant will need to be a standardized solution with known hardness, rather than an EDTA or soap solution, and that the calculations are far from obvious. Titrate a known mass of soap, and assume that soap and hardness react in a 2:1 molar ratio to form soap scum.

> **Disposal of reaction products:** The chemical waste from this experiment may be poured down the sink, unless your instructor directs otherwise.

Experiment 28

Data and Calculations: Determining the Hardness of Water

A. Preparing the calcium carbonate standard

Mass of solid $CaCO_3$ in large beaker	_____ g	Moles of $CaCO_3$ in large beaker (molar mass = 100.1 g/mol) _____ moles
Volume of Ca^{2+} standard prepared	_____ mL	Molarity of Ca^{2+} in standard _____ M
Volume of 25.0 mL of standard	_____ L	Moles of Ca^{2+} in 25.0 mL of standard _____ moles

C. Determining the blank volume

Volume of titrant needed to change the color of a blank:

Initial buret reading _____ mL	Final buret reading _____ mL	Blank volume _____ mL

D. Standardizing the EDTA titrant against the calcium carbonate standard

Titration	1	2	3
Initial buret reading	_____ mL	_____ mL	_____ mL
Final buret reading	_____ mL	_____ mL	_____ mL
Volume of EDTA dispensed	_____ mL	_____ mL	_____ mL
Volume of EDTA used to titrate blank (from Part C)	_____ mL →	_____ mL →	_____ mL
Volume of EDTA required to titrate Ca^{2+} in standard	_____ mL	_____ mL	_____ mL
	_____ L	_____ L	_____ L

Molarity of EDTA titrant =

$\dfrac{\text{mol } Ca^{2+} \text{ in 25.0 mL of standard}}{\text{liters of EDTA titrant required}}$ _____ M   _____ M   _____ M

Average (mean) molarity of EDTA titrant _____ M   (See Appendix H)

Standard deviation in molarity of EDTA titrant _____ M

(continued on following page)

E. Determining total hardness in a water sample

Unknown ID code of water sample, if applicable _____

Titration #	1	2	3
Volume of hard water titrated	_____ mL	_____ mL	_____ mL
Initial buret reading	_____ mL	_____ mL	_____ mL
Final buret reading	_____ mL	_____ mL	_____ mL
Volume of EDTA dispensed	_____ mL	_____ mL	_____ mL
Volume of EDTA used to titrate blank (from Part C)	_____ mL →	_____ mL →	_____ mL
Volume of EDTA required to titrate hardness in water sample	_____ mL	_____ mL	_____ mL
	_____ L	_____ L	_____ L
Moles of EDTA per liter of water = moles of $CaCO_3$ equivalents per liter of water	_____ mol/L	_____ mol/L	_____ mol/L
Grams of $CaCO_3$ equivalents per liter of water (molar mass = 100.1 g/mol)	_____ g/L	_____ g/L	_____ g/L
Milligrams of $CaCO_3$ per liter of water (1 g = 1000 mg)	_____ mg/L	_____ mg/L	_____ mg/L
Water hardness (1 ppm = 1 mg/L)	_____ ppm	_____ ppm	_____ ppm

Average (mean) water hardness _____ ppm

Standard deviation in water hardness _____ ppm

(See Appendix H; include only those titrations that agree within 3% when calculating the average and standard deviation of your water hardness results.)

Experiment 28

Advance Study Assignment: Determining the Hardness of Water

1. A 0.4943-g sample of solid calcium carbonate, $CaCO_3$, was dissolved in 12 M HCl and the resulting solution diluted to 250. mL in a volumetric flask to form a standard solution.

   a. How many moles of $CaCO_3$ were used? (molar mass of $CaCO_3$ = 100.1 g/mol)

   _____ moles

   b. What was the molarity of Ca^{2+} in the 250. mL of standard solution?

   _____ M

   c. How many moles of Ca^{2+} are present in a 25.0-mL sample of the standard solution in 1(b)?

   _____ moles

2. 25.0-mL samples of the standard solution prepared in Problem 1 were titrated with EDTA to the Eriochrome Black T endpoint. A blank containing a small measured amount of Mg^{2+} required 2.77 mL of the EDTA titrant to reach the endpoint. The first sample of standard solution, to which the same amount of Mg^{2+} had been added, required 49.04 mL of the EDTA titrant to reach the endpoint.

   a. How many milliliters of EDTA titrant were needed to titrate the Ca^{2+} ions in the first 25.0 mL sample of standard?

   _____ mL

   b. How many moles of EDTA were present in the volume obtained in 2(a)?

   _____ moles

   c. What is the molarity of the EDTA titrant solution?

   _____ M

(continued on following page)

3. A 100.-mL sample of hard water was titrated with the EDTA titrant solution from Problem 2. The same amount of Mg^{2+} was added, and the volume of EDTA required was 39.84 mL.

    a. What volume of EDTA was used in titrating the Ca^{2+} in the hard water?

                                                             _____ mL

    b. How many moles of EDTA are there in that volume?

        _____ moles

    c. How many moles of Ca^{2+} and Mg^{2+} (hardness) ions are there in 100. mL of the hard water?

        _____ moles

    d. If the hardness all comes from $CaCO_3$, how many moles of $CaCO_3$ would there be in one liter of this hard water? How many grams of $CaCO_3$ would be present per liter of this hard water? (The molar mass of $CaCO_3$ is 100.1 g/mol.)

        _____ mol/L

        _____ g/L

    e. If 1 ppm = 1 mg per liter, what is the water hardness, in ppm of $CaCO_3$ equivalents? (1 g = 1000 mg)

        _____ ppm $CaCO_3$

Experiment 29

Making and Analyzing a Coordination Compound

Some of the most interesting research in inorganic chemistry involves the preparation and properties of coordination compounds, sometimes called complexes. These are typically salts that contain complex ions: ions consisting of a central metal atom to which small polar molecules or negative ions, called ligands, are bonded through coordinate covalent bonds—bonds in which the shared electrons are all furnished by the ligands.

Some coordination compounds can be prepared stoichiometrically pure, which means the atoms in the compound are present in the exact ratio given by its chemical formula. Many supposedly pure compounds are not stoichiometrically pure because they contain variable amounts of water, either incorporated into their crystal structure (technical term: they are hydrates) or sticking to (technical term: adsorbed on) their surfaces. In this experiment you will be making a compound that is indeed stoichiometrically pure. It is not a hydrate and it does not tend to adsorb water on its surface.

The compound you will be preparing contains cobalt ion, ammonia, and chloride ions. Its formula is of the form $Co_x(NH_3)_yCl_z$, in which x, y, and z are counting numbers (1, 2, 3, and so on). Once you have made the compound, you may, depending on the time available, analyze it to determine how much of one or more of these "ingredients" it contains, and so determine x, y, and/or z.

Ligands attached to a metal can be labile, meaning they are easy to exchange for other ligands and will do so quickly. Most, but by no means all, complexes are labile. Some, including the one you will be making in this experiment, exchange ligands only very slowly. For such substances, the preparation reaction takes time, but once formed, the complex ion in solution is quite resistant to change. Such complexes are called inert.

In the complex ion you will be making, NH_3 and Cl^- are the possible ligands. Because the overall charge on the compound must be zero, the moles of Cl^- ion in a mole of compound will be determined by the (positive) charge on the cobalt ion. Some of the Cl^- ions may be ligands in the complex ion, while others may simply be counterions (ions that offset the positive charge of the complex ion) in the crystal. When the compound is dissolved in water, the counterions will go into solution, but the chloride ions that are ligands in the complex ion will remain firmly attached to cobalt. They will not react with a precipitating agent such as Ag^+ ion, which will form AgCl with any free Cl^- ions. All of the NH_3 molecules will be in the complex ion and will not react with added H^+ ion, as they would if they were free. To ensure that you can measure all the Cl^- and NH_3 in the compound, you will destroy the complex (break all of its coordinate covalent bonds) before attempting to analyze it.

The complete analysis of this compound involves a gravimetric (mass-measurement) procedure for chloride ion, a colorimetric (color-intensity-measurement) procedure for cobalt ion, and a volumetric (volume-measurement) procedure for ammonia. From these analyses you will be able to determine the moles of each substance present in a mole of the compound. The ratio of these quantities reflects the chemical formula of the compound. Although procedures for the analysis of all three "ingredients" in the compound are described, its formula can be determined (although with less certainty) from only two analyses: knowing the compound contains only three "ingredients" (Co, NH_3, and Cl), the amount of the third "ingredient" can be calculated by subtraction.

Making (technical term: synthesizing) the coordination compound requires one full lab period. The analyses for chloride and cobalt content can both be done in one lab period. Determining the amount of ammonia present in the complex takes most of an additional lab period.

Experimental Procedure

Carry out the procedure in either A.1 or A.2, as directed by your instructor.

A.1 Synthesizing $Co_x(NH_3)_yCl_z$ (Procedure 1)

1. Start warming a hot water bath, which you will use later. Fill a large (at least 600-mL) beaker about 20% full of deionized water. If using a hotplate, set the surface temperature to $120\,°C$ (about 10% power) and place the beaker directly on top. If using a laboratory burner, support the beaker with a wire gauze on an iron ring, above a 1-cm-tall soft blue flame. Suspend a thermometer in the water and continue with the following steps as the bath warms up.

2. Select a clean 250-mL beaker with volume markings (technical term: graduations) to use as your "reaction beaker". **Inspect it carefully** for cracks or other signs of damage. It will be exposed to big temperature changes, so you want to make sure it is in good shape.

3. Weigh 10.0 ± 0.5 g of ammonium chloride, $NH_4Cl(s)$, into your 250-mL reaction beaker.

4. Add 40. mL of deionized water to your reaction beaker and stir until most of the solid dissolves. *Keep the stirring rod in the reaction beaker even when not stirring with it.*

5. Add 8.0 ± 0.5 g of cobalt(II) chloride hexahydrate, $CoCl_2 \cdot 6\,H_2O(s)$, to your reaction beaker. Record the actual amount added on the report page.

6. Stir the reaction mixture until most of the solids dissolve.

7. **In a fume hood**, add 40. mL of 15 M ammonia, $NH_3(aq)$. **CAUTION:** **This is a concentrated base with a strong smell that can knock you unconscious**. Remaining in the hood, stir your reaction mixture with the stirring rod for about thirty seconds.

8. **If you leave the hood for this step, keep your reaction beaker covered with a watch glass when possible**. Add approximately 0.8 g of activated charcoal. This is a catalyst, so the exact amount is not critical. Its presence helps speed up the displacement of H_2O ligands in the Co complex ion by NH_3, but it is not itself chemically altered in doing so.

9. **In a hood and slowly, not more than 10 mL at a time**, add 50. mL of $10.\%_{mass}$ hydrogen peroxide, $H_2O_2(aq)$, stirring between *each* addition until bubbling slows. **CAUTION:** **Be very careful to not spill or spatter this chemical on your skin. Wear gloves**. There will be some bubbling as oxygen gas is released, and your reaction beaker will get hot. (The heat will also drive ammonia out of solution, which is why this step should be done in a hood.) Hydrogen peroxide converts Co^{2+} to Co^{3+}. Your reaction mixture volume should now be 145 ± 5 mL, and the solution deep maroon in color.

10. Once all bubbling has stopped, place your reaction beaker in your hot water bath, which should be at $60 \pm 10\,°C$. (If it is not, make adjustments to get it there.)

11. Leave the reaction beaker in the hot water bath for 35 ± 5 minutes while keeping the water bath temperature between $50\,°C$ and $70\,°C$. Stir the reaction beaker every few minutes, avoiding the fumes above it when you do. Cover the beaker with a watch glass when not stirring. The solution will appear mostly black, due to the charcoal in it, but you should be able to see that the liquid has a yellow-orange color that becomes red as you stir and heat.

12. Remove your 250-mL reaction beaker from the water bath and put it down on the lab bench. Turn off the heat source; you are done with the hot water bath (but not with the heat source).

13. Set up a cool water bath, consisting of a 400-mL beaker containing 150 mL of cold tap water. Cool the reaction beaker in this cool water bath for 2 minutes. *Then* transfer the reaction beaker to an ice bath, stirring occasionally as it cools. *Slow* cooling promotes the formation of larger crystals, less likely to clog or pass through a filtration.

14. Use a thermometer placed directly in your reaction beaker to monitor the temperature of the reaction mixture. Keep your reaction beaker in the ice bath until the temperature reads less than $10\,°C$ while the reaction mixture is being stirred.

15. Use a Büchner funnel to separate the solid out of the reaction mixture. (See Appendix D for details on how to do this.) You do not have to get all the solid out of your reaction beaker, or clean the reaction beaker, because you will be putting the solid right back into it. Pull air through the solid in the funnel for at least a minute, to pull the last of the liquid out.

16. Turn off the vacuum. Release the vacuum on the suction flask by twisting off the vacuum hose as you pull. Rotate the (cup of the) Büchner funnel to free it from the rest of the apparatus.

17. Knock (preferred) or scrape the contents of the Büchner funnel's cup, including the filter paper, back into your 250-mL reaction beaker. *You do not need to clean or rinse the reaction beaker first.*

18. Rinse the cup of the Büchner funnel, using deionized water from your squirt bottle, into the reaction beaker.

19. Pick up the filter paper with the plastic tweezers provided. Use your squirt bottle, filled with deionized water, to rinse the colored cobalt compound off the filter paper. Use a moderate amount of water: not more than 60 mL. (It is fine if some black material remains on the filter.) Discard the used filter paper in the trash.

20. Discard the liquid from the filtration (technical term: <u>filtrate</u>) in the liquid waste container. Rinse the suction flask with deionized water and reassemble your Büchner funnel apparatus.

21. Using your squirt bottle, rinse down the sides of the reaction beaker, and your stirring rod, into the reaction beaker. Add deionized water to your reaction beaker until the total solution volume is approximately 125 mL, as indicated by the volume markings on the beaker.

22. Add 5 mL of 12 M hydrochloric acid, HCl(aq), and stir. **CAUTION:** **This is a concentrated strong acid with a choking smell. Avoid contact with it**.

23. Heat the mixture to boiling while stirring to dissolve the solids. Put the reaction beaker directly on a hotplate and set it to maximum, or above a hot laboratory burner flame supported by a wire gauze on an iron ring. Stir frequently until the mixture boils, then immediately lower the temperature of the heat source to keep the beaker hot, but not boiling strongly ($1/3$ power / 5-cm quiet blue flame). If the solution volume has visibly dropped below 130 mL, add enough deionized water to get it back to that level.

24. Place a piece of filter paper in the Büchner funnel and wet it completely with a minimum amount of deionized water, but do not turn on the vacuum until you are ready to start filtration in the next step.

25. Holding the reaction beaker securely with a pair of beaker tongs or a folded paper towel, pour the hot reaction mixture, a little bit at a time, onto the moistened filter paper in the Büchner funnel apparatus, under suction. This step traps the charcoal on the filter paper. When not pouring it out, keep the reaction beaker heating. If the filtration stops, ask for help and *do not add more liquid!*

26. Squirt small amounts of deionized water at any colored material remaining in the reaction beaker, and pour the resulting colored water out into the funnel. Stop rinsing when no color remains, even if some black material is left behind.

27. What passes through the filter and ends up in the suction flask should be orange, and it *contains the dissolved product. Do not discard it!*

28. Wipe any carbon remaining in the reaction beaker out with a paper towel. Also wipe off any that remains on your stirring rod. Rinse the reaction beaker with hot tap water and then with deionized water. You do not need to dry it.

29. Transfer the contents of the suction flask to your cleaned-out 250-mL reaction beaker. Swirl the flask just before transferring portions of the liquid: try to transfer any solid present in the flask into the reaction beaker. However, *do not add any more water to the system: do not rinse!* If you leave solid behind in the flask, pour some of the liquid from your reaction beaker back into the suction flask and try again. Remember, swirl the mixture right before pouring, in order to suspend as much of the solid as possible in the liquid.

30. Add 15 mL of 12 M hydrochloric acid, HCl(aq), to your reaction beaker and stir. **CAUTION:** **This is a concentrated strong acid with a choking smell. Avoid contact with it**.

31. Place your reaction beaker in an ice bath. Periodically stir the reaction mixture as it cools, monitoring its temperature. Keep it in the ice bath until the temperature reads less than 10 °C while the reaction mixture is being stirred, and then for five minutes after that.

32. Clean the filter cup of your Büchner funnel. Throw the used filter paper (and the charcoal that was trapped on it) into the trash. Wipe out any remaining charcoal with a moist paper towel. Do this over another paper towel, because the bottom of the Büchner funnel may have product on it. Rinse the filter cup with hot water, then DI water, and then reassemble the filtering apparatus. Before you put it in and wet it, determine the mass of the filter paper you will use in the next step to ± 0.01 g.

33. Filter the contents of the cold reaction beaker through the Büchner funnel by pouring it onto moistened filter paper under suction. To transfer as much of your product as possible, stir the reaction mixture with your stirring rod just before you pour out the portions, and pour quickly, a little at a time. Some product will remain in the reaction beaker, but *do not attempt to rinse it out* or you will dissolve away more of your product from the filter cup than you end up rescuing from the reaction beaker.

34. Turn off the vacuum. Release the vacuum on the suction flask by twisting off the vacuum hose as you pull. Then reconnect the vacuum hose, but leave the vacuum turned off.

35. Pour about twenty milliliters of 95% ethanol into the Büchner funnel's cup, where the yellow-orange product crystals are. Wait about ten seconds, then turn on the vacuum to the Büchner funnel. This should remove the remaining water and HCl from the surfaces of the product crystals.

36. Pull air through the crystals for five minutes. While waiting, weigh a clean, dry watch glass to ± 0.01 g.

37. Turn off the vacuum. Release the vacuum on the suction flask by twisting off the vacuum hose as you pull. Rotate the (cup of the) Büchner funnel to free it from the rest of the apparatus.

38. Transfer the crystals *and* your filter paper to the watch glass by inverting the funnel's cup onto the watch glass. You may have to knock the filter cup a few times to get the product to fall out.

39. Weigh the watch glass, product, and filter paper to ± 0.01 g, to determine an approximate yield.

40. If you will be analyzing your product in a future lab period, leave it and the filter paper on the watch glass, in your lab drawer or locker, to dry. Otherwise do with it as directed by your instructor.

Disposal of reaction products: Discard the liquid in the suction flask in a waste container, or as directed by your instructor.

A.2 Synthesizing Co$_x$(NH$_3$)$_y$Cl$_z$ (Procedure 2)

There are several possible Co$_x$(NH$_3$)$_y$Cl$_z$ compounds. This procedure will allow you to make a different one.

1. Start warming a hot water bath, which you will use later. Fill a large (at least 600-mL) beaker about 20% full of deionized water. If using a hotplate, set the surface temperature to 185 °C (about 20% power) and place the beaker directly on top. If using a laboratory burner, support the beaker with a wire gauze on an iron ring, above a 3-cm-tall soft blue flame. Suspend a thermometer in the water and continue with the following steps as the bath warms up.

2. Select a clean 250-mL beaker with volume markings (technical term: <u>graduations</u>) to use as your "reaction beaker". **Inspect it carefully** for cracks or other signs of damage. It will be exposed to big temperature changes, so you want to make sure it is in good shape.

3. Weigh 4.0 ± 0.2 g of ammonium chloride, NH$_4$Cl(s), into your 250-mL reaction beaker.

4. **In a fume hood,** add 25 mL of 15 M ammonia, NH$_3$(aq). **CAUTION:** **This is a concentrated base with a strong smell that can knock you unconscious**.

5. **Remaining in the hood,** stir the reaction mixture until most of the solid NH$_4$Cl dissolves. *Keep the stirring rod in the reaction beaker even when not stirring with it.*

6. **If you leave the hood for this step, keep your reaction beaker covered with a watch glass when possible.** Add 8.0 ± 0.5 g of cobalt(II) chloride hexahydrate, CoCl$_2 \cdot 6$ H$_2$O(s), to your reaction beaker. Record the actual amount added on the report page.

7. Stir the contents of the reaction beaker until the CoCl$_2 \cdot 6$ H$_2$O has dissolved and magenta solid can no longer be seen through the bottom. A tan solid in a deep red to brown liquid should result.

8. **In a hood and slowly, not more than 10 mL at a time,** add 20. mL of 10.%$_{\text{mass}}$ hydrogen peroxide, H$_2$O$_2$(aq), stirring between *each* addition until bubbling slows. **CAUTION:** **Be very careful to not spill or spatter this chemical on your skin. Wear gloves.** There will be some bubbling as oxygen gas is

released, and your reaction beaker will get hot. (The heat will also drive ammonia out of solution, which is why this step should be done in a hood.) Use your stirring rod to ensure that any tan material on the walls of the beaker contacts the liquid. Hydrogen peroxide converts Co^{2+} to Co^{3+}.

9. **In a fume hood, 5 mL at a time**, add 25 mL of 12 M hydrochloric acid, HCl(aq), stirring after each addition. **CAUTION:** **This is a concentrated strong acid with a choking smell. Avoid contact with it.** The white cloud that forms is solid ammonium chloride, NH_4Cl, resulting from a gas-phase acid–base neutralization reaction between HCl and NH_3 gases in the air above the beaker. **Do not breathe this dust in.**

10. Place your reaction beaker in your hot water bath, which should be at $80 \pm 5\,°C$. (If it is not, make adjustments to get it there.)

11. Leave the reaction beaker in the hot water bath for about thirty minutes while keeping the water bath temperature between $75\,°C$ and $85\,°C$. Stir the reaction beaker every few minutes, avoiding the fumes above it when you do. Cover the beaker with a watch glass when not stirring. You should see a deep purple solid in an indigo (blue-violet) solution, with both colors changing somewhat as you stir and heat.

12. Remove your 250-mL reaction beaker from the water bath and put it down on the lab bench. Keep the water bath heating: it will be needed again soon. If needed, add more deionized water to the water bath beaker to keep it at 20% full.

13. Set up a cool water bath, consisting of a 400-mL beaker containing 150 mL of cold tap water. Cool the reaction beaker in this cool water bath for 2 minutes. *Then* transfer it to an ice bath, stirring occasionally as it cools. *Slow* cooling promotes the formation of larger crystals, less likely to clog or pass through a filtration.

14. Use a thermometer placed directly in your reaction beaker to monitor the temperature of the reaction mixture. Keep your reaction beaker in the ice bath until the temperature reads less than $10\,°C$ while the reaction mixture is being stirred.

15. Use a Büchner funnel to separate the solid out of the reaction mixture. (See Appendix D for details on how to do this.) You do not have to get all the solid out of your reaction beaker, or clean the reaction beaker, because you will be putting the solid right back into it. Pull air through the solid in the funnel for at least a minute, to pull the last of the liquid out.

16. Turn off the vacuum. Release the vacuum on the suction flask by twisting off the vacuum hose as you pull. Rotate the (cup of the) Büchner funnel to free it from the rest of the apparatus.

17. Knock (preferred) or scrape the contents of the Büchner funnel's cup, including the filter paper, back into your 250-mL reaction beaker. *You do not need to clean or rinse the reaction beaker first.*

18. Rinse the cup of the Büchner funnel, using deionized water from your squirt bottle, into the reaction beaker.

19. Pick up the filter paper with the plastic tweezers provided. Use your squirt bottle, filled with deionized water, to rinse the colored cobalt compound off the filter paper. Use a moderate amount of water: not more than 60 mL. Discard the used filter paper in the trash.

20. Discard the liquid from the filtration (technical term: <u>filtrate</u>) in the liquid waste container. Rinse the suction flask with deionized water and reassemble your Büchner funnel apparatus.

21. Using your squirt bottle, rinse down the sides of the reaction beaker, and your stirring rod, into the reaction beaker. Add deionized water to your reaction beaker until the total solution volume is approximately 75 mL, as indicated by the volume markings on the beaker.

22. **In a fume hood**, add 25 mL of 15 M ammonia, NH_3(aq), and stir. **CAUTION:** **This is a concentrated base with a strong smell that can knock you unconscious.**

23. **In a fume hood,** add 60 mL of 12 M hydrochloric acid, HCl(aq), and stir. **CAUTION:** **This is a concentrated strong acid with a choking smell. Avoid contact with it.** Keep stirring for at least thirty seconds.

24. Place your reaction beaker back into your hot water bath, ensuring the bath is still at $80 \pm 5\,°C$. (If it is not, make adjustments to get it there.)

25. Leave the reaction beaker in the hot water bath for another thirty minutes while keeping the temperature of the water bath between $75\,°C$ and $85\,°C$. Stir the reaction beaker every few minutes, avoiding the fumes above it when you do. Cover the beaker with a watch glass when not stirring.

26. Remove your 250-mL reaction beaker from the water bath and put it down on the lab bench. Remove the watch glass.

27. Set up a cool water bath, consisting of a 400-mL beaker containing 150 mL of cold tap water. Cool the reaction beaker in this cool water bath for 2 minutes. *Then* transfer the reaction beaker to an ice bath, stirring occasionally as it cools. *Slow* cooling promotes the formation of larger crystals, less likely to clog or pass through a filtration, and quicker to dry completely.

28. Use a thermometer placed directly in your reaction beaker to monitor the temperature of the reaction mixture. Keep your reaction beaker in the ice bath until the temperature reads less than 10 °C while the reaction mixture is being stirred, and then for five minutes after that.

29. Before you put it in and wet it, determine the mass of the filter paper you will use in the next step to ± 0.01 g. Record this on the report page.

30. Filter the contents of the cold reaction beaker through the Büchner funnel by pouring it onto moistened filter paper under suction. To transfer as much of your product as possible, stir the reaction mixture with your stirring rod just before you pour out the portions, and pour quickly, a little at a time. Some product will remain in the reaction beaker, but *do not attempt to rinse it out* or you will dissolve away more of your product from the filter cup than you end up rescuing from the reaction beaker.

31. Turn off the vacuum. Release the vacuum on the suction flask by twisting off the vacuum hose as you pull. Then reconnect the vacuum hose, but leave the vacuum turned off.

32. Pour about twenty milliliters of 95% ethanol into the Büchner funnel's cup, where the purple product crystals are. Wait about ten seconds, then turn on the vacuum to the Büchner funnel. This should remove the remaining water and HCl from the surfaces of the product crystals.

33. Pull air through the crystals for five minutes. While waiting, weigh a clean, dry watch glass to ± 0.01 g.

34. Turn off the vacuum. Release the vacuum on the suction flask by twisting off the vacuum hose as you pull. Rotate the (cup of the) Büchner funnel to free it from the rest of the apparatus.

35. Transfer the crystals *and* your filter paper to the watch glass by inverting the funnel's cup onto the watch glass. You may have to knock the filter cup a few times to get the product to fall out.

36. Weigh the watch glass, product, and filter paper to ± 0.01 g, to determine an approximate yield.

37. If you will be analyzing your product in a future lab period, leave it and the filter paper on the watch glass, in your lab drawer or locker, to dry. Otherwise do with it as directed by your instructor.

Disposal of reaction products: Discard the liquid in the suction flask into a waste container, or as otherwise directed by your instructor.

B. Gravimetric determination of chloride content

1. The cobalt complex you prepared previously should now be dry: any water or ethanol trapped between the crystals has had plenty of time to evaporate away into the gas phase. Because your product is dry, you need to move it about slowly so that it does not fly away! *Carefully* open your lab drawer and remove the watch glass with your product on it. Move slowly to a balance and determine and record the combined mass of your dry product, filter paper, and watch glass to ± 0.01 g.

2. Remove the filter paper from your product, using tweezers, and dispose of it as directed by your instructor. (Some product will be lost with the filter paper, but that is not a problem.)

3. Bring a clean and dry 400- or 600-mL beaker to the lab oven and use crucible tongs or a folded paper towel to carefully transfer one of the empty filter crucibles from the oven into your beaker. Close the oven door. Bring the beaker back to your lab station, cover it with a watch glass, and allow the crucible in it to cool.

4. Select a clean and dry 250-mL beaker from your lab drawer. **Inspect it carefully** for cracks or other signs of damage. It will be exposed to big temperature changes, so you want to make sure it is in good shape.

5. Place your 250-mL beaker on an *analytical* balance, and tare it out to read zero.

6. Transfer 0.30 ± 0.05 g of your compound into the beaker, recording the mass added to ± 0.0001 g.

7. Add 30. mL of 1.0 M sodium hydroxide, NaOH(aq), and stir until all the soluble material has dissolved. Avoid the fumes released as you do so. **CAUTION:** **NaOH is a strong base: if it comes in contact with your skin, wash with water until your skin no longer feels slippery. Avoid contact with it**. This may

take some time—use your stirring rod to rapidly stir the solution in a circular motion, and then crush any bits of solid that accumulate at the center of your beaker. If you have more than a few particles that refuse to dissolve, ask for help. *Keep the stirring rod in the beaker even when not stirring with it.*

8. Put the 250-mL beaker directly on a hotplate and set it to maximum, or place it above a hot laboratory burner flame supported by a wire gauze on an iron ring. Stir frequently until the mixture boils, then immediately lower the temperature of the heat source to keep the beaker hot, but not boiling strongly ($1/3$ power /5-cm quiet blue flame). Keep it there for 3 minutes, stirring occasionally. This step will destroy the complex; the mixture will turn black as the cobalt in it becomes $Co_3O_4(s)$.

9. Turn off the heat source. With a pair of beaker tongs or a folded paper towel, remove the 250-mL beaker and put it on the lab bench to cool for two minutes.

10. **In a fume hood**, add 10. mL of 6 M nitric acid, HNO_3(aq). **Remain in the hood** as you stir this solution and perform the next step—nasty fumes are released! **CAUTION:** HNO_3 **is a powerful, if slow-acting, strong acid. Avoid contact with it**.

11. **In a fume hood,** add a small spatula-full (0.2 g) of solid sodium sulfite, $Na_2SO_3(s)$, and stir.

12. Reheat the solution to boiling and boil gently for a minute or so. This will convert the cobalt from Co_3O_4 into the Co^{2+} ion. The solution should clarify and turn a rose pink color.

13. Wet your stirring rod with the solution and touch it to the walls of the beaker to wash any unreacted dark solid material back down into the solution.

14. Gradually add 50. mL of 0.10 M silver nitrate, $AgNO_3$(aq), stirring as you do so. This is an excess, enough to precipitate all the chloride ion in the solution as silver chloride, $AgCl(s)$.

15. Stir thoroughly for about a minute. This helps the AgCl crystals grow larger and clump together.

16. The filter crucible you removed from the oven earlier should now be cool. Write your name on the side of it with a lab marker and *then* weigh it to ± 0.0001 g on an analytical balance. (Note that the crucible *must be at room temperature* before you can reliably determine its mass!)

17. Turn on the vacuum to a suction flask and place the filter crucible into an adapter on it.

18. Carefully, with the suction on, transfer *all* of the AgCl precipitate into the filter crucible. Wash the AgCl out of the beaker with deionized water from your wash bottle. Use your wash bottle and rubber policeman to complete the transfer.

19. *Holding the crucible*, turn the vacuum off and release the remaining vacuum on the suction flask by twisting the vacuum hose off as you pull. Transfer the filter crucible back into the beaker you cooled it in.

20. Add 10. mL of 6 M nitric acid, HNO_3(aq), to the filter crucible and wait one minute. Then use a suction flask under vacuum to suck the nitric acid through the precipitate and out of the filter crucible. **CAUTION:** HNO_3 **is a powerful, if slow-acting, strong acid. Avoid contact with it**.

21. *Again, holding the crucible*, turn the vacuum off and release the remaining vacuum on the suction flask by twisting the vacuum hose off as you pull. Transfer the filter crucible back into the beaker you cooled it in.

22. Add about twenty milliliters of deionized water to the crucible and wait at least thirty seconds. Then use a suction flask under vacuum to suck the water through the precipitate and out of the filter crucible. This time keep the vacuum on for a minute after this water has been pulled through the filter, to help dry the solid.

23. *Again, holding the crucible*, turn the vacuum off and release the remaining vacuum on the suction flask by twisting the vacuum hose off as you pull. Transfer the filter crucible back into the beaker you cooled it in.

24. Transfer the crucible and its contents into the 150 °C oven. Allow them to dry there for at least forty minutes. You may work on Part C while you wait.

25. Once it has been in the oven *for at least forty minutes*, remove your AgCl crucible from the oven and place it into the cooling beaker again using crucible tongs or a folded paper towel. Bring the crucible back to your lab station and let it cool to room temperature.

26. Once the crucible is at room temperature, weigh it to ± 0.0001 g on an analytical balance. (Note that the crucible *must be at room temperature* before you can reliably determine its mass!)

C. Colorimetric determination of cobalt content

1. On an *analytical* balance, transfer 0.50 ± 0.05 g of your compound into a clean, dry 50-mL beaker. Determine and record the actual mass you use to ± 0.0001 g. Try to avoid having any of your product end up on the walls of the beaker, because this will make later steps harder.

2. Cover the beaker with a small watch glass and heat it directly on a hotplate set to 80% power, or supported by a wire gauze on an iron ring over a 5-cm-tall soft blue laboratory burner flame, until the solid melts, foams, and turns *completely* blue, including any on the beaker walls. In this process the complex is destroyed and the cobalt ion freed from the grip of the ligands.

3. Turn off the heat and let everything cool down a bit. Then move the beaker onto the lab bench, using a *thick* folded paper towel or *tongs NOT coated with plastic*. (The beaker will be hot enough to melt and burn plastic coatings!)

4. Let the beaker cool for 5 minutes on the lab bench. Once it has cooled, remove the watch glass and set it aside, dirty side facing upwards.

5. Add 10. mL of 6 M nitric acid, HNO_3(aq), to the beaker. **CAUTION:** **This is a powerful strong acid, especially once it is hot. Treat this solution with great care from here onward.**

6. Making sure that the watch glass is *not* on the beaker, heat the beaker until the liquid begins to boil and the solid is all dissolved. Put the 50-mL beaker directly on a hotplate and set it to maximum, or support it on a wire gauze above an iron ring over a hot laboratory burner flame. As soon as the mixture boils, lower the temperature of the heat source to keep the beaker hot, but not boiling strongly ($^1/_3$ power / 5-cm quiet blue flame).

7. Wet your stirring rod with the solution and touch it to the walls of the beaker in order to wash any undissolved solid down into the solution. Use the same technique to wash any solid off the watch glass and into the beaker, finishing with a small amount of deionized water.

8. Turn off the heat and transfer the beaker to the lab bench to cool.

9. Once the beaker can be comfortably handled with your fingers, carefully transfer its contents to a clean (wet with deionized water is fine) 25.0-mL volumetric flask. *Your results will only be accurate if all the cobalt ions in the beaker end up in the volumetric flask—do not spill!* Rinse the beaker with small amounts of deionized water, adding the rinsings to the volumetric flask.

10. Fill the volumetric flask to the mark on its neck with deionized water. (Consult Appendix D for details on how to do this.) Stopper the flask and invert it at least thirteen times, allowing the trapped air bubble to move all the way from the top of the neck to the bottom of the flask and back each time, to ensure that the solution inside becomes thoroughly mixed.

11. Determine the absorbance of the solution in the volumetric flask at 510. nm using a spectrophotometer. (See Appendix D for details on how to do this.)

12. Using the calibration curve or equation provided, determine the molarity of cobalt ion, Co^{2+}, in the solution.

Disposal of reaction products: Pour the rest of the solution in the volumetric flask into a waste container, unless your instructor directs otherwise.

D. Volumetric determination of ammonia content

In this part of the experiment you will first decompose the complex, releasing NH_3 into the solution, as you did in the chloride analysis (Part B). You will then distill off the NH_3 (force it into the gas phase and then back into a liquid) into another container, and titrate it against an acid. This procedure is called a Kjeldahl analysis, and can be used for determining the nitrogen content of many organic and inorganic substances.

Assemble the distillation apparatus shown in Figure 29.1. The details of the apparatus will vary from lab to lab, but you will need (the equivalent of) a 250-mL distilling flask, a distilling head, a condenser, and a receiver adapter, in addition to a 125-mL Erlenmeyer flask, which will serve as a receiving flask. The 250-mL flask should be on a wire gauze on an iron ring, or in a sand bath, or in a heating mantle. The flask and condenser should be held with clamps, adjusted so that all joints are tight. On the end of the receiver adapter attach a 4-inch length of flexible tubing, which should reach to the bottom of the receiving flask.

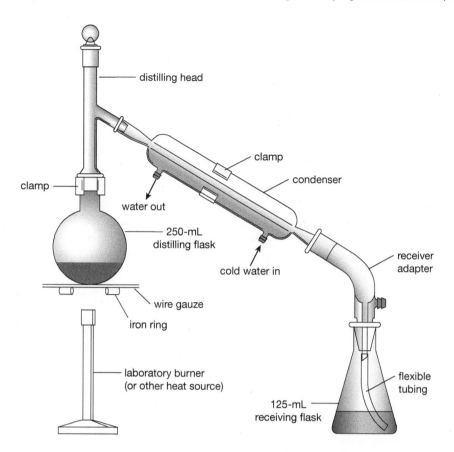

Figure 29.1 The purpose of this apparatus, used in the volumetric determination of nitrogen, is to convert the nitrogen in the distilling flask into ammonia gas, condense it into some water that co-distills with it, and finally trap that ammonia as ammonium ion in the acid in the receiving flask.

Put about fifty milliliters of saturated boric acid solution, H_3BO_3(aq), in the 125-mL receiving flask; add 5 drops of bromocresol green indicator and adjust the level of the flask, if necessary, so that the flexible tubing is at the bottom of the flask, well under the liquid surface.

Use an analytical balance to weigh out 1.0 ± 0.1 g of your compound into a clean, dry beaker, and record the mass actually used to ± 0.0001 g. Disconnect the distilling flask from the head. Transfer the sample to the distilling flask. Rinse the beaker several times with small amounts of deionized water, and add the rinsings to the flask. *All* of the product sample must end up in the flask. Swirl the distilling flask to dissolve the sample, adding as little deionized water as necessary to dissolve it completely. Add a few boiling chips and several pieces of granulated zinc. Start cold water flowing through the condenser, slowly. If you are working with standard taper glassware, lightly grease the lower joint of the distilling head.

Pour 120. mL of 1.0 M sodium hydroxide, NaOH(aq), into the distilling flask and immediately reconnect it to the distilling head. Make sure that all joints are tight, and that the top of the distilling head is stoppered. Turn on the heat source and bring the liquid in the flask to a boil. The complex will be destroyed, the liquid will turn dark as Co_3O_4 forms, and NH_3 gas will be driven from the solution into the receiving flask.

Adjust the heat as necessary to maintain smooth boiling. The color of the bromocresol green acid-base indicator in the receiving flask will change as basic NH_3 gas is absorbed into the boric acid solution. There may be a tendency for liquid to be pulled into the flexible tubing during the distillation; this is not serious, but the level can be lowered by letting in a little air by momentarily loosening one of the joints or the stopper atop the distilling head. When the volume of liquid in the receiving flask reaches about one-hundred milliliters, enough of the solution will have been distilled that essentially all of the NH_3 will have been transferred into the boric acid solution in the receiving flask.

While still heating the flask, disconnect the receiver adapter from the condenser. (If the heat is turned off first, liquid will be pulled up into the condenser.) Rinse the receiver adapter with the flexible tubing still in the receiving flask but above the boric acid solution. Use your wash bottle to rinse down the receiving adapter as well as the outer surface of the flexible tube into the receiving flask, then turn off the heat. Pour *all* of the boric acid solution in the receiving flask into a 250-mL volumetric flask, being careful not to lose any: this analysis relies on all the ammonia ending up in the volumetric flask! Rinse the receiving flask with a few small portions of deionized water, adding the rinsings to the volumetric flask. Fill the volumetric flask to the mark with deionized water. (See Appendix D for more details.) Stopper the flask and invert it at least thirteen times, allowing the trapped air bubble to move all the way from the top of the neck to the bottom of the flask and back each time, to ensure that the solution inside becomes thoroughly mixed.

Clean two burets. Rinse one of them with several small portions of the NH_3 solution in the volumetric flask, draining the rinses through the stopcock. (See the buret section of Appendix D for more details.) Fill the buret with the NH_3 solution and read and record the level. Rinse the other buret with small portions of the standardized hydrochloric acid, HCl(aq), solution from the stock supply, and then fill that buret with that solution. Record the level.

Release about forty milliliters of the NH_3 solution from the buret into a clean (wet with deionized water is not a problem) 250-mL Erlenmeyer flask, and add 5 drops of bromocresol green indicator. Titrate with HCl from the other buret to the endpoint, where the indicator changes from blue to yellow. The actual endpoint is green and can be reached by back titrating as necessary with NH_3 solution from its buret. Record the final levels in the NH_3 and HCl burets.

Repeat the titration once or twice more. The NH_3 to HCl volume ratios from your two best trials should agree within 1%, if all goes well. Use only trials whose ratios agree within 1% in your calculations.

Disposal of reaction products: The titrated solutions may be poured down the sink. The liquid remaining in the distilling flask should be discarded into a waste container, or as otherwise directed by your instructor.

Experiment 29

Data and Calculations: Making and Analyzing a Coordination Compound

A.1 or A.2 Synthesizing $Co_x(NH_3)_yCl_z$

Procedure followed (A.1 or A.2) _____

Mass of $CoCl_2 \cdot 6\ H_2O$ actually used in synthesis _____ g

Mass of filter paper on which product was collected _____ g

Mass of watch glass onto which product was transferred _____ g

Mass of damp product, filter paper, and watch glass _____ g

Mass of damp product _____ g

B. Gravimetric determination of chloride content

Mass of dry product, filter paper, and watch glass _____ g

Mass of dry product _____ g

Mass of product sample transferred to 250-mL beaker _____ g

Mass of cooled filter crucible _____ g

Mass of filter crucible containing dried silver chloride, AgCl _____ g

Mass of recovered silver chloride, AgCl _____ g

Moles of Cl^- in recovered AgCl = moles of chlorine in transferred sample _____ mol

Moles of chlorine per gram of sample _____ mol/g

C. Colorimetric determination of cobalt content

Mass of product sample transferred to 50-mL beaker _____ g

Absorbance at 510. nm of cobalt solution in 25-mL volumetric flask _____

Molarity of Co^{2+} ion in 25-mL volumetric flask _____ M

Moles of cobalt in 25-mL flask = moles of cobalt in transferred sample _____ mol

Moles of cobalt per gram of sample _____ mol/g

D. Volumetric determination of ammonia content

Mass of product sample transferred to beaker _____ g

Molarity of standardized HCl _____ M

(continued on following page)

	Trial # 1	**Trial # 2**	**Trial # 3 (if needed)**
Initial reading of NH_3 buret	_____ mL	_____ mL	_____ mL
Initial of reading HCl buret	_____ mL	_____ mL	_____ mL
Final reading of NH_3 buret	_____ mL	_____ mL	_____ mL
Final reading of HCl buret	_____ mL	_____ mL	_____ mL
Volume of NH_3 solution used	_____ mL	_____ mL	_____ mL
Volume of HCl standard used	_____ mL	_____ mL	_____ mL
Volume ratio, NH_3 to HCl	_____	_____	_____
Moles of HCl = moles of NH_3	_____ mol	_____ mol	_____ mol

Moles of NH_3 per 250.0 mL NH_3 solution = moles of NH_3 in transferred mass of product sample

	_____ mol	_____ mol	_____ mol
Moles of NH_3 per gram of product sample	_____ mol/g	_____ mol/g	_____ mol/g

Average (mean) moles of NH_3 per gram of product (based only on trials that agree within 1%) _____ mol/g

E. Determining the formula of $Co_x(NH_3)_yCl_z$

Per gram of
sample: Moles of chlorine _____   Moles of cobalt _____   Moles of NH_3 _____

Counting number ratio
(divide through by the smallest
value on the row above): $z =$ _____ $x =$ _____ $y =$ _____

Formula of product compound   _____

Molar mass of product compound (based on formula)   _____ g/mol

Moles of (dry) product obtained from synthesis   _____ mol

Moles of cobalt used in synthesis   _____ mol

Percent synthetic yield (mol product/mol cobalt used in synthesis) _____ %

Experiment 29

Advance Study Assignment: Making and Analyzing a Coordination Compound

In Part A of this experiment, a student prepared 5.65 g of $Co_x(NH_3)_yCl_z$ from 8.02 g of $CoCl_2 \cdot 6\ H_2O$. She then analyzed the product by the procedure in this experiment.

Part B. In the gravimetric determination of chloride, she weighed out a 0.3006-g sample of the product. The following data were obtained:

Mass of crucible plus AgCl	33.1627 g
Mass of empty crucible	32.6803 g

Mass of AgCl   _____ g

Moles of Cl^- in AgCl   _____ mol $MM_{AgCl} = 143.323$ g/mol
(Note that this is the moles of Cl present in the 0.3006 g of product that the student analyzed.)

Moles of Cl per gram of product   _____ mol/g

Part C. In the colorimetric determination of cobalt, she used a product sample with a mass of 0.5010 g. The molarity of cobalt ion in the solution from the 25.0-mL volumetric flask was 0.073 M.

Moles of cobalt ion in 25.0 mL solution = moles of cobalt in sample   _____ mol

Moles of Co per gram of product   _____ mol/g

Part D. In the volumetric determination of ammonia, the product sample had a mass of 0.9985 g. In the titration, 0.1000 M HCl was used. She found that 40.92 mL of the NH_3 solution required 35.97 mL of the HCl standard to reach the endpoint.

Moles of HCl used   _____ = moles of NH_3 in 40.92 mL of NH_3 solution

Moles of NH_3 in 250. mL of NH_3 solution   _____ = moles of NH_3 in 0.9985 g of product compound

Moles of NH_3 per gram of product   _____ mol/g

E. Determining the formula of $Co_x(NH_3)_yCl_z$:

In 1 g of the product there are:

moles of Co = _____ mol

moles of NH_3 = _____ mol

moles of Cl = _____ mol

Dividing each of these by smallest of them,

moles of Co = _____ mol

moles of NH_3 = _____ mol

moles of Cl = _____ mol

Rounding off each of these to the nearest whole number,

x = _____

y = _____

z = _____

Formula of the product compound, $Co_x(NH_3)_yCl_z$ _____
(This may or may not be the formula of the compound you will be making!)

Molar mass of product compound (calculated from formula) _____ g/mol

Moles of product obtained from synthesis _____ mol

Moles of $CoCl_2 \cdot 6 H_2O$ used in synthesis _____ mol

Synthetic percent yield (moles of Co in product/moles of Co added in synthesis) _____ %

Experiment 30

Determining Iron by Reaction with Permanganate—A Redox Titration

A convenient method for determining the amounts required for complete reaction with a chemical sample is called <u>titration</u>. A reactant of known concentration (the <u>titrant</u>) is added in measured amounts to the unknown sample until the reaction stops occurring, using a volume dispensing device called a <u>buret</u> (see Appendix D). Potassium permanganate, $KMnO_4$, is widely used as an oxidizing agent in titrations. <u>Oxidizing agents</u> tend to take electrons from other chemicals, gaining electrons (being <u>reduced</u>) themselves in the process. In acidic solution, MnO_4^- ion undergoes reduction to Mn^{2+} according to the following electrochemical half-reaction:

$$8\,H^+(aq) + MnO_4^-(aq) + 5\,e^- \rightarrow Mn^{2+}(aq) + 4\,H_2O(\ell) \tag{1}$$

Because the MnO_4^- ion has an intense violet color while the Mn^{2+} ion is nearly colorless, the endpoint in titrations using $KMnO_4$ as the titrant can be taken to be indicated by the first permanent pink color that appears in the solution—that is, $KMnO_4$ serves as its own <u>indicator</u>.

You will use $KMnO_4$ in this experiment to determine the mass percent of iron in an unknown solid containing iron(II) ammonium sulfate, $Fe(NH_4)_2(SO_4)_2 \cdot 6\,H_2O$, one of the few shelf-stable solid forms of iron(II). The titration, which involves the oxidation of Fe^{2+} ion to Fe^{3+} by <u>permanganate ion</u>, MnO_4^-, is carried out in acidic solution to prevent Fe^{2+} from being oxidized by oxygen in the air. The endpoint of the titration is sharpened by using phosphoric acid, H_3PO_4, for this—because the Fe^{3+} produced in the titration, which would otherwise cause the solution to turn yellow, forms an essentially colorless complex with phosphate ions.

The oxidation of Fe^{2+} to Fe^{3+} occurs by this electrochemical half-reaction:

$$Fe^{2+}(aq) \rightarrow Fe^{3+}(aq) + 1\,e^- \tag{2}$$

<u>Oxidation-reduction, or redox, reactions</u> involve the transfer of electrons from one reactant to another, rather than changes in how electrons are shared. They are the net result of one reactant being reduced (gaining electrons) and another reactant being <u>oxidized</u> (losing electrons) at the same time. Chemistry cannot create or destroy electrons, only move them around—so reduction and oxidation must happen together. Moreover, the electrons produced as a product of oxidation must *all* be consumed by a simultaneous reduction reaction. This means that when adding together two electrochemical half-reactions in order to arrive at an overall net ionic equation, the half-reactions must be multiplied by factors that cause the total number of electrons produced in the oxidation process to exactly equal the number consumed in the reduction process.

The moles of potassium permanganate used in the titration equals the product of the molarity of the $KMnO_4$ in the titrant and the volume of titrant used to reach the endpoint. The moles of iron present in the sample can be obtained from the balanced chemical reaction and the amount of MnO_4^- ion reacted. From this, the mass of iron present in the sample can be calculated, leading to the percent by mass of iron in the sample. You will calculate the average (mean) percent by mass of iron of the several duplicate (technical term: <u>replicate</u>) titrations you will carry out, along with their standard deviation. (See Appendix H for more information about that.)

Wear your safety glasses while performing this experiment.

Experimental Procedure

A. Preparing the buret

Prepare a buret for the titration by the following procedure, which is described in greater detail in the buret section of Appendix D. (The purpose of this procedure is to make sure that the solution in the buret has the same concentration as the stock solution.) Clean the buret and rinse it with deionized water. Then rinse it three times with a few milliliters of the titrant—the standard $KMnO_4(aq)$ stock solution provided in the lab—in each case thoroughly wetting the walls of the buret with the solution and then letting it out through the stopcock. Fill the buret with $KMnO_4(aq)$ stock solution; open the stopcock momentarily to fill the tip and bring the titrant level below the zero mark. Check to make sure that the buret does not leak and that there are no air bubbles in the buret tip.

B. Determining the iron content in an unknown

Obtain an unknown iron(II) sample and record its identifier on the report page. Select a clean 125- or 250-mL Erlenmeyer flask: it can be wet with deionized water on the inside, but the outside must be dry. Using an analytical balance, add 1.0 ± 0.4 g of your unknown iron(II) sample to the flask, recording the amount actually used to ± 0.0001 g. Add about twenty-five milliliters of deionized water and 3 mL of $85\%_{mass}$ phosphoric acid, H_3PO_4. CAUTION: **Corrosive substance—avoid contact**. Swirl the flask until the solid is completely dissolved.

Titrate the iron solution in your flask with the $KMnO_4$ titrant in your buret, first recording the initial volume in the buret as precisely as you can. (*Note*: Because the color of the MnO_4^- ion is so intense, it will be difficult to see the meniscus in the buret. With this titrant, it is best to read volumes by noting the position of the very top of the violet color. Whatever you choose to do, however, be consistent in how you read the initial and final readings for a given trial.)

Titrate until you obtain the first peach or pink color that no longer fades away in under 15 seconds. Record the final volume in the buret when that happens. It will be difficult to hit the endpoint exactly on your first try, but use your first titration to estimate how much titrant will be needed in later trials. Calculate the milliliters of titrant per gram of solid (the "titration ratio") in your first trial and multiply that by the mass of solid being used in a later trial. Add titrant quickly until you have added about a milliliter less than that volume, then slow down and titrate carefully to the endpoint.

You will need to conduct additional titrations to get a reliable result. Do so by refilling the buret with titrant and then preparing another sample to titrate. Follow the same procedure. Continue doing titrations until you have three in which you feel you hit the endpoint and that have titration ratios within 0.5 mL/g of each other. Use only these three when calculating the average (mean) and standard deviation on the report page.

Optional C. Standardizing the KMnO₄ titrant

The $KMnO_4$ stock titrant solution can be standardized (its concentration determined both accurately and precisely) by the same method used in this experiment. $Fe(NH_4)_2(SO_4)_2 \cdot 6H_2O$ is a primary standard with a molar mass equal to 392.2 g/mol. For standardization, use 0.8 ± 0.2 g samples of the primary standard, weighed to ± 0.0001 g. Note that the titration ratios for these titrations should differ from those obtained when titrating the unknown, but they should agree well with each other. ▨

Take it further (optional): Determine the mass percent of iron in iron supplement tablets that contain the iron(II) ("ferrous") ion, Fe^{2+}.

Disposal of reaction products: Dispose of your titrated solutions as directed by your instructor.

Experiment 30

Data and Calculations: Determining Iron by Reaction with
Permanganate—A Redox Titration

B. Determining the iron content in an unknown

Unknown ID code   ═══════════

Trial 1

Mass of unknown solid actually transferred into Erlenmeyer flask   ═══════════ g

Initial volume reading on buret (before starting titration)   ═══════════ mL

Final volume reading on buret (at titration endpoint)   ═══════════ mL

Volume of titrant used (final volume reading – initial volume reading)   ═══════════ mL

Titration ratio, $\left(\dfrac{\text{volume of titrant used in this titration}}{\text{mass of solid used in this titration}} \right)$ ═══════════ mL/g

Trial 2

Mass of unknown solid actually transferred into Erlenmeyer flask   ═══════════ g

Initial volume reading on buret (before starting titration)   ═══════════ mL

Expected titrant volume needed based on Trial 1   ═══════════ mL
(multiply the mass of solid for this trial by the titration ratio from Trial 1)

Expected final volume reading on buret   ═══════════ mL
(add the initial volume reading for this trial to the expected titrant
volume on the line above)

Actual final volume reading on buret (at titration endpoint)   ═══════════ mL

Actual volume of titrant used (actual final reading – initial reading)   ═══════════ mL

Titration ratio, $\left(\dfrac{\text{volume of titrant used in this titration}}{\text{mass of solid used in this titration}} \right)$ ═══════════ mL/g

Trial 3

Mass of unknown solid actually transferred into Erlenmeyer flask

_____ g

Initial volume reading on buret (before starting titration)

_____ mL

Expected titrant volume needed based on best previous trial
(multiply the mass of solid for this trial by the titration ratio from
your best previous trial)

_____ mL

Expected final volume reading on buret
(add the initial volume reading for this trial to the expected titrant
volume on the line above)

_____ mL

Actual final volume reading on buret (at titration endpoint)

_____ mL

Actual volume of titrant used (actual final reading – initial reading)

_____ mL

Titration ratio, $\left(\dfrac{\text{volume of titrant used in this titration}}{\text{mass of solid used in this titration}} \right)$

_____ mL / g

Trial 4 (if needed)

Mass of unknown solid actually transferred into Erlenmeyer flask

_____ g

Initial volume reading on buret (before starting titration)

_____ mL

Expected titrant volume needed based on best previous trial
(multiply the mass of solid for this trial by the titration ratio from
your best previous trial)

_____ mL

Expected final volume reading on buret
(add the initial volume reading for this trial to the expected titrant
volume on the line above)

_____ mL

Actual final volume reading on buret (at titration endpoint)

_____ mL

Actual volume of titrant used (actual final reading – initial reading)

_____ mL

Titration ratio, $\left(\dfrac{\text{volume of titrant used in this titration}}{\text{mass of solid used in this titration}} \right)$

_____ mL / g

Trial 5 (if needed)

Mass of unknown solid actually transferred into Erlenmeyer flask _____ g

Initial volume reading on buret (before starting titration) _____ mL

Expected titrant volume needed based on best previous trial _____ mL
(multiply the mass of solid for this trial by the titration ratio from your best previous trial)

Expected final volume reading on buret _____ mL
(add the initial volume reading for this trial to the expected titrant volume on the line above)

Actual final volume reading on buret (at titration endpoint) _____ mL

Actual volume of titrant used (actual final reading − initial reading) _____ mL

Titration ratio, $\left(\dfrac{\text{volume of titrant used in this titration}}{\text{mass of solid used in this titration}} \right)$ _____ mL / g

Unknown Analysis Results

Numbers of your three best unknown determination trials _____, _____, _____

Average (mean) titration ratio from those three best trials _____ mL / g

Standard deviation in titration ratio from those three best trials _____ mL / g

Relative error in titration ratio (divide standard deviation by mean) _____

Molarity of titrant (given, or from your own standardization) _____ M

Average (mean) moles of $KMnO_4$ per gram of solid unknown _____ mol / g
(multiply average titration ratio by the molarity on the line above
and divide by 1000 mL / L)

Average (mean) moles of iron per gram of solid unknown _____ mol / g
(multiply the ratio on the line above by the number of Fe^{2+}
that react with one $KMnO_4$)

Average (mean) mass of iron per gram of solid unknown _____ g / g
(multiply the ratio on the line above by the molar mass of iron, 55.847 g / mol)

Average (mean) mass percent of iron in solid unknown _____ %$_{\text{mass}}$
(express the result on the line above as a percentage)

Standard deviation in mass percent of iron in solid unknown _____ %$_{\text{mass}}$
(Multiply the line above by the relative error in titration ratio)

Optional **C. Standardizing the KMnO$_4$ titrant**

Trial A

Mass of Fe(NH$_4$)$_2$SO$_4 \cdot$ 6 H$_2$O actually transferred into Erlenmeyer flask
========== g

Initial volume reading on buret (before starting titration)
========== mL

Final volume reading on buret (at titration endpoint)
========== mL

Volume of titrant used (final volume reading − initial volume reading)
========== mL

Titration ratio, $\left(\dfrac{\text{volume of titrant used in this titration}}{\text{mass of solid used in this titration}} \right)$
========== mL/g

Trial B

Mass of Fe(NH$_4$)$_2$SO$_4 \cdot$ 6 H$_2$O actually transferred into Erlenmeyer flask
========== g

Initial volume reading on buret (before starting titration)
========== mL

Expected titrant volume needed based on Trial A
(multiply the mass of solid for this trial by the titration ratio from Trial A)
========== mL

Expected final volume reading on buret
(add the initial volume reading for this trial to the expected titrant volume
on the line above)
========== mL

Actual final volume reading on buret (at titration endpoint)
========== mL

Actual volume of titrant used (actual final reading − initial reading)
========== mL

Titration ratio, $\left(\dfrac{\text{volume of titrant used in this titration}}{\text{mass of solid used in this titration}} \right)$
========== mL/g

Trial C

Mass of $Fe(NH_4)_2SO_4 \cdot 6\,H_2O$ actually transferred into Erlenmeyer flask　　　_____ g

Initial volume reading on buret (before starting titration)　　　_____ mL

Expected titrant volume needed based on best previous trial　　　_____ mL
(multiply the mass of solid for this trial by the titration ratio from your
best previous trial)

Expected final volume reading on buret　　　_____ mL
(add the initial volume reading for this trial to the expected titrant volume
on the line above)

Actual final volume reading on buret (at titration endpoint)　　　_____ mL

Actual volume of titrant used (actual final reading – initial reading)　　　_____ mL

Titration ratio, $\left(\dfrac{\text{volume of titrant used in this titration}}{\text{mass of solid used in this titration}} \right)$　　　_____ mL/g

Trial D (if needed)

Mass of $Fe(NH_4)_2SO_4 \cdot 6\,H_2O$ actually transferred into Erlenmeyer flask　　　_____ g

Initial volume reading on buret (before starting titration)　　　_____ mL

Expected titrant volume needed based on best previous trial　　　_____ mL
(multiply the mass of solid for this trial by the titration ratio from
your best previous trial)

Expected final volume reading on buret　　　_____ mL
(add the initial volume reading for this trial to the expected titrant
volume on the line above)

Actual final volume reading on buret (at titration endpoint)　　　_____ mL

Actual volume of titrant used (actual final reading – initial reading)　　　_____ mL

Titration ratio, $\left(\dfrac{\text{volume of titrant used in this titration}}{\text{mass of solid used in this titration}} \right)$　　　_____ mL/g

(continued on following page)

Trial E (if needed)

Mass of $Fe(NH_4)_2SO_4 \cdot 6\ H_2O$ actually transferred into Erlenmeyer flask

＿＿＿＿＿＿ g

Initial volume reading on buret (before starting titration)

＿＿＿＿＿＿ mL

Expected titrant volume needed based on best previous trial (multiply the mass of solid for this trial by the titration ratio from your best previous trial)

＿＿＿＿＿＿ mL

Expected final volume reading on buret (add the initial volume reading for this trial to the expected titrant volume on the line above)

＿＿＿＿＿＿ mL

Actual final volume reading on buret (at titration endpoint)

＿＿＿＿＿＿ mL

Actual volume of titrant used (actual final reading – initial reading)

＿＿＿＿＿＿ mL

Titration ratio, $\left(\dfrac{\text{volume of titrant used in this titration}}{\text{mass of solid used in this titration}} \right)$

＿＿＿＿＿＿ mL / g

Standardization Results

Letters of your three best standardization trials

＿＿＿ , ＿＿＿ , ＿＿＿

Average (mean) titration ratio from those three best trials

＿＿＿＿＿＿ mL / g

Standard deviation in titration ratio from those three best trials

＿＿＿＿＿＿ mL / g

Relative error in titration ratio (divide standard deviation by mean)

＿＿＿＿＿＿

Mass of $Fe(NH_4)_2SO_4 \cdot 6\ H_2O$ required to react with one mole of $KMnO_4$

＿＿＿＿＿＿ g / mol

Average (mean) volume of titrant containing one mole of $KMnO_4$
(multiply average titration ratio by the mass on the line above)

＿＿＿＿＿＿ mL / mol

Average (mean) molarity of $KMnO_4$ in titrant
(Divide <u>1000</u> mL / L by the volume on the line above)

＿＿＿＿＿＿ M

Standard deviation in molarity of $KMnO_4$ in titrant
(Multiply the line above by the relative error in titration ratio)

＿＿＿＿＿＿ M

Experiment 30

Advance Study Assignment: Determining Iron by Reaction with Permanganate—A Redox Titration

1. Write the balanced net ionic equation for the reaction between MnO_4^- ion and Fe^{2+} ion in acidic solution. Remember that when you add electrochemical half-reactions, the electrons must cancel out: chemistry can not create or destroy electrons! In this case, in order to meet this requirement, you will need to multiply Reaction 2 by a factor of five (or add it five times as it is written) when adding it to Reaction 1 in order to obtain the net ionic reaction for the titration in this experiment.

2. How many moles of Fe^{2+} ion can be oxidized by 1.85×10^{-4} moles of MnO_4^- ion? (Base this on the balanced reaction you got as your answer to Question 1.)

_____ moles

(continued on following page)

3. A solid sample containing some Fe^{2+} ion had a total mass of 0.9791 g. It required 18.2 mL of 0.01495 M $KMnO_4$ to titrate the Fe^{2+} in the dissolved sample to a peach endpoint.

   a. How many moles of MnO_4^- ion were required?

   _____ moles

   b. How many moles of Fe^{2+} were present in the sample?

   _____ moles

   c. How many grams of iron were present in the sample?

   _____ g

   d. What was the mass percent of iron in the sample?

   _____ %$_{mass}$

4. What is the mass percent of iron in iron(II) ammonium sulfate hexahydrate, $Fe(NH_4)_2(SO_4)_2 \cdot 6H_2O$?

   _____ %$_{mass}$

Experiment 31

Determining an Equivalent Mass by Electrolysis

In the process called electrolysis, electricity is used to force a chemical reaction—one that would not otherwise occur—to happen. Electrons are continually pulled out of a liquid containing ions (technical term: an electrolyte) at a positive electrode (technical term: anode) and pushed into it at a negative electrode (technical term: cathode); both electrodes are made of electrically conductive materials, usually metals, that can support a constant flow of electrons in the same direction without undergoing any chemical change. Ion-containing liquids (electrolytes) cannot do this: they can support a continuing alternating current (flow of electrons back and forth), but electrochemical half-reactions must occur at each electrode to support a continuous direct current (flow of electrons in a single direction), the kind used in electrolysis. Where the positive electrode contacts the liquid, an oxidation half-reaction must take place: something must lose electrons, taking on a more positive charge. At the negative electrode, a reduction half-reaction must take place: something must gain electrons, taking on a more negative charge. Many reactions are possible at each electrode, but what actually happens is what most easily occurs under the conditions at each electrode. If those conditions are controlled, a specific reaction can be made to dominate at each electrode (rather than having a mixture of multiple reactions take place).

In this experiment, the liquid subjected to electrolysis will be an aqueous solution of sodium sulfate, Na_2SO_4, and acetic acid, CH_3CHOOH, chosen such that the most favorable reaction at the negative electrode will be the reduction of hydrogen ion, $H^+(aq)$, to hydrogen gas, $H_2(g)$:

$$2\,H^+(aq) + 2\,e^- \rightarrow H_2(g) \tag{1}$$

Reaction 1 is an example of an electrochemical reduction half-reaction; it indicates that for every H^+ ion reduced *one* electron is required, and for every molecule of H_2 formed *two* electrons are needed.

Ordinarily chemistry deals not with individual ions or molecules, but with moles of substances. In terms of moles, by Equation 1:

The reduction of one mole of H^+ ion requires one mole of electrons.

The production of one mole of $H_2(g)$ requires two moles of electrons.

A mole of electrons is a fundamental amount of electricity in the same way that a mole of pure substance is a fundamental amount of matter, at least from a chemical point of view. A mole of electrons is called a faraday, after Michael Faraday, who discovered the basic laws of electrolysis. The mass of a substance that reacts with a mole of electrons, or one faraday, is defined as the equivalent mass of that substance. Because one faraday will reduce one mole of H^+ ion, we say that the equivalent mass of hydrogen is 1.008 g/mol, equal to the mass of one mole of H^+ ion [or $1/2$ mole of $H_2(g)$]. To form one mole of $H_2(g)$, we would have to pass two faradays through the electrolysis system.

In the electrolysis you will carry out for this experiment, you will measure the volume of hydrogen gas produced under known conditions of temperature and pressure. Using the ideal gas law, you will be able to calculate how many moles of H_2 were formed, and therefore how many faradays of electricity passed through the system.

At the positive electrode, electrons are taken from whatever they can most easily be pulled away from. The solvent itself (water, in this case) might be oxidized (forming O_2 gas), or something dissolved in the solution might be oxidized (for example, in this case acetic acid might be oxidized to CO_2); but there is also a third possibility, that being that the electrode itself can undergo oxidation. If the electrode material is very resistant to oxidation—for example, if it is made of platinum—something in the solution will be oxidized. However, if the electrode is made of a more easily oxidized metal, the oxidation of that metal can be more favorable than oxidation of either the solvent or anything dissolved in it, and is what will take place. That is the desired situation in this case, because the goal of this experiment is to determine the equivalent mass of a metal: the metal used as the positive electrode. Because the metal the electrode is made from will be relatively easy to oxidize, the most favorable reaction at the positive electrode will be the oxidation of the metal, M:

$$M(s) \rightarrow M^{n+}(aq) + n\,e^- \qquad (2)$$

As electrolysis proceeds the atoms in the metal electrode are converted into positively charged ions and dissolve into the solution. The mass of the metal electrode itself decreases, by an amount related to the number of electrons forced through the system and the nature of the metal. To oxidize one mole, or one molar mass, of the metal, requires n faradays, where n is the charge on the metal ion formed in Reaction 2. By definition, one faraday would cause one equivalent mass, EM, of metal to go into solution. The molar mass, MM, and the equivalent mass of the metal are related by Equation 3:

$$MM = EM \times n \qquad (3)$$

In an electrolysis experiment, because n is not determined independently, it is not possible to find the molar mass of a metal directly. It is possible, however, to find the equivalent mass of an easily oxidized metal, and that will be your goal. You will oxidize a sample of an unknown metal at the positive electrode, weighing the metal before and after the electrolysis and thereby determining its loss in mass. The same number of electrons removed from the metal will go into reducing hydrogen ions by Reaction 1, at the negative electrode. (A power source is not an electron source! It can only push or pull electrons around, not create or destroy them—so every electron that enters the positive electrode has to exit at the negative electrode.) From the volume of H_2 gas produced at the negative electrode, you can calculate the moles of H_2 formed, and from that the number of faradays that passed through the system. The equivalent mass of the metal is then calculated as the amount of metal that would be oxidized if one faraday were used. In an optional part of the experiment, you may use Equation 3 to determine possible identities of your unknown metal.

Experimental Procedure

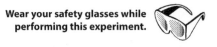

Wear your safety glasses while performing this experiment.

A. Preparing for electrolysis

Obtain a 50-mL buret and an unknown metal electrode. If necessary, lightly sand the metal to clean it. Rinse the metal electrode with water and then with acetone. Let the acetone evaporate. When the electrode is thoroughly dry, weigh it on an analytical balance to ±0.0001 g and record the result on the report page.

Set up the electrolysis apparatus shown in Figure 31.1. There should be about one hundred milliliters of 0.5 M CH_3COOH in 0.5 M Na_2SO_4 in the beaker to serve as the conducting solution. Insert the bare, tightly coiled end of the heavy copper wire into the open end of the buret, then lower this combination into the beaker containing the conducting solution until the buret opening is about one centimeter above the bottom of the beaker. Attach a length of flexible tubing to the upper end (the tip) of the buret. Open the stopcock of the buret and, with suction, carefully pull the conducting solution up to near the top of (but not above) the volume markings (technical term: <u>graduations</u>) in the buret. Close the stopcock. Ensure that all of the exposed heavy copper wire is above the bottom lip of the buret opening; any wire contacting the conducting solution but not inside the buret must be covered with watertight insulation. Check the solution level after a few minutes to make sure the stopcock does not leak. Record the initial buret reading; think carefully as you do this, because the buret is upside down!

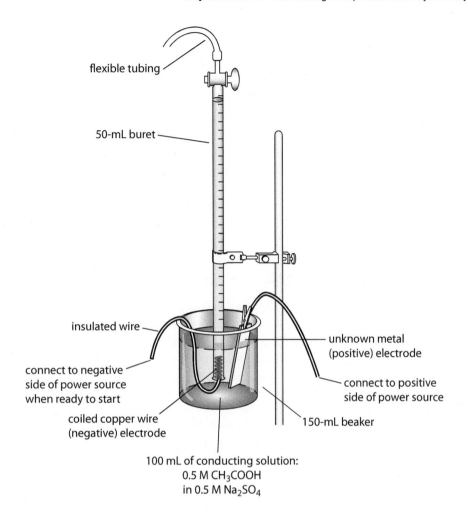

flexible tubing

50-mL buret

insulated wire

connect to negative
side of power source
when ready to start

coiled copper wire
(negative) electrode

unknown metal
(positive) electrode

connect to positive
side of power source

150-mL beaker

100 mL of conducting solution:
0.5 M CH₃COOH
in 0.5 M Na₂SO₄

Figure 31.1 Obtaining accurate data with the apparatus used in this experiment requires all of the hydrogen that bubbles off of the bare portion of the copper wire negative electrode to be collected in the inverted buret.

B. First electrolysis cycle

The metal unknown will serve as the positive electrode in your electrolysis cell. Connect the unknown metal electrode to the positive (+) side of the power source with an alligator clip and immerse the electrode (but not the clip!) in the conducting solution. The heavy copper wire electrode will be the negative electrode. Connect the dry end of that electrode to the negative (−) side of the power source. When you do, hydrogen gas should immediately begin to bubble from the copper wire (negative) electrode, but no bubbles should be seen at the unknown metal (positive) electrode. (In some cases the solution may become cloudy during the electrolysis. This is caused by the formation of a metal hydroxide and will not hurt your results.) Collect the hydrogen gas generated at the negative electrode until almost fifty milliliters have been produced—but *before* the conducting solution level drops below the final volume marking in the buret. At that point, stop the electrolysis by disconnecting the copper wire (negative) electrode from the power source if it cannot be turned off with a switch. Measure and record the temperature and the atmospheric pressure in the laboratory. Once the gas bubbles in the buret have risen to the top, record the final level of the liquid. Again, keep in mind that the buret markings are upside down.

C. Second electrolysis cycle

The mass loss of the positive electrode is still too small for a reliable equivalent mass measurement, but there are enough metal ions in the solution that they will soon start to plate out (be reduced back to the metal) at the negative electrode, in competition with hydrogen generation; therefore, you should now replace the conducting solution and then restart the electrolysis. Ensure you have recorded the final volume correctly, then drain the buret by opening the stopcock. Raise the buret in its clamp, above the beaker. Remove your electrodes from the electrolyte solution and stand them up in a small beaker, disturbing them as little as possible—you want to put them back "as they were" after you replace the conducting solution. Do not rinse or attempt to clean the electrodes in any way, at this point. You do not need to disconnect the electrical connections or alligator clips: it is better if you do not! (However, do not let the two electrodes touch each other.) Discard the used conducting solution in a waste container, then rinse the 150-mL beaker with tap water and refill it with one hundred milliliters of fresh conducting solution. Gently transfer the electrodes back into the conducting solution beaker and lower the buret over the coiled end of the copper wire (negative) electrode again. Pull conducting solution back up into the buret and record the second initial volume reading. Then repeat the electrolysis, again generating almost fifty milliliters of H_2 and recording the final liquid level in the buret on the report page.

D. Determining the mass lost by the positive electrode

Take the alligator clip off the unknown metal (positive) electrode and drip 0.10 M acetic acid, CH_3COOH, onto all sides of the unknown metal electrode while holding it over the 150-mL conducting solution beaker. Rinse all sides of the unknown metal electrode in hot tap water, then in deionized water, and finally with ethanol or acetone. Let the ethanol or acetone evaporate completely; do not dry the electrode with anything else. Weigh the dry unknown metal electrode to the nearest $\pm$ 0.0001 g. Record the ID code of your unknown, as well as detailed observations of it, on the report page.

> **Disposal of reaction products:** When you are finished with this experiment, return the metal electrodes and discard the conducting solution in a waste container, unless your instructor directs otherwise.

Experiment 31

Data and Calculations: Determining an Equivalent Mass
by Electrolysis

Initial mass of dry unknown metal (positive) electrode   _____ g

Initial buret reading (Be careful, the scale is upside down!)   _____ mL

Buret reading after first electrolysis cycle   _____ mL

Ambient (atmospheric) pressure, $P_{ambient}$ _____ mm Hg

Room temperature, t _____ °C

Buret reading after refilling   _____ mL

Buret reading after second electrolysis cycle   _____ mL

Mass of dry unknown metal (positive) electrode after electrolysis   _____ g

Vapor pressure of water at t (consult Appendix A), VP_{H_2O} _____ mm Hg

Total volume of (water-saturated) H_2 gas produced, V_{gas}

                                                                              _____ mL

Room temperature measured on an absolute temperature scale, T

                                                                              _____ K

Pressure exerted by dry H_2, $P_{H_2} = P_{ambient} - VP_{H_2O}$
(Ignore any pressure effect due to liquid levels in the buret.)

                                                                              _____ mm Hg

(continued on following page)

Moles of H_2 produced, n_{H_2} (Use the ideal gas law, $PV = nRT$; in this case V will be V_{gas} but P will be P_{H_2} and n will be n_{H_2}.)

_____ moles

Faradays of charge passed (moles of electrons)

_____ F

Mass lost by unknown metal (positive) electrode

_____ g

Equivalent mass of unknown metal $\left(EM = \dfrac{\text{grams of metal lost}}{\text{faradays passed}} \right)$

_____ g/mol e$^-$

Unknown metal ID code   ════════

Observations of unknown metal:

Optional If for this metal	$n = +1$	$n = +2$	$n = +3$
Molar mass of metal, MM {g/mol}	_____	_____	_____
Possible metal(s)	_____	_____	_____

Experiment 31

Advance Study Assignment: Determining an Equivalent Mass by Electrolysis

1. In an electrolysis experiment similar to the one in this experiment, a student observed that their unknown metal anode lost 0.242 g when a total volume of 98.24 mL of water-saturated hydrogen gas was produced. The temperature in the laboratory was 23 °C and the ambient (atmospheric) pressure was 729 mm Hg. You can find the vapor pressure of water in Appendix A. Fill in the following series of blanks to calculate the equivalent mass of the metal.

   VP_{H_2O} (from Appendix A) = _____ mm Hg

   $P_{H_2} = P_{ambient} - VP_{H_2O}$ = _____ mm Hg = _____ atm

   V_{gas} = _____ mL = _____ L

   T = _____ K

   n_{H_2} = _____ moles $n_{H_2} = \dfrac{PV}{RT}$ (where $P = P_{H_2}$ and $V = V_{gas}$; show your work below)

   Producing 1 mole of H_2 requires passing _____ faraday(s) through the electrochemical cell

   Faradays passed = _____ mol of e^-

   Mass lost by metal anode = _____ g

   Grams of metal lost per faraday passed = $\dfrac{\text{grams lost}}{\text{faradays passed}}$ = _____ g/mol e^- = EM

(continued on following page)

The student was told that their metal anode was made of copper.

$MM_{Cu} =$ _____ g/mol. The charge n on the Cu ion is therefore _____. (See Equation 3; show your work below.)

2. The coulomb (C) is the SI base unit for electrical charge. One faraday (F) has a total charge of 96480 coulombs, while a mole (N_A) consists of 6.02214×10^{23} things. Based on this information, calculate the charge on a *single* electron, in coulombs. (*Hint*: It will be a very small number!)

_____ coulombs per electron

Experiment 32

Voltaic Cell Measurements

When a piece of copper wire, Cu(s), is placed in an aqueous solution of silver nitrate, $AgNO_3$(aq), a remarkable transformation occurs. Needles of silver metal, Ag(s), form on the copper, and the originally colorless solution gradually turns the blue color characteristic of aqueous copper(II) ion, Cu^{2+}(aq). The following net ionic equation describes this change:

$$Cu(s) + 2\ Ag^+(aq) \rightarrow Cu^{2+}(aq) + 2\ Ag(s) \tag{1}$$

The nitrate ion, NO_3^-, introduced with the $AgNO_3$, remains in solution but is not changed in the reaction: it is called a <u>spectator ion</u>. Spectator ions are not included in <u>net ionic equations</u> such as Reaction 1.

Reaction 1 occurs because electrons will move from copper metal to Ag^+ ions if given the chance; it is the net result of Ag^+ being <u>reduced</u> (gaining electrons) and Cu(s) being <u>oxidized</u> (losing electrons) at the same time. Chemistry cannot create or destroy electrons, only move them around, so reduction and oxidation *must* happen together; reactions of this sort are called <u>oxidation–reduction, or "redox", reactions</u>. Reaction 1 can be written as the sum of two <u>electrochemical half-reactions</u>, one representing the reduction of the Ag^+ ion and the other the oxidation of metallic copper, Cu(s):

$$\text{Oxidation (electron loss): } Cu(s) \rightarrow Cu^{2+}(aq) + 2\ e^- \tag{2}$$

$$\text{Reduction (electron gain): } 2\ Ag^+(aq) + 2\ e^- \rightarrow 2\ Ag(s) \tag{3}$$

Though both half-reactions must occur at the same time, it is possible to make the reduction (electron gain) half-reaction and the oxidation (electron loss) half-reaction happen in different places, and to force the electrons to flow through a wire to get from one half-reaction to the other. This is called a <u>voltaic cell</u>, and such cells make up the batteries we use every day: they convert chemical energy into electrical energy. Electrons enter and leave electrochemical cells through <u>electrodes</u> made of electrically conductive materials, usually solid metals. Oxidation occurs at one electrode, called the <u>anode,</u> and reduction occurs at the other electrode, called the <u>cathode,</u> of an electrochemical cell. How much electron "pressure" to flow exists between the two electrodes of a voltaic cell is called the <u>cell potential</u> and is measured in volts, abbreviated V. The cell potential reflects how energetically favorable the redox reaction is, which is related to the reaction's equilibrium constant and how far away from equilibrium the cell currently is.

Making a useful voltaic cell requires physically separating the reactants from one another, yet preventing them from becoming electrically charged as they react. The key to accomplishing this is allowing spectator ions to flow between the electrodes, while keeping the reactants themselves separate. Taking Reaction 1 as an example, as silver is reduced from Ag^+ to silver metal at the cathode the amount of positive charge there decreases, but the nitrate ions that were associated with the positively charged Ag^+ ions (which are now neutral Ag metal atoms) are still around. They will lead to a buildup of negative charge at the cathode if they do not move. At the same time, at the anode, copper metal is being oxidized to Cu^{2+} ions. Without negative ions around to compensate, a positive charge would quickly build up there. The resulting ionic charge imbalance in the cell would quickly grow to offset the cell potential completely, causing the electron pressure to flow through the wire (the voltage) to drop to zero, unless nitrate ions move from the cathode to the anode...but they cannot flow through the wire like the electrons do. There must be some sort of ionic path between the anode and cathode, but it must be one that prevents the reactants from mixing. One way of accomplishing this is shown in Figure 32.1, in which the flow of ions between the electrodes is made possible by a thin layer of ceramic filled with tiny holes (technical term: a <u>porous frit</u>). Ions cannot flow through this layer very quickly,

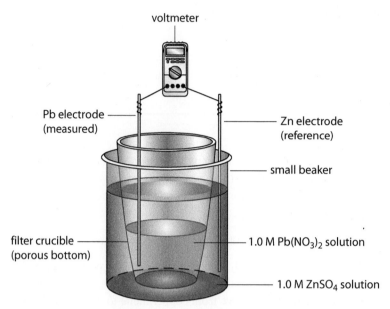

Figure 32.1 The voltaic cell configuration used in this experiment cannot support much current, but works well for measuring cell potentials provided that the bottom of the crucible has very small holes, preventing significant liquid flow between the chambers, and the meter does not allow many electrons to flow through it.

so this cell design can produce very little power: if a large number of electrons actually flows between the electrodes, ion flow will not be able to keep up, the cell will develop an internal charge, and electrons will quickly stop flowing. This cell design is useful, however, for measuring the cell potential when (almost) no electrons are flowing, and that is what you will be doing in this experiment.[1]

The voltaic cell in Figure 32.1 is set up to study the following (net ionic) redox reaction involving zinc, Zn, and lead, Pb, by having all of the reactants and products present at known concentrations:

$$Zn(s) + Pb^{+2}(aq) \rightarrow Zn^{2+}(aq) + Pb(s) \tag{4}$$

The voltmeter is designed to allow very few electrons to flow through it in the process of measuring voltage, and its reference is connected to the zinc electrode. The voltmeter displays the potential of the measuring (in this case, Pb) electrode relative to the potential of the reference (Zn) electrode. If this is positive, the reference (Zn) electrode is the anode, zinc is being oxidized, and Reaction 4 goes in the direction it is written. If negative, the zinc electrode is the cathode, zinc is being reduced, and Reaction 4 goes in reverse. Reaction 4 will go from whichever metal and ion combination is less stable to the pair that is more stable. The cell potential is given the symbol E_{cell}, and its sign must be recorded along with which electrode was used as the reference. In this experiment a zinc electrode will always be used as the reference, and it will always be the anode: that is, the cell potentials, E_{cell}, that you measure in this experiment should always be positive. But understand that it is possible for cell potentials to be negative and the reference to be the cathode, where reduction takes place.

In this experiment you will work with seven electrode systems: one metal ion/metal ion system, three halogen(Cl_2, Br_2, I_2)/halide(Cl^-, Br^-, I^-) systems, and three metal ion/solid metal systems:

Row 1: $Fe^{3+}(aq), Fe^{2+}(aq)|Pt$ $Cl_2(g), Cl^-(aq)|Pt$ $Br_2(\ell), Br^-(aq)|Pt$ $I_2(s), I^-(aq)|Pt$

Row 2: $Ag^+(aq)|Ag(s)$ $Cu^{2+}(aq)|Cu(s)$ $Zn^{2+}(aq)|Zn(s)$

In every case, the oxidized form of the reactive substance is written before the reduced form. This means that in a reduction, the substance indicated first is converted to the substance written second. The three reactions in Row 2 involve solid metals, which are present as *active* electrodes. In the four reactions in Row 1, solid

---

[1]Cell designs with large anodes and cathodes separated by much smaller distances and a barrier able to provide lots of mobile ions can maintain their voltage even when electrons are flowing rapidly between the electrodes; these can provide significant electrical power, but are harder to set up and to take measurements on.

electrodes are necessary to serve as the reaction site and to connect the wire that completes the circuit, but the electrodes do not chemically participate in the reaction. A strip of platinum, Pt, is used in these cases: it is an *inactive* electrode. The platinum will not react: it serves as a chemically inert path for electrons. The oxidized-form metal ions in Row 2 will be made available to you as solutions of soluble salts, while the solutions used for the reactions in Row 1 will contain both the oxidized and reduced forms of the reactive substance.

Voltaic cells are important in our everyday lives, but they are also useful to chemists for measuring very low concentrations. For example, pH meters use a voltaic cell to measure hydrogen ion concentrations as low as 10^{-14} M, something difficult to do in other ways. This requires knowing how concentrations change cell potentials, which is described by the <u>Nernst equation</u>. In a simplified form valid at 25 °C, it is:

$$E_{cell} = E_{cell}^0 - \frac{0.0591}{n} \log Q \tag{5}$$

where n is the number of electrons transferred in the balanced redox reaction and Q is the <u>reaction quotient</u> for the same reaction, which is the ratio or product of concentrations that its equilibrium constant is equal to. Taking Reaction 4 as an example, its equilibrium expression would be

$$K_{eq} = \frac{[\underline{Zn^{2+}}]}{[\underline{Pb^{2+}}]} = Q$$

(*Note*: Underlined chemicals in square brackets indicate concentrations measured in moles per liter, but used without units.)

Reaction 4 is the sum of these two electrochemical half-reactions:

Oxidation (electron loss): $\quad Zn(s) \rightarrow Zn^{2+}(aq) + 2\ e^-$ \hfill **(6)**

Reduction (electron gain): $\quad Pb^+(aq) + 2\ e^- \rightarrow Pb(s)$ \hfill **(7)**

These indicate that $n = 2$ for Reaction 4, because 2 e⁻ cancels out when the half-reactions are added together: 2 electrons transfer from Zn to Pb^{2+} each time Reaction 4 takes place as written. E_{cell}^0 is the cell potential measured under <u>standard conditions</u>, which are concentrations of 1 M, gas pressures of 1 atm, and a temperature of 25 °C, while E_{cell} is the cell potential when the concentrations are equal to the values plugged into Q.

In an optional part of this experiment, you will use a voltaic cell to determine very low equilibrium concentrations, making it possible to estimate some equilibrium constants. Specifically, you will study the formation constant, K_f, of the copper(II)-ammonia complex ion $Cu(NH_3)_4^{2+}$(aq) and the solubility product, K_{sp}, of silver chloride, AgCl:

copper(II)–ammonia
complex ion formation: $\quad Cu^{2+}$ (aq) $\quad + \quad 4\ NH_3$ (aq) $\rightleftharpoons Cu(NH_3)_4^{2+}$ (aq) $\qquad K_f = \dfrac{[Cu(NH_3)_4^{2+}]}{[Cu^{2+}]\ [NH_3]^4}$ \hfill **(8)**
$\qquad\qquad\qquad\qquad$ (blue) $\qquad\qquad$ (colorless) $\qquad$ (deep blue-violet)

silver chloride solubility: $\qquad\qquad\qquad\qquad\qquad AgCl(s) \rightleftharpoons Ag^+(aq) + Cl^-(aq) \qquad K_{sp} = [\underline{Ag^+}][\underline{Cl^-}]$ \hfill **(9)**

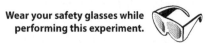

Wear your safety glasses while performing this experiment.

Experimental Procedure

A. Cell potentials

Obtain a voltmeter, a set of metal electrodes, and a filter crucible. The metal electrodes in this experiment can be difficult to tell apart, so they have been cut to different sizes to help you identify them. Unless your instructor tells you otherwise, the zinc electrode is the longest, copper the second longest, silver the third longest, and the platinum electrode is the shortest.

Pick the smallest beaker from your lab drawer that allows you to have your zinc electrode next to the crucible while the crucible still drops all the way to the bottom of the beaker without you having to push it down. Fill this beaker about one-quarter (¼) full with 1.0 M zinc sulfate, $ZnSO_4$(aq), solution and put the zinc electrode into it as shown in Figure 32.1. Next, put the filter crucible into the beaker next to the zinc electrode. Connect the reference side of the voltmeter to the zinc electrode, as shown in Figure 32.1.

(On most meters, the reference port is labelled COM, for "common reference", and is connected to a black wire.) You will now set up and measure the cell potential of a series of voltaic cells like the one shown in Figure 32.1, but with the lead (Pb) electrode and solution replaced with an electrode and solution pair from Row 1 or 2. Start by connecting the platinum electrode to the measuring side of the voltmeter (usually a red wire) and placing the end of the platinum electrode in the crucible, touching the bottom. Make sure the two electrodes and their electrical connections are not touching one another. Now add 0.10 M Fe^{2+}/0.10 M Fe^{3+} solution to the crucible until it reaches a depth of about two millimeters. Switch the voltmeter on and set it to an appropriate voltage range setting (likely 2 V). Wait for the reading on the voltmeter to stabilize, then record it on the report page. In many cases, the cell potential will gradually change (technical term: <u>drift</u>) over a long period of time: you do not need to wait for it to stabilize completely—it may never do so. Instead, record the cell potential you observe after about twenty seconds, with an appropriate number of significant figures—reflecting how much the value is changing. (See Appendix E.)

Remove the platinum electrode from the crucible and dry it with a paper towel. Then remove the crucible from the beaker and pour out its contents into a waste container. Rinse the crucible a few times with deionized water and drain it. Put the crucible back into the beaker, with the same zinc solution and zinc electrode still in place, and put the platinum electrode back in the crucible. Now add 1.0 M sodium chloride, NaCl, saturated with chlorine gas, Cl_2, to the crucible, again to a depth of about two millimeters. If you are using a voltmeter with a 2 V range, it might now be reading "OL", for "overload", because this cell potential is large. Change to the next highest range (likely 20 V), for this reading only. Record the reading from the voltmeter on the report page, once it stabilizes. Return the voltmeter to its original, more sensitive range and repeat this procedure for the other combinations in Row 1 and the first two in Row 2. When working on Row 2, you will replace the platinum electrode with the appropriate metal: silver or copper. The measurements can be taken in any order, but if doing the optional part of this experiment it will be most efficient to collect data for the copper cell last, counting how many drops of copper solution you add to the crucible and not disassembling the cell after the measurement is taken.

B. The effect of concentration on cell potentials

1. Determining the formation constant of the $Cu(NH_3)_4^{2+}$ complex ion If you do not already have it, set up a voltaic cell using $Zn^{2+}(aq)\,|\,Zn(s)$ as the reference and $Cu^{2+}(aq)\,|\,Cu(s)$ in the crucible. Carefully *count* how many drops of copper solution you add to the crucible to get to a depth of about two milliliters. Record the cell potential on the report page. Then add the same number of drops of 6 M ammonia, $NH_3(aq)$, to the crucible and stir well using the copper electrode. The solution in the crucible should turn a deep blue-violet color and the cell potential should change appreciably. Record the new cell potential once it stabilizes (or after about twenty seconds, with an appropriate number of significant figures).

2. Determining the solubility product of silver chloride, AgCl Empty the crucible from the measurement above into a waste container and rinse it well with deionized water, several times. Drain it and put it back into the beaker, then add 1.0 M potassium chloride, KCl, solution to it, to a depth of about two millimeters. Connect the measuring side of the voltmeter to the silver electrode and place it into the crucible. Notice, but do not yet record, the cell potential indicated on the voltmeter. If your instructor tells you to do so, add 1 drop of 1.0 M silver nitrate, $AgNO_3(aq)$, solution to the crucible and stir well with the silver electrode. (This step is not actually necessary, but does make it more obvious what is going on. Even without adding Ag^+ ions explicitly, there are enough around the silver metal electrode to establish the tiny equilibrium concentration needed to reach the solubility product of AgCl.) Wait for the reading on the voltmeter to stabilize and record it on the report page. (Did it change when you added a drop of $AgNO_3$ solution, if you did so?) ▨

Disposal of reaction products: The solutions used in this experiment are quite concentrated, and should not be poured down the sink, unless your instructor directs otherwise. If you did the optional part of this experiment and any AgCl remains in the crucible, it can be dissolved out using 6 M ammonia, $NH_3(aq)$. **CAUTION:** **Aqueous ammonia is a base with a strong smell that can knock you unconscious. Avoid its fumes, as well as contact with it!**

Experiment 32

Data and Calculations: Voltaic Cell Measurements

A. Cell potentials

Record your cell potential measurements in this table. You do not need to measure them in the order they are listed.

Reference electrode	Measured electrode	Cell potential, E^0_{cell} {Volts}	Oxidation half-reaction	Reduction half-reaction
$Zn^{2+}(aq)\,\|\,Zn(s)$	$Fe^{3+}(aq),\ Fe^{2+}(aq)\,\|\,Pt$	══════	$Zn(s) \rightarrow Zn^{2+}(aq) + 2\,e^-$	$Fe^{3+}(aq) + 1\,e^- \rightarrow Fe^{2+}(aq)$
$Zn^{2+}(aq)\,\|\,Zn(s)$	$Cl_2(g),\ Cl^-(aq)\,\|\,Pt$	══════	$Zn(s) \rightarrow Zn^{2+}(aq) + 2\,e^-$	$Cl_2(g) + 2\,e^- \rightarrow 2\,Cl^-(aq)$
$Zn^{2+}(aq)\,\|\,Zn(s)$	$Br_2(\ell),\ Br^-(aq)\,\|\,Pt$	══════	$Zn(s) \rightarrow Zn^{2+}(aq) + 2\,e^-$	_____
$Zn^{2+}(aq)\,\|\,Zn(s)$	$I_2(s),\ I^-(aq)\,\|\,Pt$	══════	$Zn(s) \rightarrow Zn^{2+}(aq) + 2\,e^-$	_____
$Zn^{2+}(aq)\,\|\,Zn(s)$	$Ag^+(aq)\,\|\,Ag(s)$	══════	$Zn(s) \rightarrow Zn^{2+}(aq) + 2\,e^-$	_____
$Zn^{2+}(aq)\,\|\,Zn(s)$	$Cu^{2+}(aq)\,\|\,Cu(s)$	══════	$Zn(s) \rightarrow Zn^{2+}(aq) + 2\,e^-$	

(continued on following page)

Standard reduction potentials

The (standard) cell potentials you measured all use $Zn^{2+}(aq)\,|\,Zn(s)$ as their reference, and each involves a different reduction half-reaction. If you were to put $Zn^{2+}(aq)\,|\,Zn(s)$ in the crucible as well as outside of it, electrons would have no reason to flow (they would be leaving one chemical environment for an identical one), and the cell potential would be zero. This means your measurements are actually standard reduction potentials measured using the reduction of $Zn^{2+}(aq)$ to $Zn(s)$ as their reference. By sorting these you can construct your own table of standard reduction potentials. This is similar to such tables found elsewhere, except that the $H^+(aq)$, $H_2(g)\,|\,Pt$ electrode is used as the reference for most tables of standard electrode potentials—because it is very reproducible. However, for day-to-day laboratory work, the hydrogen gas electrode is relatively impractical: hydrogen is a flammable gas, and very good results can be obtained using other, safer and more convenient, references, as you have done in lab today. In this table, sort your results from the previous page by cell potential, largest to smallest. Then look up the reduction potentials of the associated reduction half-reactions in a table of standard reduction potentials, and calculate the differences between those values and your measurements.

Standard reduction potential {Volts, $Zn^{2+}(aq)\,|\,Zn(s) = 0$}	Reduction half-reaction	Standard reduction potential {Volts, $H^+(aq)$, $H_2(g)\,|\,Pt = 0$}	Difference {Volts}
___	___	___	___
___	___	___	___
___	___	___	___
___	___	___	___
___	___	___	___
0	$Zn(s) \rightarrow Zn^{2+}(aq) + 2\,e^-$	−0.76	0.76

Optional **B. The effect of concentration on cell potentials**

1. Determining the formation constant of the Cu(NH₃)₄²⁺ complex ion

Potential, E^0_{cell}, before adding 6 M ammonia, $NH_3(aq)$ ═══════════ V

Potential, E_{cell}, after $Cu(NH_3)_4^{2+}$ formed ═══════════ V

Applying Equation 5 to this situation gives

$$E_{cell} = E^0_{cell} - \frac{0.0592\ V}{2} \log \frac{[Cu^{2+}]}{[Zn^{2+}]} \qquad \text{(5a)}$$

Use Equation 5a to calculate the concentration of free Cu^{2+} ion in equilibrium with $Cu(NH_3)_4^{2+}$ in the solution in the crucible.

$$[Cu^{2+}] = \underline{\hspace{5cm}}\ M$$

Hopefully you have obtained an incredibly low concentration of free Cu^{2+}. This indicates that almost all the copper ions were pulled into the $Cu(NH_3)_4^{2+}$ complex ion (and that its formation constant, K_f, is very large). As a result, the concentration of $Cu(NH_3)_4^{2+}$ in the crucible can be taken to be 0.50 M (because the 1.0 M Cu^{2+} that was present before the 6 M ammonia solution was added was diluted by an equal volume of ammonia solution). The concentration of NH_3 in the crucible would have been 3 M if no complex formed, but because 0.50 M complex formed, its concentration has dropped to 1 M (because each complex ion contains 4 NH_3 molecules). Use these values to estimate the formation constant, K_f, of the $Cu(NH_3)_4^{2+}$ complex ion, using Equation 8.

$$K_f = \underline{\hspace{5cm}}$$

(continued on following page)

2. Determining the solubility product of silver chloride, AgCl

Potential, E_{cell}^0, of the $Zn(s)\,|\,Zn^{2+}(aq)\,\|\,Ag^+(aq)\,|\,Ag(s)$ cell (from Part A)   ═══════════ V

Potential, E_{cell}, with 1.0 M KCl present   ═══════════ V

Using Equation 5, calculate $[Ag^+]$ in the cell, where it is in equilibrium with 1.0 M Cl^- ion. (This time *you* get to figure out Q and n!)

$[Ag^+] = $ _____ M

Because the Ag^+ and Cl^- in the crucible are in equilibrium with solid AgCl, you can estimate K_{sp} for AgCl from the concentrations of Ag^+ and Cl^-, both of which you now know. Use Equation 9.

$K_{sp} = $ _____

Experiment 32

Advance Study Assignment: Voltaic Cell Measurements

1. A student measures the potential of a voltaic cell similar to that in Figure 32.1, but with 1.0 M $Cu(NO_3)_2$ in the beaker and 1.0 M $AgNO_3$ in the crucible. A metallic copper (Cu) electrode is in the $Cu(NO_3)_2$, replacing the Zn electrode in Figure 32.1, and a metallic silver (Ag) electrode is in the $AgNO_3$ in the crucible, replacing the Pb electrode. She finds that the standard cell potential, or voltage, of the cell, E^0_{cell}, is +0.44 V, with the silver (measured) electrode positive relative to the copper (reference) electrode.

   a. At which electrode is oxidation occurring?

   b. Write the equation for the oxidation half-reaction in this cell.

   c. Write the equation for the reduction half-reaction in this cell.

   d. Write the net ionic equation for the oxidation–reduction reaction that occurs in this cell. Remember that when you add half-reactions, the electrons must cancel out: chemistry cannot create or destroy electrons! In this case, in order to meet this requirement, you will need to add your answer to 1(c) *twice* (or double it before adding) when combining it with your answer to 1(b).

   e. Do nitrate, NO_3^-(aq), spectator ions need to flow into or out of the crucible in order to compensate for the flow of electrons in this cell?

2. In another voltaic cell, like the one in Question 1 but with 1.0 M $Ni(NO_3)_2$ and a nickel (Ni) electrode in the crucible, the standard potential of the nickel metal│nickel(II) ion electrode was found to be −0.11 V relative to the copper metal│copper(II) ion electrode. (The copper electrode is the reference.)

   a. At which electrode is oxidation occurring?

   b. Write the equation for the oxidation half-reaction in this cell.

   c. Write the equation for the reduction half-reaction in this cell.

(continued on following page)

d. Write the net ionic equation for the oxidation–reduction reaction that occurs in this cell. Remember that when you add half-reactions, the electrons must cancel out: chemistry cannot create or destroy electrons!

e. Do nitrate, $NO_3^-(aq)$, spectator ions need to flow into or out of the crucible in order to compensate for the flow of electrons in this cell?

3. Optional All of the solutions used in this experiment are 1.0 M, with one exception: the $Fe^{3+}(aq)$, $Fe^{2+}(aq)$ solution, which has 0.10 M concentrations. We can get away with this because as long as the concentrations of the two iron ions are the same we will still measure standard cell potentials. The Nernst equation will show you why!

a. Write the electrochemical half-reaction in which iron(III) ion, $Fe^{3+}(aq)$, is reduced to iron(II) ion, $Fe^{2+}(aq)$. (If you are at a loss, look for it on the report pages.)

b. Combine this with the half-reaction for the oxidation of zinc, Reaction 6, to get the net ionic equation for the redox reaction that occurs in a voltaic cell with $Zn^{2+}(aq)|Zn(s)$ as the reference electrode and $Fe^{3+}(aq)$, $Fe^{2+}(aq)|Pt$ as the measured electrode. Remember that the electrons must cancel out when you add the half-reactions.

c. What is n for this redox reaction? [How many electrons are transferred from Zn to Fe every time the balanced redox reaction occurs as written? This will be the number of electrons cancelled out in arriving at your answer to 3(b).]

$n =$ _____

d. What would the equilibrium expression be for the redox reaction in 3(b)?

$$Q = K_{eq} =$$

e. Write the Nernst equation (Equation 5) for the redox reaction in 3(b).

Experiment 33

Making Copper(I) Chloride

Oxidation–reduction, or "redox", reactions involve the transfer of electrons from one reactant to another, rather than changes in how electrons are shared. They are the net result of one reactant being reduced (gaining electrons) and another reactant being oxidized (losing electrons) at the same time. Chemistry cannot create or destroy electrons, only move them around, so reduction and oxidation *must* happen together. In this experiment you will use a series of redox reactions to make one of the less common salts of copper: copper(I) chloride, CuCl.

Your first step in making (technical term: synthesizing) CuCl will be to dissolve copper metal in oxidizing (electron-removing) nitric acid, $HNO_3(aq)$. These react according to the following net ionic equation—in which copper metal, $Cu(s)$, is oxidized to blue copper(II) ions, $Cu^{2+}(aq)$, and colorless nitrate ions, $NO_3^-(aq)$, are reduced to orange-brown nitrogen dioxide gas, $NO_2(g)$:

$$Cu(s) \;+\; 4\,H^+(aq) \;+\; 2\,NO_3^-(aq) \;\rightarrow\; Cu^{2+}(aq) \;+\; 2\,NO_2(g) \;+\; 2\,H_2O(\ell) \qquad (1)$$
(colorless)   (colorless)   (blue)   (orange-brown)

You will then add an excess of sodium carbonate, Na_2CO_3, to neutralize the remaining acid (generating bubbles of colorless carbon dioxide gas, CO_2) by Reaction 2 and precipitate the blue copper(II), Cu^{2+}, ions in solution as the blue-green solid copper(II) carbonate, $CuCO_3$, by Reaction 3:

$$2\,H^+(aq) \;+\; CO_3^{2-}(aq) \;\rightleftharpoons\; (H_2CO_3)(aq) \;\rightleftharpoons\; CO_2(g) \;+\; H_2O(\ell) \qquad (2)$$
(colorless)   (colorless)   (colorless)   (colorless)

$$Cu^{2+}(aq) \;+\; CO_3^{2-}(aq) \;\rightleftharpoons\; CuCO_3(s) \qquad (3)$$
(blue)   (colorless)   (blue-green)

After isolating the $CuCO_3$ by filtration, and washing it to improve its purity, you will dissolve the $CuCO_3$ in hydrochloric acid. Copper metal then added to the highly acidic solution reduces the Cu(II) to Cu(I) by being oxidized itself (losing an electron) to form Cu(I). In the presence of excess chloride ion, Cl^-, the copper(I) will be present as a yellow $CuCl_4^{3-}$ complex ion. Diluting this solution with water destroys the complex, and white CuCl precipitates:

$$CuCO_3(s) \;+\; 2\,H^+(aq) \;+\; 4\,Cl^-(aq) \;\rightarrow\; CuCl_4^{2-}(aq) \;+\; CO_2(g) \;+\; H_2O(\ell) \qquad (4)$$
(blue-green)   (colorless)   (colorless)   (color varies)   (colorless)

$$CuCl_4^{2-}(aq) \;+\; Cu(s) \;+\; 4\,Cl^-(aq) \;\rightarrow\; 2\,CuCl_4^{3-}(aq) \qquad (5)$$
(color varies)   (colorless)   (yellow)

$$CuCl_4^{3-}(aq) \xrightarrow{\;H_2O\;} CuCl(s) \;+\; 3\,Cl^-(aq) \qquad (6)$$
(yellow)   (white)   (colorless)

Because CuCl is quickly oxidized by oxygen, you must minimize its exposure to air while you are making it.

Determining the amount of CuCl theoretically obtainable from this experiment's reaction sequence (the theoretical yield) is a bit tricky, because the chemical completely used up in the reaction sequence, and that limits how much product can result (the limiting reagent)—copper metal—is added in two different steps.

The key is understanding that the copper metal added for Reaction 5 can only react with as much Cu^{2+} as is present in the solution it is added to. That is controlled by how much copper metal was oxidized in Reaction 1. If the amount of copper metal used in Reaction 1 is less than the amount used for Reaction 5, the amount used for Reaction 5 is <u>an excess</u> (it does not limit the theoretical yield) and the theoretical yield is controlled by the amount of copper metal used in Reaction 1.

An optional part of this experiment proves that the formula of the product compound is really CuCl by converting the copper in it back into the Cu^{2+} ion, forming an intensely colored copper(II) ammonia complex, and determining the amount of complex present based on the amount of light it absorbs (technical term: <u>spectroscopically</u>). Knowing the product contains only copper and chlorine, the mass of chlorine in it is determined by subtracting the mass of copper in the product from the total mass of product used in the analysis. It can then be determined whether the mole ratio of copper to chlorine in the product is 1:1.

Experimental Procedure

Wear your safety glasses while performing this experiment.

A. Oxidizing copper metal to copper(II) ions using nitric acid

Obtain a 1.0-g sample of copper metal turnings, a Büchner funnel, and a filter flask. Weigh the copper metal to ± 0.01 g in a tared 150-mL beaker (see Appendix D), and record the mass on the report page.

In a fume hood, add 5 mL of concentrated nitric acid, 15 M HNO_3, to the 150-mL beaker. **CAUTION:** HNO_3 **is a powerful oxidizing acid, especially once it is hot. Treat it with great care.** Brown nitrogen dioxide gas, NO_2, will bubble out of the liquid as the solution turns blue and gets hot. **CAUTION: Avoid exposure to NO_2 gas, which is toxic.** Nitric acid is relatively slow-acting unless hot, but the heat released by Reaction 1 is usually enough to quickly turn all of the copper into ions. However, *if this proves not to be the case and some solid copper remains,* you may **carefully and gently** warm the beaker to speed it up. When all of the copper has gone into solution, add 50 mL of deionized water to the 150-mL beaker and allow it to cool.

B. Precipitating the copper(II) ions as copper(II) carbonate

Weigh out about five grams of sodium carbonate, Na_2CO_3, in a small beaker. With your spatula, add small amounts of this solid to the solution, adding more only as the bubbling caused by the release of carbon dioxide gas, $CO_2(g)$, in Reaction 2 slows down. Stir the solution to expose it to the Na_2CO_3. Once the acid is neutralized, the bubbling will stop and a blue-green precipitate of copper(II) carbonate, $CuCO_3$, will begin to form by Reaction 3. At that point, add the rest of the Na_2CO_3 all at once, stirring the mixture well to ensure complete precipitation of $CuCO_3$.

C. Isolating the solid copper(II) carbonate

Transfer the precipitate to the Büchner funnel, using suction to remove the excess liquid. (See Appendix D and Figure 33.1 for more details on how to do this.) To transfer as much of the solid as possible, stir the contents of the 150-mL beaker with your stirring rod just before pouring it out, and pour quickly, a bit at a time. Use your rubber policeman and a spray of deionized water from your wash bottle to ensure complete transfer of the solid. Wash the precipitate well with deionized water, with the suction still turned on; then let the solid remain on the filter paper with the suction on for a minute or two in order to dry the $CuCO_3$ (which is stable in air). Turn off the vacuum line, break the suction by twisting the vacuum hose as you pull it off the suction flask, and rotate the (cup of the) Büchner funnel to free it from the rest of the apparatus.

D. Reacting copper(II) ions with copper metal to obtain copper(I) as the chloride

Transfer the solid $CuCO_3$ back into the 150-mL beaker, removing the filter paper and rinsing any solid remaining on it into the beaker with about ten milliliters of deionized water. Then, *slowly and a bit at a time*, add 30 mL of 6 M hydrochloric acid, HCl(aq), to the solid, stirring continuously. When the $CuCO_3$ has all dissolved and the bubbling has stopped, add 1.5 g of copper metal turnings to the beaker and cover it with a watch glass.

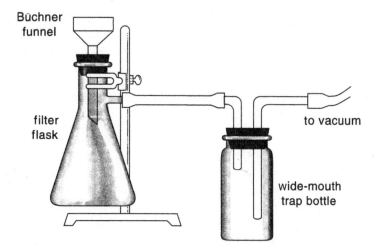

Büchner
funnel

filter
flask

to vacuum

wide-mouth
trap bottle

Figure 33.1 To separate a solid from a liquid using a Büchner funnel, put a piece of circular filter paper in the funnel. Thoroughly moisten the filter paper with deionized water from a wash bottle, but do not add more water than the filter itself will absorb. Just before pouring the mixture of solid and liquid into the filter, turn on the suction. Keep the suction on while filtering the mixture. Consult Appendix D for more details.

Heat the mixture in the 150-mL beaker to boiling and keep it just barely boiling. (See Appendix D for details.) Put 150 mL of deionized water in a 400-mL beaker and put the 400-mL beaker in an ice bath. After you have heated the acidic Cu-CuCl$_2$ mixture in the 150-mL beaker for ten minutes, or as soon as it is no longer cloudy and becomes yellow, carefully pour off the hot liquid into the 400-mL beaker of cold water, taking care not to transfer any of the excess copper metal to the beaker. White crystals of CuCl should form. Cover the 400-mL beaker with a watch glass and continue to cool it in the ice bath to promote crystallization and to increase the yield of solid CuCl.

E. Isolating the air-sensitive solid copper(I) chloride

In an ice bath, cool 25 mL of deionized water to which you have added 5 drops of 6 M hydrochloric acid, HCl(aq). Put 20 mL of acetone into a small beaker. Place a piece of filter paper that will fit in your Büchner funnel onto a watch glass, then determine their combined mass to ± 0.01 g. Record this combined mass on the report page, then place the filter paper into your Büchner funnel.

Read over the rest of this paragraph before going further, because *you will want to carry out the following steps with minimal delay.* Filter out the product crystals from the mixture in your 400-mL beaker using the Büchner funnel under suction. As before, stir the contents of the beaker just before pouring, to aid in transferring the solid to the funnel. Just as the last of the liquid is being pulled through, wash the product crystals with one-third of the acidified cold water. Rinse the last of the product from the beaker into the funnel with another portion of the cold acidic water, then use the final third to wash the solid in the funnel again. Turn off the vacuum line, break the suction by twisting the vacuum hose as you pull it off the suction flask, and then add half of the acetone to the funnel; wait about ten seconds and then turn the vacuum line back on. Repeat this operation with the other half of the acetone. Pull air through the product for a few minutes to dry it. If everything goes well, your product should be pure white: if the wet compound is allowed to come into contact with air, it will tend to turn pale green, due to oxidation of Cu(I) to Cu(II), so you want to *minimize delays while carrying out these filtration steps.*

Turn off the vacuum line, then break the suction by twisting the vacuum hose as you pull it off the suction flask. Rotate the (cup of the) Büchner funnel to free it from the rest of the apparatus. Transfer the crystals *and* your filter paper to the watch glass you weighed earlier, by inverting the filter's cup onto the watch glass. You may have to knock the filter cup a few times to get the product to fall out. Weigh the watch glass, product, and filter paper to ± 0.01 g, to determine an approximate yield. Show your sample to your instructor for evaluation.

Disposal of reaction products: The contents of the suction flask after the first filtration (of CuCO$_3$) can be discarded down the sink, unless your instructor directs otherwise. The contents after the second filtration include an appreciable amount of copper, so they should be put in a waste container, unless your instructor directs otherwise. Solid copper and used filter paper can be placed in the trash.

Optional **F. Confirming the product is copper(I) chloride**

There are many ways to prove that the formula of the copper compound you have prepared is indeed CuCl. The following procedure usually gives good results.

1. Preparing a product sample solution in which all the copper has been oxidized to Cu^{2+}

Dry the copper compound by putting it under a heat lamp or in a drying oven at 110°C for 10. minutes. Using an analytical balance, accurately determine the mass of a 0.1 ± 0.04-g sample of the product compound transferred into a tared 50-mL beaker. Dissolve the sample in 5 mL of 6 M nitric acid, HNO_3, which will oxidize all the copper present to the copper(II) ion, Cu^{2+}. Add about ten milliliters of deionized water, stir, and transfer *all* of the resulting solution to a clean (wet with deionized water is fine) 100.-mL volumetric flask. (See Appendix D for more details.) Use deionized water from your wash bottle to rinse the remaining solution from the beaker into the flask, then fill the flask to the mark with deionized water. Stopper the flask and invert it at least thirteen times, allowing the air bubble inside to travel all the way from one end of the flask to the other each time, to ensure that the solution in the flask is thoroughly mixed.

2. Converting the copper(II) ions from the product sample to the ammonia complex $Cu(NH_3)_4^{2+}$

Use a clean 10-mL pipet to transfer a 10.0-mL sample of the solution from the volumetric flask to a clean and *dry* 50-mL beaker. Add 10.0 mL of 6 M ammonia, $NH_3(aq)$, to the beaker, using the 10-mL pipet, after you have rinsed it with 6 M NH_3. The ammonia will bind to any Cu^{2+} in the solution to form the deep blue-violet complex ion $Cu(NH_3)_4^{2+}$.

3. Spectroscopic analysis

Method 1 After mixing well, measure the absorbance of the $Cu(NH_3)_4^{2+}$ solution at a wavelength of 575 nm, using deionized water as your blank. (See Appendix D for more details.) Determine the concentration of the copper(II) ammonia complex ion, $[Cu(NH_3)_4^{2+}]$, from a graph or equation provided to you, or by comparison with a standard solution, prepared as described in the next subsection. (Measure the absorbance of the standard at 575 nm, then calculate $[Cu(NH_3)_4^{2+}]$ in the solution of the compound by assuming that Beer's law holds.)

Method 2 Place identical test tubes next to one another, one containing the product sample solution and the other the standard solution prepared as described in the next subsection. Look down both test tubes toward a well-illuminated piece of white paper on the laboratory bench. Transfer standard solution to or from its test tube until the color intensity you see when looking down the tube containing the standard matches the intensity you see when looking down the tube containing the product sample solution. Record the depth of the liquid in each tube on the report page.

4. Preparing a $Cu(NH_3)_4^{2+}$ standard solution (if needed)

You can prepare a standard by accurately weighing 0.06 ± 0.02 g of copper turnings and putting them into a 50-mL beaker. Dissolve the copper in 5 mL of 6 M HNO_3, and proceed as you did with the solution of the product compound in parts F.1 and F.2.

5. Calculating the mole ratio

Knowing the concentration of Cu^{2+} in the 100. mL sample solution, $[Cu^{2+}]_{product\ sample\ solution}$, calculate the moles of copper in the product sample you used to make that solution. Then calculate the number of grams of copper in the product sample and subtract that from the mass of the entire product sample to determine the grams of chlorine in it, by difference. Use that value to find the moles of chlorine in the product sample, and then the mole ratio Cu:Cl in the product compound you made.

Experiment 33

Data and Calculations: Making Copper(I) Chloride

Mass of copper metal sample in 150-mL beaker _____ g

Mass of empty watch glass and filter paper _____ g

Mass of watch glass, filter paper, and product compound _____ g

Mass of product actually prepared = experimental yield _____ g

Theoretical yield of CuCl (show your work in the space below) _____ g

$$\text{Percent yield} = \frac{\text{experimental yield}}{\text{theoretical yield}} \text{ (show your work in the space below)} \quad \underline{\hspace{2cm}} \%$$

Optional **F. Confirming the product is copper(I) chloride**

Mass of copper chloride product sample used _____ g

Absorbance of $Cu(NH_3)_4^{2+}$ solution (Method 1) or depth of product sample
solution (mixed with ammonia) in tube (Method 2) _____

Mass of copper turnings used in making standard _____ g

Absorbance of standard solution in ammonia (Method 1) or depth of standard
solution (mixed with ammonia) in tube (Method 2) _____

Moles of copper in turnings used in making standard

_____ moles

Copper concentration in standard solution, $[Cu^{2+}]_{standard}$
(All of the copper in the turnings is converted to Cu^{2+} in a standard solution whose total volume is 100. mL.)

_____ M

(continued on following page)

Assume that Beer's law holds, so that

$$[Cu^{2+}]_{\text{product sample solution}} = [Cu^{2+}]_{\text{standard}} \times ratio \qquad (7)$$

where $ratio = \dfrac{\text{absorbance of sample}}{\text{absorbance of standard}}$ if using Method 1, or

$ratio = \dfrac{\text{depth of standard solution in tube}}{\text{depth of sample solution in tube}}$ if using Method 2.

Copper concentration in solution of product sample, $[Cu^{2+}]_{\text{product sample solution}}$

_____ M

Moles of copper in product sample (All of the copper in the product sample was converted to Cu^{2+} in a solution with a total volume of 100. mL)

_____ moles

Grams of copper in product sample

_____ g

Grams of chlorine in product sample (sample mass = mass Cu + mass Cl)

_____ g

Moles of chlorine (as Cl, not Cl_2) in product sample

_____ moles

Mole ratio in sample (and in product), Cu:Cl   _____

Cu:Cl mole ratio in product (rounded to nearest counting numbers)   _____

Experiment 33

Advance Study Assignment: Making Copper(I) Chloride

1. Suppose the Cu^{2+} ions in Part A of this experiment are produced by the reaction of 1.16 g of copper turnings with excess nitric acid. How many moles of Cu^{2+} are produced?

_____ mol

2. Why is hydrochloric acid not used in a direct reaction with copper to prepare the $CuCl_2$ solution?

3. How many grams of metallic copper are required to react with the moles of Cu^{2+} calculated in Problem 1 to form the CuCl? The overall reaction can be taken to be:

$$Cu^{2+}(aq) + 2\,Cl^-(aq) + Cu(s) \rightarrow 2\,CuCl(s)$$

_____ g

4. What is the theoretical maximum mass of CuCl that can be prepared from the reaction sequence of this experiment, if 1.16 g of Cu turnings are used to prepare the Cu^{2+} solution at the very start of the procedure?

_____ g

(continued on following page)

5. Optional   A sample of the copper chloride product compound prepared in this experiment, weighing 0.0989 g, is dissolved in HNO_3 and diluted to a volume of 100. mL. A 10.0-mL sample of that solution is mixed with 10.0 mL of 6 M NH_3 and analyzed. The concentration of Cu^{2+}, $[Cu^{2+}]$, in the product sample solution, *before* it was mixed with ammonia, is found to be 0.0100_6 M.

   a. How many moles of copper were in the original product sample, which had been dissolved into a total volume of 100. mL of product sample solution?

   _____ moles

   b. How many grams of copper were in the product sample?

   _____ g

   c. The dry product sample contained *only* copper and chlorine atoms. How many grams of chlorine were in the product sample? How many moles?

   _____ g _____ moles

   d. What is the mole ratio, Cu:Cl, in the product compound?

Experiment 34

Developing a Scheme for Qualitative Analysis

In previous experiments you have determined how much of something was present in a sample. You may have determined the concentration of chloride in an unknown salt mixture, the molarity of a sodium hydroxide solution, or the amount of calcium ion in a sample of hard water. All of these experiments fall into the part of chemistry called quantitative analysis.

Sometimes chemists are more interested in whether or not something is present in a sample than in how much there is. This is called qualitative analysis. For example, in forensic (crime investigation) chemistry work it might be important to determine which metal ions are present in a solution. This experiment is part of a series that deals with the qualitative analysis of solutions containing various anions (negative ions) and metallic cations (positive ions).

This experiment uses precipitation reactions to identify and remove the cations from a mixture in a carefully chosen order. If a precipitate can contain only one cation under a specific set of conditions, the formation of that precipitate proves the presence of that cation. If the precipitate may contain several cations, it can be dissolved and further identified in a series of steps that may include acid–base, complex ion formation, redox, or other precipitation reactions. The ultimate result is separation of the sample into portions that can contain only one cation, whose identity and presence are established by the formation of a characteristic precipitate or colored complex ion.

In this experiment you will be asked to develop a scheme for the qualitative analysis of four cations, using such a systematic approach. The differing behavior of the cations toward a set of common test reagents (chemicals) provides the basis for their separation and identification.

The cations you will study in this experiment are Ba^{2+}, Mg^{2+}, Cd^{2+}, and Al^{3+}. The test reagents you will use are 1.0 M Na_2SO_4, 1.0 M Na_2CO_3, 6 M NaOH, and 6 M NH_3. These reagents furnish anions or molecules that will precipitate or form complex ions with the cations, so you may observe the formation of insoluble sulfates, carbonates, or hydroxides (hydroxides with either NaOH or NH_3). In addition, you may form complexes when larger amounts of the same test reagent are added. These complexes may be quite stable: stable enough to prevent the precipitation of an otherwise insoluble salt upon adding a particular anion. The complexes, however, are all unstable in excess acid and can be broken down by adding 6 M HNO_3, releasing the cation for further reactions.

In the first part of this experiment you will observe the behavior of the four cations in the presence of one or more of the test reagents. On the basis of your observations, you will develop a scheme for identifying the cations present in a mixture. After testing your scheme with a known containing all of the cations, you will be given an unknown to analyze.

Experimental Procedure

Wear your safety glasses while performing this experiment.

To four small (13 × 100 mm) test tubes add about one milliliter (in these test tubes, a depth of about one centimeter) of 0.10 M solutions of the nitrate or chloride of Ba^{2+}, Mg^{2+}, Cd^{2+}, and Al^{3+}, one solution to a test tube. Add 1 drop of 1.0 M sodium sulfate, Na_2SO_4, to each test tube. If a precipitate forms in a test tube, write the formula of the precipitate formed by that cation–SO_4^{2-} pair on the top line in the corresponding box on the report page, indicating that the sulfate of that cation is insoluble in water. If no precipitate forms, put a dash (—) on that top line. Then add about one additional milliliter of 1.0 M Na_2SO_4 to each test tube and stir with your stirring rod. Rinse or store the rod in a 400-mL beaker of deionized water between *each* use. If a precipitate dissolves after adding additional Na_2SO_4, it indicates that the cation forms a complex with sulfate ion. If that happens, you should write "complex" on the second line in the corresponding box; otherwise put a dash there.

To any tubes containing precipitates or complex ions, add 6 M nitric acid, HNO_3, one drop at a time (though not necessarily slowly), observing carefully as you do so, until the solution is acidic to litmus (blue litmus turns red). If a precipitate dissolves, note that on the third line with an "A", meaning that the precipitate dissolves in acid. In the case of complexes, the precipitate that originally formed may reprecipitate if the ligands react with acid, and then dissolve again when the solution becomes acidic. If you observe reprecipitation, indicate that on the third line with a "P", followed by an arrow leading to an "A" ("P $\rightarrow$ A") if the precipitate dissolves when the solution becomes acidic. If the initial precipitate or complex remains unchanged upon acidification, put a dash on the third line.

Pour the contents of the four test tubes into a beaker and rinse the tubes with deionized water. Conduct similar tests, first with 1.0 M sodium carbonate, Na_2CO_3, then with 6 M sodium hydroxide, NaOH, and finally with 6 M aqueous ammonia, NH_3(aq). Although in most cases you will not observe formation of complex ions, you will see a few, and it is important not to miss them. One drop of reagent will typically produce a precipitate if the neutral cation–anion salt is insoluble in water. If a complex forms, it will do so when you add excess reagent, and the solution will become clear as you do so (possibly after you stir). A precipitate may or may not form as you acidify a complex with HNO_3, but to avoid missing it be sure to add the acid one drop at a time, without stirring, and watch the test tube carefully as you do so. The appearance of a persistent cloudiness in the liquid is evidence of reprecipitation.

When your Table of Solubility Properties is complete you will know whether the sulfate, carbonate, and hydroxide of each of the cations is insoluble in water, and whether it dissolves in acid. You should also know whether the cation forms a complex with SO_4^{2-}, CO_3^{2-}, OH^-, and/or NH_3, and whether the complex can be turned back into a solid by carefully adding acid.

Now, given the solubility and reaction data you have obtained, your challenge is to devise a step-by-step procedure for deciding whether or not each metal cation is present in an unknown mixture. There will be more than one way to do this, but you will need to think through whatever procedure you come up with to make sure it gives clear answers. As you construct your scheme, here are a few things to keep in mind besides the solubility and reaction data:

1. To separate a precipitate from a solution, you can use a centrifuge. Pour off (technical term: <u>decant</u>) the solution into a test tube for use in further steps, leaving the solid behind. Consult the "Separating Solids from Liquids" section of Appendix D for more information.

2. If a precipitate can only be explained by the presence of one cation, observing that precipitate proves that this metal cation is present. If it may involve more than one cation, it must be further analyzed. In that case, the precipitate must be washed free of any cations that did not precipitate in that step, because those cations could interfere with later steps. To clean a precipitate, wash it twice with a 1:1 mixture of water and the precipitating reagent, stirring before centrifuging and pouring off the wash liquid each time.

3. pH is important. Your original tests were made with the cations in a neutral solution. If you want to obtain a precipitate, the solution must have a pH in which that precipitate can form. 6 M HNO_3 or 6 M NaOH can be used to bring a mixture to a pH of about seven. (Washing a precipitate with a neutral solution and pouring off the wash is also sufficient.) You can test for neutrality using litmus paper.

When you have your separation scheme in mind, describe it by constructing a flow diagram. The design of a flow diagram is discussed in the Advance Study Assignment. In a flow diagram, the formulas of all precipitates involving the cations should be indicated. The reagents used for a specific step are shown next to the arrow connecting reactants to products.

When you have completed your flow diagram, test your scheme with a known solution containing all four cations. If your scheme works, show your flow diagram to your instructor, who will then give you an unknown to analyze.

Disposal of reaction products: On completing this experiment, pour the contents of the beaker used for reaction products into a waste container, unless your instructor directs otherwise.

Experiment 34

Observations and Analysis: Developing a Scheme
for Qualitative Analysis

Table of Solubility Properties

	Ba^{2+} barium(II)	Mg^{2+} magnesium(II)	Cd^{2+} cadmium(II)	Al^{3+} aluminum(III)
1.0 M Na_2SO_4 (sulfuric acid)				
1.0 M Na_2CO_3 (sodium carbonate)				
6 M NaOH (sodium hydroxide)				
6 M NH_3 (aqueous ammonia)				

Top line: Formula of insoluble salt,
　　　or a dash (—) if one drop of test reagent does not form a precipitate

Middle line: "complex" if an insoluble salt dissolves into a complex upon adding more of the same test reagent,
　　　or a dash (—) if the solid remains

Bottom line: "A" if the end result upon acidification with 6 M nitric acid, HNO_3, is a clear solution;
　　　"P → A" if a precipitate is seen reforming from a complex as the acidification is being carried out;
　　　or a dash (—) if adding acid has no effect

Flow diagram for separation scheme:

Indicate on this flow diagram the actual observations made on the known and the unknown. For the known, only note any deviations from the expected results indicated in the flow diagram, labeling them with "KNOWN". For the unknown, clearly indicate the path through the flow diagram that you followed to reach your conclusions.

Unknown # _____ Ion(s) present in unknown: _____ _____ _____ _____

Experiment 34

Advance Study Assignment: Developing a Scheme
for Qualitative Analysis

Qualitative analysis schemes can be summarized by flow diagrams. A flow diagram for a scheme that might be used to analyze a mixture that could contain Cu^{2+}, Pb^{2+}, and/or Sn^{2+} is shown here:

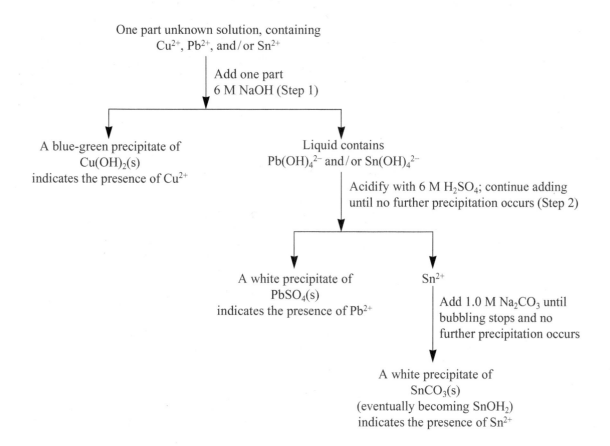

A flow diagram summarizes the steps in a scheme and the observations expected if a given cation is present. It is important that it include amounts, though they may be relative (one part this, two parts that, or add this reagent until the solution becomes basic). In the first step of this scheme, 6 M NaOH is added <u>in excess</u> (meaning that more moles of NaOH are added than there are moles of things for it to react with—note that while the volumes used are the same, the concentration of NaOH is much higher than the concentration of any of the metal ions). $Cu(OH)_2$ precipitates (if Cu^{2+} is present in the sample), and Pb^{2+} and Sn^{2+} remain in solution because of the formation of the complex ions $Pb(OH)_4{}^{2-}$ and $Sn(OH)_4{}^{2-}$. Any $Cu(OH)_2$ precipitate formed is removed by centrifuging the tube and then pouring off the clear liquid, which still contains any complex ions present. In Step 2, the poured-off liquid is made acidic with H_2SO_4, and $PbSO_4$ precipitates after the $Pb(OH)_4{}^{2-}$ complex is destroyed by H^+ ions from the acid (if Pb is present in the sample). Any $PbSO_4$ that precipitates is removed by centrifuging, and the clear liquid is poured off into a new test tube. In Step 3, adding Na_2CO_3 to the liquid neutralizes the acid and then precipitates $SnCO_3$ (if Sn is present in the sample). (*Note*: As indicated in the flow diagram, $SnCO_3$ eventually becomes $SnOH_2$, which is also a white solid.)

(*continued on following page*)

1. In the space below, construct a flow diagram for the following separation scheme, applicable to a solution containing Ag^+, Ni^{2+}, and/or Zn^{2+}:

Step 1. Add 6 M hydrochloric acid, HCl(aq), to precipitate any Ag^+ as white AgCl(s). Ni^{2+} and Zn^{2+} remain in solution, unchanged. Centrifuge out the AgCl and carry out further analysis on the liquid.

Step 2. Add one part 6 M sodium hydroxide solution, NaOH(aq), (an excess), precipitating green $Ni(OH)_2(s)$ if Ni is present and converting any Zn^{2+} to the $Zn(OH)_4^{2-}$ complex ion. Centrifuge out the $Ni(OH)_2$ and carry out further analysis on the liquid.

Step 3. Acidify the solution with 6 M HCl, then add 1.0 M Na_2CO_3 until bubbling stops, neutralizing the acidic solution and precipitating white $ZnCO_3(s)$ if Zn is present.

Experiment 35

Spot Tests for Some Common Anions

There are two broad categories of problems in <u>analytical chemistry</u> (or just chemical <u>analysis</u>), the chemistry subfield focused on determining what is in a sample. <u>Quantitative analysis</u> involves determining how much of something is present in a sample. The other area of analysis, called <u>qualitative analysis</u>, has a more limited purpose: establishing whether given substances are or are not present in detectable amounts in a sample. This is a qualitative analysis experiment.

Qualitative analysis can be carried out in various ways. The simplest approach, and the one you will use in this experiment, is to test for the presence of each possible substance of interest by adding a chemical that will cause that substance, if it is present in the sample, to react in a characteristic way. This method involves a series of "spot" tests, one for each substance, carried out on separate samples of the unknown. The difficulty with this way of doing qualitative analysis is that frequently, particularly in complex mixtures, the presence of one substance can interfere with the analytical test for another. Although interferences are common, there are many ions that can, under optimum conditions at least, be identified in mixtures by chemical spot tests.

In this experiment you will use spot tests for the analysis of a solution that may contain the following commonly encountered <u>anions</u> (negative ions):

CO_3^{2-}	PO_4^{3-}	Cl^-	SCN^-
(carbonate)	(phosphate)	(chloride)	(thiocyanate)
SO_4^{2-}	SO_3^{2-}	CH_3COO^-	NO_3^-
(sulfate)	(sulfite)	(acetate)	(nitrate)

The procedures in this experiment involve acid–base, precipitation, complex ion formation, and redox (oxidation–reduction) reactions. At each reaction step in the procedure, you should try to recognize the kind of reaction that occurs so that you can write the net ionic equation that describes it.

Experimental Procedure

Wear your safety glasses while performing this experiment.

Carry out the spot test for each of the anions, as indicated in the following instructions. Then repeat each test using a more dilute sample solution made by combining 2 drops of the anion stock solution with 18 drops of deionized water, which will provide a total volume of about one milliliter. (If using Pasteur pipets, which provide smaller drops, use 3 drops of stock solution in 27 drops of water.) Make sure to mix these well before continuing with the analysis. In some of the tests, a boiling-water bath containing about one hundred milliliters of water in a 150-mL beaker will be needed, so set that up now, before you start. When performing a test, if no reaction is immediately visible, stir the mixture with a stirring rod. Rinse or store your stirring rod(s) in a 400-mL beaker of deionized water between *each* use. The tests in this experiment can detect the anions at concentrations of 0.02 M or greater, but in dilute solutions approaching this limit careful observation may be required. These instructions assume the use of small (13 × 100 mL) test tubes, in which a depth of about one centimeter corresponds to a volume of about one milliliter. Using this depth-to-volume relationship is sufficient for all of the volume measurements required in this experiment. These instructions also assume the use of normal dropper bulbs; if Pasteur pipets are used for drop measurements, increase specified drop counts by 50%. (For example, you would use 3 drops instead of 2 and 9 instead of 6; where one drop is specified, you should still use a single drop.)

A. Test for the presence of carbonate ion, CO_3^{2-}

Gradually add about one milliliter of 6 M hydrochloric acid, HCl(aq), to about one milliliter of 1.0 M sodium carbonate, Na_2CO_3, in a small (13×100 mm) test tube. With concentrated solutions, bubbles of carbon dioxide gas are immediately formed (and will cause the solution to bubble out of the test tube if the acid is added too quickly). With dilute solutions, the bubbling will be less obvious. Warming in the boiling-water bath, with stirring, will increase the rate of bubble formation. Carbon dioxide is colorless and has no smell.

B. Test for the presence of sulfate ion, SO_4^{2-}

Add about one milliliter of 6 M hydrochloric acid, HCl(aq), to about one milliliter of 0.5 M sodium sulfate, Na_2SO_4. Add a few drops of 1.0 M barium chloride, $BaCl_2$. A white, finely divided precipitate of the insoluble salt barium sulfate, $BaSO_4$, indicates the presence of the sulfate ion, SO_4^{2-}. [*Note*: $BaCO_3$, $Ba_3(PO_4)_2$, and $BaSO_3$ are all reasonably insoluble in neutral solutions, but dissolve completely in acidic solutions such as this one. (See Appendix B1.) Thus the CO_3^{2-}, and PO_4^{3-} ions do not interfere with this test. SO_3^{2-} does not interfere with this test directly, but because sulfite ion is oxidized to sulfate ion by the air (see Spot Test D), its presence can provide a false positive for sulfate. Use the odor or oxidation after precipitation tests described in Spot Test D to rule out sulfite as the cause of a white precipitate observed with this spot test.]

C. Test for the presence of phosphate ion, PO_4^{3-}

Add about one milliliter of 6 M nitric acid, HNO_3, to about one milliliter of 0.5 M sodium hydrogenphosphate, Na_2HPO_4. Then add about one milliliter of 0.5 M ammonium molybdate, $(NH_4)_2MoO_4$, and stir thoroughly. A yellow precipitate of ammonium phosphomolybdate, $(NH_4)_3PO_4 \cdot 12\,MoO_3$, indicates the presence of the phosphate ion, PO_4^{3-}. The precipitate may form slowly, particularly in more dilute solutions; if it does not appear right away, put the test tube in a boiling-water bath for a few minutes.

D. Test for the presence of sulfite ion, SO_3^{2-}

Sulfite ion, SO_3^{2-}, in acidic solutions tends to form sulfur dioxide gas, SO_2, which can be detected by its smell—even at low concentrations. Sulfite ion is oxidized to sulfate, SO_4^{2-}, if exposed to oxygen or other oxidizing agents, so sulfite-containing solutions exposed to the air will slowly accumulate—and usually test positive for—sulfate ion.

To about one milliliter of 0.5 M sodium sulfite, Na_2SO_3, add about one milliliter of 6 M hydrochloric acid, HCl(aq), and mix with your stirring rod. Cautiously smell the rod to try to detect the "burning matches" smell of SO_2, which is a good test for sulfite. If the smell is too faint to detect, put the test tube in the boiling-water bath for 10. seconds and smell the rod again.

To do a visual test for sulfite, add about one milliliter of 1.0 M barium chloride, $BaCl_2$, to the acidified solution in the tube from the odor test. Stir, and centrifuge out any precipitate of white barium sulfate, $BaSO_4$. Decant (pour off—see Appendix D) the clear solution into a test tube and add about one milliliter of $3\%_{mass}$ hydrogen peroxide, H_2O_2, an oxidizing (electron removing) agent. Stir the solution and let it stand for a few seconds. If sulfite is present, its oxidation to sulfate by the hydrogen peroxide will cause more white barium sulfate, $BaSO_4$, to precipitate out of solution.

If thiocyanate ion, SCN^-, is present, it may interfere with the visual test because it too can be oxidized by hydrogen peroxide to form sulfate ion. If thiosulfate may be present, add about one milliliter of 1.0 M barium chloride, $BaCl_2$, to about one milliliter of the (neutral, *not* acidified) sample. Centrifuge out any precipitate, which will contain some of any sulfite that is present as white barium sulfite, $BaSO_3$, but will not contain any thiocyanate. ($BaSO_3$ is very soluble in acid, but much less so in neutral solutions; see Appendix B1.) Wash the solid with about two milliliters of water, stir, centrifuge, and discard the wash. To the precipitate add about one milliliter of 6 M hydrochloric acid, HCl(aq), and about two milliliters of water, and stir. Centrifuge out any white barium sulfate, $BaSO_4$, that remains undissolved and decant the clear liquid into a new test tube. To the liquid add about one milliliter of $3\%_{mass}$ hydrogen peroxide, H_2O_2. If sulfite is present, you will observe a small amount of new white $BaSO_4$ precipitating out of the solution within a few seconds.

E. Test for the presence of thiocyanate ion, SCN⁻

Add about one milliliter of 6 M acetic acid, CH_3COOH, to about one milliliter of 0.5 M potassium thiocyanate, KSCN, and stir. Add one or two drops of 0.10 M iron(III) nitrate, $Fe(NO_3)_3$. A red-orange to deep red color resulting from the formation of the complex ion $FeSCN^{2+}$ indicates the thiocyanate ion, SCN^-, is present.

F. Test for the presence of chloride ion, Cl⁻

Add about one milliliter of 6 M nitric acid, HNO_3, to about one milliliter of 0.5 M sodium chloride, NaCl. Add two or three drops of 0.10 M silver nitrate, $AgNO_3$. A thick white precipitate of AgCl will quickly form if chloride ion is present.

Several other anions interfere with this test, because under these conditions they too form white precipitates with the silver(I) ion, Ag^+, present in $AgNO_3$. In this experiment only the thiocyanate ion, SCN^-, will interfere. [*Note*: Ag_3PO_4, Ag_2SO_3, and Ag_2CO_3 and are all reasonably insoluble in neutral solutions, but dissolve completely under acidic conditions such as those established for this spot test (see Appendix B1). Thus the PO_4^{3-}, SO_3^{2-}, and CO_3^{2-} ions do not interfere with this test. Ag_2SO_4 is only slightly soluble, and a small amount of it may slowly precipitate out as a white solid if sulfate ion is present—but it will not appear immediately the way AgCl and AgSCN do.] If the sample contains SCN^- ion, put about one milliliter of the solution into a 30- or 50-mL beaker and add about one milliliter of 6 M nitric acid, HNO_3. Boil the solution gently until its volume is decreased by half. This will destroy most of the thiocyanate. Transfer the solution to a small test tube and add about one milliliter of 6 M nitric acid, HNO_3, and a few drops of 0.10 M silver nitrate, $AgNO_3$. If Cl^- is present you will get a thick white precipitate. If Cl^- is absent, you may see a slight cloudiness due to remaining trace amounts of SCN^-.

G. Test for the presence of acetate ion, CH₃COO⁻

Generally you begin this test by ensuring the solution is acidic as indicated by litmus—adding 6 M nitric acid, HNO_3, if needed, to reach that point. To demonstrate the test, use about one milliliter of 0.5 M sodium acetate, $NaCH_3COO$. This is guaranteed to start out basic (because acetate ion, CH_3COO^-, is the conjugate base of the weak acid acetic acid, CH_3COOH), so add 6 M nitric acid, HNO_3, until the solution turns blue litmus red, indicating that it has been made acidic. Then add 6 M aqueous ammonia, $NH_3(aq)$, until the solution is just basic to litmus, thereby assuring a nearly neutral solution. Add $\underline{1}$ drop of 1.0 M barium chloride, $BaCl_2$. If a precipitate forms, add about one milliliter of the $BaCl_2$ solution to precipitate anions that would interfere with the next step. Stir, centrifuge, and decant the clear liquid into a test tube. Add $\underline{1}$ drop of $BaCl_2$ to make sure that precipitation is complete. To about one milliliter of the liquid add 0.10 M potassium triiodide, KI_3, drop by drop, until the solution takes on a fairly strong rust-orange color. Add about half a milliliter (ten drops) of 0.10 M lanthanum nitrate, $La(NO_3)_3$, and 6 drops of 6 M NH_3. Stir, and put the test tube in the boiling-water bath. If acetate ion, CH_3COO^-, is present, the mixture will darken to nearly black in a few minutes. The color is due to elemental iodine, I_2, sticking to (technical term: <u>adsorbed on</u>) a precipitate of (very difficult to otherwise see) lanthanum acetate.

H. Test for the presence of nitrate ion, NO₃⁻

To about one milliliter of 0.5 M sodium nitrate, $NaNO_3$, add about one milliliter of 6 M sodium hydroxide, NaOH. Then add a few granules of aluminum, Al, metal, using your spatula, and put the test tube in the boiling-water bath. In a few seconds, the reaction between aluminum metal and sodium hydroxide will produce H_2 gas, which will reduce the nitrate ion, NO_3^-, to ammonia, NH_3, which will come off as a gas. To detect the NH_3, hold a piece of moistened red litmus paper just above the end of the test tube. If the sample contains nitrate ion, the litmus paper will gradually (over a period of about two minutes) turn blue because ammonia gas acts as a base. Do not confuse blue spots on the litmus caused by spattering of the liquid with a change in the overall color of the litmus to blue, which is what indicates the presence of NH_3 gas and thus of nitrate ion in the original sample. Cautiously, and at first indirectly, smell the top of the tube; you may be able to detect the distinct and sharp smell of ammonia. **CAUTION:** **Inhaling too much ammonia at once can knock you unconscious!**

If thiocyanate ion, SCN^-, is present, it will interfere with this test. In that case, first add about one milliliter of 1.0 M copper sulfate, $CuSO_4$, to about one milliliter of the sample and put the test tube in the boiling-water bath for a minute or two. Centrifuge out the copper thiocyanate, $Cu(SCN)_2$, precipitate, and decant the clear liquid into a test tube. Add about one milliliter of 1.0 M sodium carbonate, Na_2CO_3, to remove excess copper(II) ion, Cu^{2+}, as blue-green copper(II) carbonate, $CuCO_3$. Centrifuge out the precipitate, and decant the clear liquid into a test tube. To about one milliliter of the liquid add about one milliliter of 6 M sodium hydroxide, NaOH, and proceed, starting with the second sentence of this spot test's procedure.

I. Analysis of an unknown

Once you have familiarized yourself with all of the spot tests, and have *run them again with diluted stock solution samples*, obtain an unknown from your instructor and analyze it by applying the spot tests to separate portions of about one milliliter of it. The unknown will contain three or four of the anions for which spot tests are introduced in this experiment, so your test for a given ion may be affected by the presence of others. When you think you have properly analyzed your unknown, you may, if you wish, make a "known" with the composition you found and test it to see if it behaves as your unknown did. Be sure to record the ID number of your unknown on the report page.

> **Disposal of reaction products:** As you complete each test, pour the waste materials into a beaker. When you have finished the experiment, pour the contents of the beaker into a waste container, unless your instructor directs otherwise.

Name _____ Section _____

Experiment 35

Observations and Analysis: Spot Tests for Some Common Anions

Table of Spot Test Results

Anion	Result with stock solution	Result with diluted stock solution	Result with unknown
CO_3^{2-} (carbonate)			
SO_4^{2-} (sulfate)			
PO_4^{3-} (phosphate)			
SO_3^{2-} (sulfite)			

(continued on following page)

Table of Spot Test Results (*continued*)

Anion	Result with stock solution	Result with diluted stock solution	Result with unknown
SCN⁻ (thiocyanate)			
Cl⁻ (chloride)			
CH₃COO⁻ (acetate)			
NO₃⁻ (nitrate)			

Unknown # _____ contains _____

Experiment 35

Advance Study Assignment: Spot Tests for Some Common Anions

1. Each of the observations in the following list was made on a *different* solution containing a single anion from this experiment. Given each observation, state which anion is present. If the test is not definitive, indicate that with a question mark and explain why. (*Hint*: There will be *two* of those!)

   a. Adding 6 M NaOH and Al metal to the solution produces a gas that turns red litmus blue.
      Ion present:

   b. Adding 6 M HCl produces a gas with the smell of burning matches.
      Ion present:

   c. Adding 6 M HCl produces bubbles.
      Ion present:

   d. Adding 6 M HNO_3 plus 0.10 M $AgNO_3$ produces a white precipitate.
      Ion present:

   e. Adding the oxidizing acid 6 M HNO_3 plus 1.0 M $BaCl_2$ produces a white precipitate.
      Ion present:

   f. Adding 6 M HNO_3 plus 0.5 M $(NH_4)_2MoO_4$ produces a yellow precipitate.
      Ion present:

2. A *single* unknown solution containing one or more of the anions studied in this experiment has the following properties:

   a. No effect or odor is observed upon adding 6 M HCl to an equal volume of the unknown.

   b. A white precipitate is observed upon adding 1.0 M $BaCl_2$ to the solution from 2(a). The liquid above this precipitate does *not* form more precipitate when $3\%_{mass}$ H_2O_2 is added to it.

   c. No immediate effect is observed upon dripping 0.10 M $AgNO_3$ into equal volumes of 6 M HNO_3 and the unknown.

   d. No yellow precipitate is observed upon adding 0.5 M $(NH_4)_2MoO_4$ to equal volumes of 6 M HNO_3 and the unknown.

   On the basis of this information, which ions are present, which are absent, and which remain undetermined?

Present	Absent	Undetermined

3. The chemical reactions used in the anion spot tests in this experiment are for the most part precipitation or acid–base reactions. Given the information in each test procedure, try to write the net ionic equation for the key reaction(s) in each test.

A. CO_3^{2-}

B. SO_4^{2-}

C. PO_4^{3-} (Reactants are HPO_4^{2-}, NH_4^+, MoO_4^{2-}, and H^+; products are $(NH_4)_3PO_4 \cdot 12\,MoO_3$ and H_2O; no oxidation or reduction occurs.)

D. SO_3^{2-} (Can be written as two or three reactions, one of them a redox reaction involving H_2O_2.)

E. SCN^-

F. Cl^-

H. NO_3^- [Take $Al(s)$ and $NO_3^-(aq)$ to be the reactants, and $NH_3(g)$ and $Al(OH)_4^-(aq)$ to be the products; the final reaction also contains OH^- and H_2O as reactants. Note that this is a redox reaction, and is most easily balanced as such: the electrons released in the oxidation must be balanced by the electrons taken in by the reduction. The half-reactions involved are

$$NO_3^-(aq) + 6\,H_2O(\ell) + 8\,e^- \rightarrow NH_3(g) + 9\,OH^-(aq)$$
and
$$Al(s) + 4\,OH^-(aq) \rightarrow Al(OH)_4^-(aq) + 3\,e^-.]$$

Experiment 36

Qualitative Analysis of Group I Cations

T he aim of (inorganic) <u>qualitative chemical analysis</u> is to determine which ions are present in an unknown (rather than in what amount).

There are a small number of <u>cations</u> (positive ions) that form insoluble <u>chlorides</u> (Cl^- salts): these are called the <u>Group I cations</u>, and the three most commonly encountered are lead(II), Pb^{2+}; mercury(I), which typically exists as Hg_2^{2+}; and silver(I), Ag^+. Their chlorides are all insoluble in cold water, so they can be removed as a group from a solution by adding hydrochloric acid, $HCl(aq)$. The (net ionic) reactions that occur are precipitation reactions:

$$Ag^+(aq) + Cl^-(aq) \rightarrow AgCl(s) \tag{1}$$

$$Pb^{2+}(aq) + 2\,Cl^-(aq) \rightarrow PbCl_2(s) \tag{2}$$

$$Hg_2^{2+}(aq) + 2\,Cl^-(aq) \rightarrow Hg_2Cl_2(s) \tag{3}$$

It is important to add enough HCl to ensure complete precipitation, but not too large an excess. In concentrated HCl solutions these chlorides tend to dissolve into Cl^- complexes, such as $AgCl_2^-$.

Lead chloride is separated from the other two solid chlorides by heating the solids in water. Any $PbCl_2$ that is present will dissolve in hot water by the reverse of Reaction 2:

$$PbCl_2(s) \rightarrow Pb^{2+}(aq) + 2\,Cl^-(aq) \tag{4}$$

Once the hot liquid has been separated from the solid, the presence of Pb^{2+} can be confirmed by adding a solution of potassium chromate, K_2CrO_4. The chromate ion, CrO_4^{2-}, gives a yellow precipitate with Pb^{2+}:

$$Pb^{2+}(aq) + CrO_4^{2-}(aq) \rightarrow PbCrO_4(s) \tag{5}$$
$$\text{colorless liquid} \quad \text{yellow liquid} \quad \text{yellow solid}$$

The other two insoluble chlorides, AgCl and Hg_2Cl_2, can be separated from one another by adding aqueous ammonia, $NH_3(aq)$. Any silver chloride that is present dissolves, forming the complex ion $Ag(NH_3)_2^+$:

$$AgCl(s) + 2\,NH_3(aq) \rightarrow Ag(NH_3)_2^+(aq) + Cl^-(aq) \tag{6}$$

Ammonia also reacts with Hg_2Cl_2, through a rather unusual redox (oxidation–reduction) reaction. The products include finely divided metallic mercury, $Hg(\ell)$, which is black, and a white solid called mercury(I) amidochloride, $HgNH_2Cl$:

$$Hg_2Cl_2(s) + 2\,NH_3(aq) \rightarrow Hg(\ell) + HgNH_2Cl(s) + NH_4^+(aq) + Cl^-(aq) \tag{7}$$
$$\text{white} \qquad\qquad \text{black} \qquad \text{white}$$

As this reaction occurs, the solid appears to change color: from white to black or gray.

The liquid potentially containing $Ag(NH_3)_2^+$ is isolated and further tested to determine whether silver is present in it. Adding a strong acid (nitric acid, HNO_3) to the solution destroys the complex ion and reprecipitates white silver chloride, $AgCl(s)$. This reaction can be thought of as occurring in two steps:

$$Ag(NH_3)_2^+(aq) + 2\,H^+(aq) \rightarrow Ag^+(aq) + 2\,NH_4^+(aq)$$
$$\underline{Ag^+(aq) + Cl^-(aq) \rightarrow AgCl(s)}$$
$$Ag(NH_3)_2^+(aq) + 2\,H^+(aq) + Cl^-(aq) \rightarrow AgCl(s) + 2\,NH_4^+(aq) \qquad \textbf{(8)}$$
$$\text{white}$$

Experimental Procedure

Wear your safety glasses while performing this experiment.

See Appendix D for some suggestions regarding procedures in qualitative analysis. The following instructions assume the use of small (13 × 100 mL) test tubes, in which a depth of about one centimeter corresponds to a volume of about one milliliter. Using this depth-to-volume relationship is sufficient for all of the volume measurements required in this experiment. These instructions also assume the use of eye droppers; if Pasteur pipets are used for drop measurements, increase the specified drop counts by 50%. (For example, you would use 3 drops instead of 2 and 7 or 8 instead of 5; where one drop is specified, you should still use a single drop.) Be sure to rinse or store your stirring rod(s) in a 400-mL beaker of deionized water between *each* use, to avoid cross-contamination. A boiling-water bath will be needed in this procedure, so you should set one up now (in a 250-mL beaker), before starting Step 1.

Step 1. Precipitating the Group I cations To gain familiarity with the procedures used in qualitative analysis, you will first analyze a known Group I solution made by mixing equal volumes of 0.10 M $AgNO_3$, 0.20 M $Pb(NO_3)_2$, and 0.10 M $Hg_2(NO_3)_2$. **CAUTION:** **Mercury—in any form—is very toxic. Avoid all contact with its compounds, and make sure you wash your hands at the conclusion of this experiment.**

Add 2 drops of 6 M hydrochloric acid, $HCl(aq)$, to about one milliliter of the known solution in a small (13 × 100 mm) test tube (1 mL ≈ 1 cm depth in the tube). Mix with your stirring rod. Centrifuge the mixture, making sure there is a blank test tube containing about the same amount of water in the opposite opening in the centrifuge. (See Appendix D for more information.) Add one more drop of the 6 M HCl to test for completeness of precipitation. If more precipitate forms, centrifuge again; repeat until no new precipitate forms. Decant (pour off) the clear liquid (technical term: supernatant) into another test tube and save it for further tests if cations from other groups may be present. The precipitate will be white and will contain the chlorides of all the Group I cations present in the sample.

Step 2. Isolating Pb²⁺ Wash the precipitate with one or two milliliters of deionized water. Stir, centrifuge, and decant off the wash liquid, which may be discarded.

Add about two milliliters of deionized water from your wash bottle to the precipitate in the test tube, then place the test tube in your boiling-water bath for a minute or two, stirring occasionally with a glass rod. This will dissolve most of the $PbCl_2$, but not the other two chlorides. Without delay, centrifuge the hot mixture and decant the liquid into a test tube while still hot. Set aside the remaining precipitate for use in Step 4 (right now, in Step 3, you will work with the liquid).

Step 3. Identifying Pb²⁺ Add 1 drop of 6 M acetic acid, CH_3COOH, and a few drops of 0.10 M potassium chromate, K_2CrO_4, to the *liquid* decanted off in Step 2. If Pb^{2+} is present, a bright yellow precipitate of lead chromate, $PbCrO_4$, will form.

Step 4. Isolating and identifying Hg$_2$$^{2+}$ To the *precipitate* from Step 2, add about one milliliter of 6 M aqueous ammonia, NH$_3$(aq), and stir thoroughly. Centrifuge the mixture and decant the clear liquid into a test tube. A gray or black precipitate left behind, produced by the reaction of Hg$_2$Cl$_2$ with ammonia, indicates the presence of Hg$_2$$^{2+}$.

Step 5. Identifying Ag$^+$ Add 6 M nitric acid, HNO$_3$, to the liquid from Step 4 until it is acidic to litmus paper, turning it red. (This will require about one milliliter of 6 M HNO$_3$.) Test for acidity by dipping the end of your stirring rod in the solution, then touching it to a piece of blue litmus paper (which will turn red in acidic solution). If Ag$^+$ is present, it will precipitate as white AgCl as the solution is acidified and any silver–ammonia complex in it is destroyed.

Step 6. Analyzing an unknown When you have finished analyzing the known solution, obtain an unknown and analyze it for the possible presence of Ag$^+$, Pb^{2+}, and/or Hg$_2$$^{2+}$. Be sure to record the ID number of your unknown on the report page.

Disposal of reaction products: All reaction products in this experiment should be dealt with as directed by your instructor. Mercury, Hg, in any form, must **never** go down a sink drain.

Flow Diagrams

It is possible to summarize the directions for analysis of the Group I cations in a <u>flow diagram</u>. In such a diagram, successive steps in the procedure are linked with arrows. Reactant cations or reactant substances containing the cations are at one end of each arrow and products formed are at the other end. Chemicals and conditions used to carry out each step are placed next to the arrows. Complete the flow diagram for the Group I ions below, by filling in the blanks.

Group I Flow Diagram

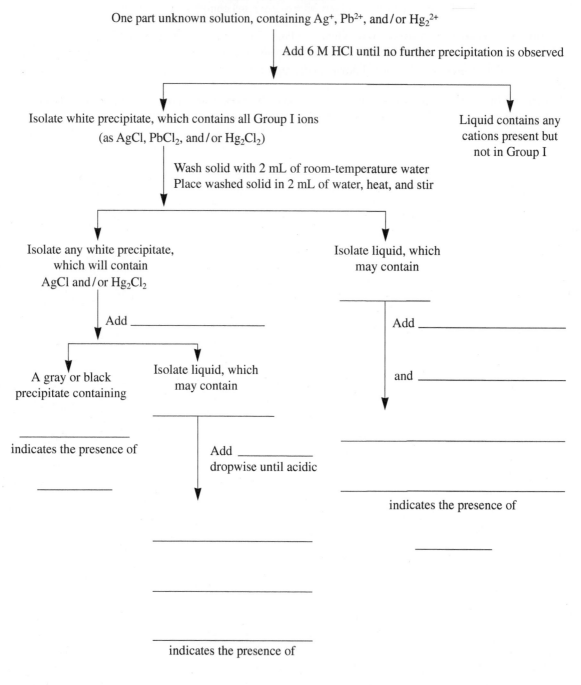

One part unknown solution, containing Ag^+, Pb^{2+}, and/or Hg_2^{2+}

Add 6 M HCl until no further precipitation is observed

Isolate white precipitate, which contains all Group I ions
(as $AgCl$, $PbCl_2$, and/or Hg_2Cl_2)

Liquid contains any cations present but not in Group I

Wash solid with 2 mL of room-temperature water
Place washed solid in 2 mL of water, heat, and stir

Isolate any white precipitate, which will contain $AgCl$ and/or Hg_2Cl_2

Isolate liquid, which may contain _____

Add _____

A gray or black precipitate containing

Isolate liquid, which may contain _____

Add _____ and _____

_____ indicates the presence of

_____

Add _____ dropwise until acidic

_____ indicates the presence of

_____

_____ indicates the presence of

_____

You will find it useful to construct flow diagrams for qualitative analysis schemes. You can use such diagrams in the laboratory as brief guides to procedure, and you can use them to record your observations on your known and unknown solutions.

Experiment 36

Observations and Analysis: Qualitative Analysis of Group I Cations
Group I Flow Diagram

Indicate on this flow diagram the actual observations made on the known and the unknown. For the known, only note any deviations from the expected results indicated in the flow diagram, labeling them with "KNOWN". For the unknown, clearly indicate the path through the flow diagram that you followed to reach your conclusions.

Unknown # _____   Ion(s) present in unknown: _____ _____ _____

Experiment 36

Advance Study Assignment: Qualitative Analysis of Group I Cations

1. On the report page for this experiment, draw a complete flow diagram for the separation and identification of the ions in Group I. (Copy and complete the partial flow diagram provided at the end of the instructions.)

2. Why would it be unwise to just add 10 drops of 6 M HCl immediately in Step 1? What could go wrong?

3. A solution may contain Ag^+, Pb^{2+}, and/or Hg_2^{2+}. A white precipitate forms upon adding 6 M HCl. The isolated precipitate does not turn gray or dissolve completely upon adding ammonia to it. Heating the resulting mixture causes the precipitate to dissolve completely. Which of the Group I ions are present, which are absent, and which remain undetermined? State your reasoning. *Note*: For "paper unknowns" such as this one, confirmatory tests are usually not included, and you do not need to provide any. The information provided here, without confirmatory tests, is sufficient to clearly indicate the presence or absence of some of the Group I ions studied in this experiment, while possibly leaving others undetermined.

Present _____

Absent _____

Undetermined _____

(continued on following page)

4. You are given an unknown solution that contains *only one* of the Group I cations, and no other cations. Develop the simplest procedure you can think of to determine which cation is present. Draw a flow diagram showing the procedure, and the observations expected at each step with each of the possible cations. The information in Appendices B1 and B2 may be helpful.

Experiment 37

Qualitative Analysis of Group II Cations

T he aim of (inorganic) <u>qualitative chemical analysis</u> is to determine which ions are present in an unknown (rather than in what amount).

The sulfides (S^{2-} salts) of the <u>Group II cations</u> are insoluble at a pH of 0.5. They include the <u>cations</u> (positive ions) bismuth(III), Bi^{3+}; tin(IV), Sn^{4+}; antimony(III), Sb^{3+}; and copper(II), Cu^{2+}. If a solution containing these ions is adjusted to this very acidic pH and then saturated with hydrogen sulfide, H_2S, then the sulfides Bi_2S_3, SnS_2, Sb_2S_3, and CuS will precipitate. The reaction with Bi^{3+} is typical:

$$2\ Bi^{3+}(aq) + 3\ H_2S(aq) \rightarrow Bi_2S_3(s) + 6\ H^+(aq) \tag{1}$$
$$\text{brown}$$

Saturation with H_2S could be achieved by bubbling the gas itself through the liquid. A more convenient method, however, is to heat the acid solution after adding a small amount of thioacetamide, CH_3CSNH_2, which reacts with water (technical term: undergoes <u>hydrolysis</u>) when heated in aqueous solution to produce hydrogen sulfide dissolved in the solution, $H_2S(aq)$:

$$CH_3CSNH_2(aq) + 2\ H_2O(\ell) \rightarrow H_2S(aq) + CH_3COO^-(aq) + NH_4^+(aq) \tag{2}$$

Using thioacetamide minimizes odor problems (H_2S gas stinks of rotten eggs!) and offers the advantage of producing denser precipitates, which tend to sink and are easier to centrifuge out.

The four insoluble sulfides of Group II can be separated into two subgroups by extracting with a solution of sodium hydroxide, NaOH. The sulfides of tin, Sn, and antimony, Sb, dissolve, forming OH^- complexes:

$$SnS_2(s) + 6\ OH^-(aq) \rightarrow Sn(OH)_6^{2-}(aq) + 2\ S^{2-}(aq) \tag{3}$$

$$Sb_2S_3(s) + 8\ OH^-(aq) \rightarrow 2\ Sb(OH)_4^-(aq) + 3\ S^{2-}(aq) \tag{4}$$

Because Cu^{2+} and Bi^{3+} do not easily form OH^- complexes, CuS and Bi_2S_3 do not dissolve in solutions of NaOH.

The solution containing the $Sb(OH)_4^-$ and $Sn(OH)_6^{2-}$ complex ions is treated with hydrochloric acid, HCl(aq), and thioacetamide. The H^+ ions of the strong acid HCl destroy the OH^- complexes; the free cations then reprecipitate as the sulfides. The reaction with $Sn(OH)_6^{2-}$ may be written as follows:

$$Sn(OH)_6^{2-}(aq) + 6\ H^+(aq) \rightarrow Sn^{4+}(aq) + 6\ H_2O(\ell)$$
$$\underline{Sn^{4+}(aq) + 2\ H_2S(aq) \rightarrow SnS_2(s) + 4\ H^+(aq)}$$
$$Sn(OH)_6^{2-}(aq) + 2\ H^+(aq) + 2\ H_2S(aq) \rightarrow SnS_2(s) + 6\ H_2O(\ell) \tag{5}$$
$$\text{yellow}$$

The $Sb(OH)_4^-$ ion behaves in a similar manner, being converted first to Sb^{3+} and then to bright orange Sb_2S_3. The Sb_2S_3 and SnS_2 are then dissolved as Cl^- complexes, in additional hydrochloric acid, and their presence is confirmed by appropriate tests.

The confirmatory test for tin, Sn, takes advantage of its two oxidation states, $+2$ and $+4$. Aluminum metal, Al(s), is added to <u>reduce</u> (add electrons to) Sn^{4+} to (produce) Sn^{2+}:

$$2\ Al(s) + 3\ Sn^{4+}(aq) \rightarrow 2\ Al^{3+}(aq) + 3\ Sn^{2+}(aq) \tag{6}$$

The classic test for the presence of tin(II), Sn^{2+}, in this solution is to add a solution of mercury(II) chloride, $HgCl_2$, which brings about another redox (oxidation–reduction) process:

$$Sn^{2+}(aq) + 2\ Hg^{2+}(aq) + 2\ Cl^-(aq) \rightarrow Sn^{4+}(aq) + Hg_2Cl_2(s) \tag{7}$$
$$\text{white}$$

The appearance of a white precipitate of insoluble Hg_2Cl_2 confirms the presence of tin. We no longer use this test due to the hazards and disposal challenges associated with mercury compounds. Instead, we use an organic compound called Janus Green (abbreviated JG). This compound is easily reduced by a strong reducing agent, such as Sn^{2+}, and a color change from blue to either violet-red or colorless (which one depends on the relative concentrations) indicates the presence of Sn^{2+}:

$$Sn^{2+}(aq) + 2\ JG(aq) \rightarrow Sn^{4+}(aq) + 2\ JG^-(aq) \qquad (8)$$
$$\text{blue} \qquad\qquad\qquad\qquad \text{violet-red}$$
$$\text{or}$$
$$Sn^{2+}(aq) + JG(aq) \rightarrow Sn^{4+}(aq) + JG^{2-}(aq) \qquad (9)$$
$$\text{blue} \qquad\qquad\qquad\qquad \text{colorless}$$

The antimony(III), Sb^{3+}, ion is difficult to confirm in the presence of Sn^{4+}; the colors of the sulfides of these two ions are too similar. To prevent interference by Sn^{4+}, the solution to be tested for Sb^{3+} is first treated with oxalic acid, $H_2C_2O_4$. This forms a very stable oxalate, $C_2O_4^{2-}$, complex with Sn^{4+}, $Sn(C_2O_4)_3^{2-}$. Treatment with H_2S then gives a bright red-orange precipitate of Sb_2S_3 only if antimony, Sb, is present:

$$2\ Sb^{3+}(aq) + 3\ H_2S(aq) \rightarrow Sb_2S_3(s) + 6\ H^+(aq) \qquad (10)$$
$$\text{bright}$$
$$\text{red-orange}$$

As pointed out earlier, CuS and Bi_2S_3 are insoluble in NaOH solutions, and they do not dissolve in hydrochloric acid. However, these two sulfides can be brought into solution by treatment with an oxidizing acid such as nitric acid, HNO_3. The reactions that occur are redox (oxidation–reduction) reactions. The NO_3^- ion is reduced, usually to NO_2; S^{2-} ions are <u>oxidized</u> (have electrons taken away from them) to (produce) bright yellow solid elemental sulfur, $S(s)$, and the cation, Cu^{2+} or Bi^{3+}, is brought into solution. The reactions are:

$$CuS(s) + 4\ H^+(aq) + 2\ NO_3^-(aq) \rightarrow Cu^{2+}(aq) + S(s) + 2\ NO_2(g) + 2\ H_2O(\ell) \qquad (11)$$
$$\text{black} \qquad\qquad\qquad\qquad\qquad \text{aqua} \quad \text{yellow}$$

$$Bi_2S_3(s) + 12\ H^+(aq) + 6\ NO_3^-(aq) \rightarrow 2\ Bi^{3+}(aq) + 3\ S(s) + 6\ NO_2(g) + 6\ H_2O(\ell) \qquad (12)$$
$$\text{brown} \qquad\qquad\qquad\qquad\qquad\qquad\qquad \text{yellow}$$

The two ions, Cu^{2+} and Bi^{3+}, can then be separated by adding aqueous ammonia, $NH_3(aq)$. The Cu^{2+} ion is converted to the deep-blue-colored complex ion $Cu(NH_3)_4^{2+}$:

$$Cu^{2+}(aq) + 4\ NH_3(aq) \rightarrow Cu(NH_3)_4^{2+}(aq) \qquad (13)$$
$$\text{aqua} \qquad\qquad\qquad\qquad \text{deep blue}$$

The reaction of ammonia with Bi^{3+} is quite different. The hydroxide ions, OH^-, produced by the reaction of NH_3 with water precipitate Bi^{3+} as $Bi(OH)_3$. We can consider this reaction as occurring in two steps:

$$3\ NH_3(aq) + 3\ H_2O(\ell) \rightleftharpoons 3\ NH_4^+(aq) + 3\ OH^-(aq)$$
$$\underline{Bi^{3+}(aq) + 3\ OH^-(aq) \rightarrow Bi(OH)_3(s)}$$
$$Bi^{3+}(aq) + 3\ NH_3(aq) + 3\ H_2O(\ell) \rightarrow Bi(OH)_3(s) + 3\ NH_4^+(aq) \qquad (14)$$
$$\text{white}$$

To confirm the presence of Bi^{3+}, the $Bi(OH)_3$ precipitate is first dissolved in hydrochloric acid:

$$Bi(OH)_3(s) + 3\ H^+(aq) \rightarrow Bi^{3+}(aq) + 3\ H_2O(\ell) \qquad (15)$$
$$\text{white}$$

The clear and colorless solution formed is then poured into deionized water. If Bi^{3+} is present, a white precipitate of bismuth oxychloride, BiClO, will form:

$$Bi^{3+}(aq) + H_2O(\ell) + Cl^-(aq) \rightarrow BiClO(s) + 2\ H^+(aq) \qquad (16)$$
$$\text{white}$$

Wear your safety glasses while performing this experiment.

Experimental Procedure

See Appendix D for some suggestions regarding procedures in qualitative analysis. The following instructions assume the use of small (13 × 100 mL) test tubes, in which a depth of about one centimeter corresponds to a volume of about one milliliter. Using this depth-to-volume relationship is sufficient for all of the volume measurements required in this experiment. These instructions also assume the use of eye droppers; if Pasteur pipets are used for drop measurements, increase the specified drop counts by 50%. (For example, you would use 3 drops instead of 2 and 7 or 8 instead of 5; where one drop is specified, you should still use a single drop.) Be sure to rinse or store your stirring rod(s) in a 400-mL beaker of deionized water between *each* use, to avoid cross-contamination. A boiling-water bath will be needed in this procedure, so you should set one up now (in a 250-mL beaker), before starting Step 1.

Step 1. Adjusting pH prior to precipitation Prepare a pH test paper by spotting methyl violet indicator on a piece of filter paper, about 1 drop to a spot, and letting the paper dry completely in the air.

Pour about one milliliter of the "known" solution for Group II, containing equal volumes of 0.10 M solutions of the nitrates or chlorides of Sn^{4+}, Sb^{3+}, Cu^{2+}, and Bi^{3+}, into a small (13×100 mm) test tube (1 mL ≈ 1 cm depth in the tube). The Group II known will be very acidic because of the presence of hydrochloric acid, HCl(aq), which is necessary to keep the salts of Bi^{3+}, Sn^{4+}, and Sb^{3+} in solution. Add 6 M aqueous ammonia, NH_3(aq), drop by drop, until the solution, after stirring, produces a violet spot on the pH test paper; test for pH by dipping your stirring rod in the solution and touching it to the paper. (A precipitate of the insoluble salts of the Group II cations will probably form during this step.) Then add 1 drop of 6 M hydrochloric acid, HCl(aq), for each milliliter of solution. This should bring the pH of the solution close to 0.5. Test the pH, using the test paper. At a pH of 0.5, methyl violet will have a blue-green color. Compare your spot with that made by putting 1 drop of 0.3 M HCl (pH = 0.5) on the test paper. Adjust the pH of your solution as necessary by adding HCl or NH_3, until the pH test gives about the right color on the paper. If you have trouble deciding on the color, centrifuging the precipitate to the bottom, so you can test clear liquid, may help. When you have established the proper pH, add about one milliliter of 1.0 M thioacetamide, CH_3CSNH_2, to the solution and stir.

Step 2. Precipitating the Group II sulfides Heat the test tube in the boiling-water bath for at least 5 minutes. **CAUTION:** **Small amounts of H_2S will be liberated; this gas is toxic (and smells intensely of rotten eggs), so avoid inhaling it unnecessarily.**

In the presence of Group II ions, a precipitate will form; typically, its color will initially be light, but it will gradually darken and finally approach black. Continue to heat the tube for at least two minutes after the color has stopped changing. Cool the test tube under cold tap water and let it stand for a minute or so. Centrifuge out the precipitate and <u>decant</u> (pour off) the solution, which will contain any Group III ions that may be present, into a test tube. (See Appendix D for more details on decanting.) Test the solution for completeness of precipitation by adding 2 drops of thioacetamide and letting it stand for a minute. If a precipitate forms, add a few more drops of thioacetamide and heat again in the boiling-water bath; then combine the two batches of precipitate. Wash the precipitate with about two milliliters of 1.0 M ammonium chloride, NH_4Cl, and stir thoroughly. Centrifuge, then discard the wash into a waste beaker.

Step 3. Separating the Group II sulfides into two subgroups To the precipitate from Step 2, add about two milliliters of 1.0 M sodium hydroxide, NaOH. Heat in the boiling-water bath, with stirring, for two minutes. Any SnS_2 and/or Sb_2S_3 should dissolve. The undissolved solid will typically be dark and may contain CuS and/or Bi_2S_3. Centrifuge and decant off the liquid into a test tube to be used in Step 4. Wash the precipitate twice, each time using about two milliliters of deionized water and a few drops of 1.0 M NaOH. Stir, centrifuge, and decant, discarding the liquid each time into a waste beaker. To the precipitate add about two milliliters of 6 M nitric acid, HNO_3, and put the test tube aside (for Step 8). At this point any copper and bismuth are in this test tube and any tin and antimony are in the test tube you previously set aside, as complex ions.

Step 4. Reprecipitating SnS_2 and Sb_2S_3 To the liquid from Step 3, add 6 M hydrochloric acid, HCl(aq), drop by drop, until the mixture is just acidic to litmus. Upon acidification, any tin and/or antimony present will again precipitate as yellow to red-orange sulfides. Add 5 drops of 1.0 M thioacetamide, CH_3CSNH_2, and heat in the boiling-water bath for two minutes to complete the precipitation. Centrifuge the tube and decant off the liquid, which may be discarded.

Step 5. **Dissolving SnS₂ and Sb₂S₃ in acid** Add about two milliliters of 6 M hydrochloric acid, HCl(aq), to the precipitate from Step 4 and heat in the boiling-water bath for a minute or two to dissolve the precipitate. Transfer the solution to your smallest beaker and bring it to a gentle boil for about a minute to drive out the H_2S. Add about one more milliliter of 6 M HCl and about one milliliter of water, then pour the liquid back into a small test tube. If there is any insoluble solid remaining, centrifuge it out and decant the liquid into a test tube.

Step 6. **Confirming the presence of tin** Pour half of the liquid from Step 5 into a different small test tube and add about two milliliters of 6 M hydrochloric acid, HCl(aq), and a piece of 24-gauge aluminum, Al, wire about one centimeter long. Heat the test tube in the boiling-water bath to promote reaction of the Al with the acid and the production of elemental hydrogen, H_2. In this reducing (electron-giving) environment, any tin present (as Sn^{4+}) will be converted to Sn^{2+} and any antimony to the metal, which will appear as black specks. Heat the tube for two minutes after *all* of the wire has reacted. Centrifuge out any solid and decant the liquid into a test tube. To the liquid, add 1 drop of Janus Green (JG) indicator solution. The presence of tin will be established by the blue indicator just added either turning a violet/red color or yielding a colorless solution, depending on the amount of tin present. The color change reaction is evidence of the reducing (electron-giving) power of the tin(II) ion, Sn^{2+}.

Step 7. **Confirming the presence of antimony** To the other half of the solution from Step 5, add 1.0 M sodium hydroxide, NaOH, to bring the pH to 0.5, as indicated by a blue-green color on methyl violet indicator paper. Add about two milliliters of water and about half a gram (a small spatula worth) of oxalic acid, $H_2C_2O_4$, then stir until no more crystals dissolve. The oxalate ion, $C_2O_4{}^{2-}$, forms a very stable complex with the tin(IV) ion, Sn^{4+}. Add about one milliliter of 1.0 M thioacetamide, CH_3CSNH_2, and put the test tube in the boiling-water bath. The formation of a bright red-orange precipitate of Sb_2S_3 confirms the presence of antimony.

Step 8. **Dissolving CuS and Bi₂S₃** Heat the test tube containing the precipitate from Step 3 in the boiling-water bath. Any sulfides that have not already dissolved should go into solution in a minute or two, possibly leaving some insoluble (yellow) solid sulfur, S(s), behind. Continue heating until no further reaction is visible—at least two minutes after the initial changes. Centrifuge and decant the liquid, which may contain Cu^{2+} and/or Bi^{3+}, into a test tube. Discard any solid.

Step 9. **Confirming the presence of copper** To the liquid from Step 8 add 6 M aqueous ammonia, NH_3(aq), one drop at a time until the mixture is just basic to litmus. Add about half a milliliter more. Centrifuge out any white precipitate that forms, and decant the liquid into a test tube. If the liquid is deep blue, the color is due to the $Cu(NH_3)_4{}^{2+}$ ion, and copper is present.

Step 10. **Confirming the presence of bismuth** Wash the precipitate from Step 9, which will contain bismuth if it was present, with about one milliliter of water and about half a milliliter of 6 M aqueous ammonia, NH_3(aq). Stir, centrifuge, and discard the wash. To the precipitate add about half a milliliter of 6 M hydrochloric acid, HCl(aq), and about half a milliliter of water. Stir to dissolve any white precipitate of $Bi(OH)_3$ that is present. Add the resulting liquid, drop by drop, to 400 mL of water in a 600-mL beaker. A white cloudiness, caused by the slow precipitation of bismuth oxychloride, BiClO, confirms the presence of bismuth.

Step 11. **Analyzing an unknown** When you have finished analyzing the known, obtain a Group II unknown and test it for the possible presence of Sn^{4+}, Sb^{3+}, Cu^{2+}, and/or Bi^{3+}. Be sure to record the ID number of your unknown on the report page.

Take it further (optional): Experimentally establish the presence of copper in a mineral supplement tablet. Estimate the amount of copper present by comparison with a suitable standard.

Disposal of reaction products: As you complete each part of this experiment, put the waste products in a beaker. At the end of the experiment, pour the contents of the beaker into a waste container, unless your instructor directs otherwise.

Experiment 37

Observations and Analysis: Qualitative Analysis of Group II Cations

Group II Flow Diagram

Indicate on this flow diagram the actual observations made on the known and the unknown. For the known, only note any deviations from the expected results indicated in the flow diagram, labeling them with "KNOWN". For the unknown, clearly indicate the path through the flow diagram that you followed to reach your conclusions.

Unknown # _____   Ion(s) present in unknown: _____ _____ _____ _____

Experiment 37

Advance Study Assignment: Qualitative Analysis of Group II Cations

1. Prepare a complete flow diagram for the separation and identification of the Group II cations and put it on the report page for this experiment.

2. Write balanced net ionic equations for the following reactions:

   a. The precipitation of tin(IV) sulfide with hydrogen sulfide, H_2S

   b. The confirmatory test for antimony

   c. The dissolution of bismuth(III) sulfide in hot nitric acid, HNO_3

   d. The confirmatory test for bismuth

(continued on following page)

3. A solution that may contain Cu^{2+}, Bi^{3+}, Sn^{4+}, and/or Sb^{3+} ions is treated with thioacetamide at pH 0.5. The dark precipitate that forms is partly soluble in strongly basic solution. The precipitate that remains is soluble in 6 M nitric acid, HNO_3, and gives only a white solid—no blue color—upon treatment with excess aqueous ammonia, $NH_3(aq)$. The basic liquid obtained from the initial dark precipitate, when acidified, produces an orange precipitate. On the basis of this information, which ions are present, which are absent, and which are still undetermined? State your reasoning. *Note*: For "paper unknowns" such as this one, confirmatory tests are usually not included, and you do not need to provide any. The information provided here, without confirmatory tests, is sufficient to clearly indicate the presence or absence of some of the Group II ions studied in this experiment, while possibly leaving others undetermined.

Present _____

Absent _____

Undetermined _____

Experiment 38

Qualitative Analysis of Group III Cations

The aim of (inorganic) <u>qualitative chemical analysis</u> is to determine which ions are present in an unknown (rather than in what amount).

The <u>Group III cations</u> form insoluble hydroxides at pH 9—just barely basic. The four <u>cations</u> (positive ions) from Group III that will be considered in this experiment are Cr^{3+}, Al^{3+}, Fe^{3+}, and Ni^{2+}. The first step in their analysis involves treating a solution containing them with sodium hydroxide, NaOH, and the <u>oxidizing</u> (electron-removing) agent sodium hypochlorite, NaClO, which is commonly referred to as "chlorine bleach". The hypochlorite ion, ClO^-, <u>oxidizes</u> (takes electrons away from) the chromium(III) ion, Cr^{3+}, to (produce) the yellow chromate ion, CrO_4^{2-}:

$$2\ Cr^{3+}(aq) + 3\ ClO^-(aq) + 10\ OH^-(aq) \rightarrow 2\ CrO_4^{2-}(aq) + 3\ Cl^-(aq) + 5\ H_2O(\ell) \qquad (1)$$
$$\text{violet}^\ddagger \qquad\qquad\qquad\qquad\qquad \text{yellow}$$

The chromate ion, CrO_4^{2-}, stays in solution. The same is true of the OH^- complex ion $Al(OH)_4^-$, formed by the reaction of Al^{3+} with excess hydroxide ion, OH^-:

$$Al^{3+}(aq) + 4\ OH^-(aq) \rightarrow Al(OH)_4^-(aq) \qquad (2)$$

In contrast, the other two ions in the group form insoluble hydroxides under these conditions:

$$Ni^{2+}(aq) + 2\ OH^-(aq) \rightarrow Ni(OH)_2(s) \qquad (3)$$
$$\text{green} \qquad\qquad\qquad \text{green}$$

$$Fe^{3+}(aq) + 3\ OH^-(aq) \rightarrow Fe(OH)_3(s) \qquad (4)$$
$$\text{yellow} \qquad\qquad\qquad \text{rust orange}$$

The Ni^{2+} and Fe^{3+} ions, unlike Al^{3+}, do not easily form complexes with OH^-. Unlike Cr^{3+}, they do not have a more oxidized form that is stable, so they are not oxidized by ClO^-.

To separate aluminum, Al, from chromium, Cr, the liquid containing CrO_4^{2-} and $Al(OH)_4^-$ is first acidified. This destroys the OH^- complex of aluminum:

$$Al(OH)_4^-(aq) + 4\ H^+(aq) \rightarrow Al^{3+}(aq) + 4\ H_2O(\ell) \qquad (5)$$

Treatment with aqueous ammonia, $NH_3(aq)$, then gives a floating, jelly-like, and white-to-almost-transparent precipitate containing aluminum(III) hydroxide, $Al(OH)_3$. This reaction can be thought of as occurring in two steps:

$$3\ NH_3(aq) + 3\ H_2O(\ell) \rightleftharpoons NH_4^+(aq) + 3\ OH^-(aq)$$
$$\underline{Al^{3+}(aq) + 3\ OH^-(aq) \rightarrow Al(OH)_3(s)}$$
$$Al^{3+}(aq) + 3\ NH_3(aq) + 3\ H_2O(\ell) \rightarrow Al(OH)_3(s) + 3\ NH_4^+(aq) \qquad (6)$$
$$\text{white}$$

The concentration of OH^- in dilute NH_3 is too low to form the $Al(OH)_4^-$ complex ion by Reaction 2, and aluminum does not form a complex ion with ammonia. (See Appendix B1.)

Equation 6 represents the classic confirmatory test for aluminum. Because the precipitate is sometimes difficult to see, a few drops of a solution of the organic dye aluminon could be added. The aluminon absorbs onto any $Al(OH)_3$ present and gives it a reddish color. This test is not always reliable, however—it sometimes gives false positives. An improved confirmatory test involves dissolving the $Al(OH)_3$ precipitate in dilute acetic acid, CH_3COOH, and then adding a few drops of a solution of the organic dye catechol violet, a weak

---

‡ Blue-gray in natural light; other colors are often observed—particularly green—in the presence of chloride, sulphate, and most other anions that can act as ligands, though ligand substitution is slow at room temperature.

acid that can be represented as HCV. With the improved confirmatory test, the Al^{3+} ions, if present, interact with the deprotonated and negatively charged (technical term: <u>anionic</u>) form of catechol violet, CV^-, to form a distinctive blue complex ion, $AlCV^{2+}$:

$$Al(OH)_3(s) + 3\ CH_3COOH(aq) \rightarrow Al^{3+}(aq) + 3\ CH_3COO^-(aq) + 3\ H_2O(\ell) \tag{7}$$

$$Al^{3+}(aq) + HCV(aq) \rightarrow AlCV^{2+}(aq) + H^+(aq) \tag{8}$$
blue

The CrO_4^{2-} ion remains in solution after Al^{3+} has been precipitated. It can be tested for by precipitation as yellow barium(II) chromate, $BaCrO_4$, by adding a solution of barium(II) chloride, $BaCl_2$:

$$Ba^{2+}(aq) + CrO_4^{2-}(aq) \rightarrow BaCrO_4(s) \tag{9}$$
yellow   light yellow

To confirm the presence of chromium, the $BaCrO_4$ precipitate is dissolved in acid and then treated with hydrogen peroxide, H_2O_2. A short-lived deep blue color results from the formation of a soluble and unstable chromium compound, probably CrO_5. The reaction may be represented by the following overall equation:

$$2\ BaCrO_4(s) + 4\ H^+(aq) + 4\ H_2O_2(aq) \rightarrow 2\ Ba^{2+}(aq) + 2\ CrO_5(aq) + 6\ H_2O(\ell) \tag{10}$$
light yellow   deep blue

The mixed precipitate of nickel(II) hydroxide, $Ni(OH)_2$, and iron(III) hydroxide, $Fe(OH)_3$, formed by Reactions 3 and 4 is dissolved by adding a strong acid: in this case nitric acid, HNO_3. An acid–base reaction occurs with each hydroxide precipitate:

$$Ni(OH)_2(s) + 2\ H^+(aq) \rightarrow Ni^{2+}(aq) + 2\ H_2O(\ell) \tag{11}$$
green   green

$$Fe(OH)_3(s) + 3\ H^+(aq) \rightarrow Fe^{3+}(aq) + 3\ H_2O(\ell) \tag{12}$$
rust orange   yellow

At this point, the Ni^{2+} and Fe^{3+} ions are separated by adding aqueous ammonia, $NH_3(aq)$. The nickel(II) ion, Ni^{2+}, is converted to the deep blue complex ion $Ni(NH_3)_6^{2+}$, which stays in solution,

$$Ni^{2+}(aq) + 6\ NH_3(aq) \rightarrow Ni(NH_3)_6^{2+}(aq) \tag{13}$$
green   deep blue

while the iron(III) ion, Fe^{3+}, which does not tend to form a complex with NH_3, is reprecipitated as rust orange $Fe(OH)_3$:

$$3\ NH_3(aq) + 3\ H_2O(\ell) \rightleftharpoons 3\ NH_4^+(aq) + 3\ OH^-(aq)$$
$$\underline{Fe^{3+}(aq) + 3\ OH^-(aq) \rightarrow Fe(OH)_3(s)}$$
$$Fe^{3+}(aq) + 3\ NH_3(aq) + 3\ H_2O(\ell) \rightarrow Fe(OH)_3(s) + 3\ NH_4^+(aq) \tag{14}$$
yellow   rust orange

The confirmatory test for nickel in the liquid is made by adding the organic compound dimethylglyoxime, $C_4H_8N_2O_2$. This produces a deep-rose-colored precipitate with nickel:

$$Ni^{2+}(aq) + 2\ C_4H_8N_2O_2(aq) \rightarrow Ni(C_4H_7N_2O_2)_2(s) + 2\ H^+(aq) \tag{15}$$
green   deep rose

The presence of the iron(III) ion, Fe^{3+}, can be confirmed by dissolving the precipitate of $Fe(OH)_3$ in hydrochloric acid, $HCl(aq)$, by Reaction 12 and adding a solution of potassium thiocyanate, $KSCN$. If iron(III) is present, the red-orange (blood red at high concentrations) $FeSCN^{2+}$ complex ion will form:

$$Fe^{3+}(aq) + SCN^-(aq) \rightarrow FeSCN^{2+}(aq) \tag{16}$$
yellow   red-orange
to blood red

Experimental Procedure

See Appendix D for some suggestions regarding procedures in qualitative analysis. The following instructions assume the use of small (13 × 100 mL) test tubes, in which a depth of about one centimeter corresponds to a volume of about one milliliter. Using this depth-to-volume relationship is sufficient for all of the volume measurements required in this experiment. These instructions also assume the use of eye droppers; if Pasteur pipets are used for drop measurements, increase the specified drop counts by 50%. (For example, you would use 3 drops instead of 2 and 7 or 8 instead of 5; where one drop is specified, you should still use a single drop.) Be sure to rinse or store your stirring rod(s) in a 400-mL beaker of deionized water between *each* use, to avoid cross-contamination. A boiling-water bath will be needed in this procedure, so you should set one up now (in a 250-mL beaker), before starting Step 1.

Step 1. Setting the stage If you are testing a solution from which the Group II ions have been precipitated, remove the excess H_2S and excess acid by boiling the solution until the volume is reduced to about one milliliter. Remove any bright yellow solid sulfur by centrifuging the solution and pouring off (technical term: <u>decanting</u> off) the liquid. (See Appendix D for more details on decanting.)

If you are working on the analysis of Group III cations only, prepare a known solution containing Fe^{3+}, Al^{3+}, Cr^{3+}, and Ni^{2+} by mixing together equal volumes of each of the appropriate 0.10 M solutions containing those cations. (About one milliliter of solution is required for this analysis.)

Step 2. Oxidizing Cr(III) to Cr(VI) and isolating insoluble hydroxides In your smallest beaker, combine about one milliliter of 6 M sodium hydroxide, NaOH, with about one milliliter of the solution to be analyzed. Boil very gently for about one minute while stirring or swirling the beaker. Remove the beaker from the heat and slowly add about one milliliter of 1.0 M sodium hypochlorite, NaClO. Swirl the beaker for about thirty seconds, using tongs if necessary. Then boil the mixture gently for a minute. Add about half a milliliter of 6 M aqueous ammonia, $NH_3(aq)$, and let stand for about thirty seconds. Then boil for another minute. Transfer the mixture to a test tube and centrifuge out any remaining solid, which contains iron and/or nickel hydroxides. Pour off the liquid, which contains any chromium and aluminum in the sample [as CrO_4^{2-} and $Al(OH)_4^-$ ions], into a test tube (for use in Step 3). Wash the solid twice, with about two milliliters of water and about half a milliliter of 6 M NaOH each time; after mixing, centrifuge each time, discarding the wash. Add about one milliliter of water and about one milliliter of 6 M sulfuric acid, H_2SO_4, to the solid and put the test tube aside (for Step 6).

Step 3. Separating aluminum from chromium Acidify the liquid from Step 2 by slowly adding 6 M acetic acid, CH_3COOH, until (after stirring) the mixture is definitely acidic to litmus (it turns blue litmus red). If necessary, transfer the solution to a small beaker and boil it to reduce its volume to about three milliliters. Pour the solution into a test tube. Add 6 M aqueous ammonia, $NH_3(aq)$, drop by drop until the solution is basic to litmus, and then add about half a milliliter more. Stir the mixture for a minute or so to bring the system to equilibrium. If aluminum is present, a light (not very dense), jelly-like, and white-to-almost-transparent solid containing $Al(OH)_3$ should be floating in the clear (possibly yellow) liquid. Centrifuge out the solid and transfer the liquid, which may contain CrO_4^{2-}, into a test tube (for use in Step 5).

Step 4. Confirming the presence of aluminum Wash the precipitate from Step 3 with about three milliliters of water once or twice, while warming the test tube in a boiling-water bath and stirring well. Centrifuge and discard the wash each time. Dissolve the precipitate in <u>2</u> drops of 6 M acetic acid, CH_3COOH: *no more, no less*. Add 3 mL of water and 2 drops of catechol violet solution, then stir. If Al^{3+} is present, the liquid will turn blue.

Step 5. **Confirming the presence of chromium** If the liquid from Step 3 is yellow, chromium is probably present; if it is colorless, chromium is absent. To the liquid add about half a milliliter of 1.0 M barium(II) chloride, $BaCl_2$. In the presence of chromium you will obtain a finely divided light yellow precipitate of barium(II) chromate, $BaCrO_4$, which may be mixed with a white precipitate of barium(II) sulfate, $BaSO_4$. Put the test tube in a boiling-water bath for a few minutes, then centrifuge out the solid and discard the liquid. Wash the solid with about two milliliters of water; centrifuge and discard the wash. To the solid add about half a milliliter of 6 M nitric acid, HNO_3, and stir to dissolve the $BaCrO_4$. Add about one milliliter of water, stir the orange solution, and add 2 drops of $3\%_{mass}$ hydrogen peroxide, H_2O_2. A deep blue color, which may fade rapidly, confirms the presence of chromium.

Step 6. **Separating iron from nickel** Returning to the precipitate from Step 2, stir to dissolve the solid in the H_2SO_4. If necessary, warm the test tube in a boiling-water bath to complete the dissolution process. Then add 6 M aqueous ammonia, $NH_3(aq)$, until the liquid is basic to litmus (turns red litmus blue). At that point any iron present will precipitate as rust-orange $Fe(OH)_3$. Add about one milliliter more of the NH_3 and stir to bring the nickel into solution as the $Ni(NH_3)_6^{2+}$ complex ion. Centrifuge and pour off the liquid into a test tube. Save the precipitate (for Step 8).

Step 7. **Confirming the presence of nickel** If the liquid from Step 6 is blue, nickel is probably present. Add about half a milliliter of a $1\%_{mass}$ solution of dimethylglyoxime, $C_4H_8N_2O_2$, to it. Formation of a rose-red precipitate confirms the presence of nickel.

Step 8. **Confirming the presence of iron** Dissolve the precipitate from Step 6 in about half a milliliter of 6 M hydrochloric acid, $HCl(aq)$. Add about two milliliters of water and stir. Then add 2 drops of 0.5 M potassium thiocyanate, KSCN. Formation of a red-orange to blood-red solution of $FeSCN^{2+}$ confirms the presence of iron.

Step 9. **Analyzing an unknown** When you have finished analyzing the known solution, obtain a Group III unknown and test it for the possible presence of Fe^{3+}, Al^{3+}, Cr^{3+}, and/or Ni^{2+}. Be sure to record the ID number of your unknown on the report page.

Disposal of reaction products: As you complete each step in the procedure, put the waste products into a beaker. When you are finished with the experiment, pour the contents of the beaker into a waste container, unless your instructor directs otherwise.

Experiment 38

Observations and Analysis: Qualitative Analysis of Group III Cations

Group III Flow Diagram

Indicate on this flow diagram the actual observations made on the known and the unknown. For the known, only note any deviations from the expected results indicated in the flow diagram, labeling them with "KNOWN". For the unknown, clearly indicate the path through the flow diagram that you followed to reach your conclusions.

Unknown # _____ Ion(s) present in unknown: _____ _____ _____ _____

Name ______________________________ Section ______________________________

Experiment 38

Advance Study Assignment: Qualitative Analysis of Group III Cations

1. Prepare a complete flow diagram for the separation and identification of the ions in Group III and put it on the report page for this experiment.

2. Write balanced net ionic equations for the following reactions:

   a. The oxidation of Cr^{3+} to CrO_4^{2-} by ClO^- in basic solution (ClO^- is converted to Cl^-)

   b. The dissolution of iron(III) hydroxide in nitric acid, HNO_3

   c. The confirmatory test for Ni^{2+}

   d. The confirmatory test for Fe^{3+}

(continued on following page)

3. A solution may contain any of the four Group III cations considered in this experiment. Treatment of the solution with hypochlorite ion, ClO^-, in base yields a yellow liquid and a colored precipitate. The acidified liquid is unaffected by treatment with aqueous ammonia, $NH_3(aq)$. The colored precipitate dissolves in nitric acid, HNO_3; adding excess NH_3 to this acidic solution produces a rust-orange precipitate and does not cause the liquid to turn a blue color. On the basis of this information, which Group III cations are present, absent, or still undetermined? State your reasoning. *Note*: For "paper unknowns" such as this one, confirmatory tests are usually not included, and you do not need to provide any. The information provided here, without confirmatory tests, is sufficient to clearly indicate the presence or absence of some of the Group III ions studied in this experiment, while possibly leaving others undetermined.

Present _____

Absent _____

Undetermined _____

Experiment 39

Identifying a Pure Ionic Solid

In this experiment you will be asked to determine the <u>cation</u> (positive ion) and anion (negative ion) in a solid sample of a pure ionic salt. The possible cations in this experiment are:

$$Ag^+ \quad Pb^{2+} \quad Hg_2^{2+} \quad Sn^{2+} \quad Sb^{3+}$$
$$Cu^{2+} \quad Bi^{3+} \quad Cr^{3+} \quad Al^{3+} \quad Ni^{2+} \quad Fe^{3+}$$

The possible anions are:

$$CO_3^{2-} \quad PO_4^{3-} \quad Cl^- \quad SCN^- \quad SO_4^{2-} \quad SO_3^{2-} \quad CH_3COO^- \quad NO_3^-$$

There are therefore 88 possible salts that you might be given to identify. It might seem that the best approach would be to carry out the procedures for analysis of Groups I, II, and III, in succession, and stop when you get to the cation in your unknown. That approach should work, but it would take longer than necessary, and would require that you have a solution of the salt to work with. Given knowledge of the properties of the possible salts, an experienced chemist would carry out some preliminary tests to determine the solubility properties of the salt, and try to use those to significantly narrow down the number of possibilities. The solubility of your unknown salt in various solvents, plus its color, and its odor (should it have one), may be very useful information.

Most of the possible salts are not soluble in water but may dissolve in strong acids, such as 6 M hydrochloric acid, HCl(aq); in strong bases, such as 6 M sodium hydroxide, NaOH; or in solutions containing <u>ligands</u> (electron pair donors) that form stable complex ions with the cation in the salt. Some important complexing ligands that may dissolve otherwise insoluble solids containing the possible cations include:

NH_3	OH^-	Cl^-	CH_3COO^-
(ammonia)	(hydroxide ion)	(chloride ion)	(acetate ion)

Nearly every one of the 88 possible salts will dissolve in, or react with, at least one of these ligands.

Before beginning the analysis, it will be helpful to note some of the properties of the cations that distinguish them from one another. Those properties are listed in Appendix B2. You should read that appendix and complete the Advance Study Assignment before coming to lab. You will need the information in Appendices B1 and B2 to complete this experiment.

Wear your safety glasses while performing this experiment.

Experimental Procedure

In this experiment you will be given two unknowns to identify. The cation in your first sample will be colored or will produce colored solutions. The second unknown will be white or colorless, and somewhat more difficult to identify. In both cases the analysis will involve determining the solubility of the solid in water and in some acidic or basic solutions. You should have about a half gram of finely divided solid available for you to work with. Read the section on qualitative analysis at the end of Appendix D if you have not already done so. The following instructions assume the use of small (13 × 100 mL) test tubes, in which a depth of about one centimeter corresponds to a volume of about one milliliter. Using this depth-to-volume relationship is sufficient for all of the volume measurements required in this experiment. Be sure to rinse or store your stirring rod(s) in a 400-mL beaker of deionized water between *each* use, to avoid cross-contamination. A boiling-water bath will be needed, so you should set one up now (in a 250-mL beaker). Set up a 250-mL beaker to serve as a waste container, and pour out your test tubes there as you finish working with their contents.

A. Identifying the cation in a colored ionic solid

The following solvents are available for use in this experiment:

A) Water   B) 6 M nitric acid, HNO_3

C) 6 M hydrochloric acid, $HCl(aq)$ D) 6 M sodium hydroxide, $NaOH$

E) 6 M aqueous ammonia, $NH_3(aq)$ F) 6 M sulfuric acid, H_2SO_4

CAUTION: **B is a powerful, if slow-acting, strong acid. C is a strong acid with a choking smell. D is a strong base: if it comes in contact with your skin, wash with water until your skin no longer feels slippery. E is a base with a strong smell that can knock you unconscious. F is a strong acid. Avoid contact with all of these.**

Test the solubility of your colored solid in as many of the solvents as you need to find one in which it will dissolve. Use a small sample of the solid: about fifty milligrams, enough to cover a two-mm-diameter circle on the end of a small spatula. Put the solid in a small (13×100 mm) test tube and add about one milliliter of water (1 mL ≈ 1 cm depth in the tube). Stir the solid with a stirring rod for a minute or so, noting any color changes in the solid or the solution. If the solid does not dissolve, put the tube in your boiling-water bath for a few minutes. If at this point you have a solution, add about five more milliliters of water and stir. From the color of the solution you should be able to make a good guess as to which cation is in your unknown. (The information in Appendices B1 and B2 should prove helpful.) Use about one milliliter of your solution to carry out the confirmatory test for that cation from the Group II or Group III qualitative analysis schemes. (None of the possible Group I cations are colored.) Make sure that you set the pH and any other conditions required for the confirmatory test properly before making a decision. If your guess is confirmed, you may start analyzing for the anion. If your guess proves incorrect, perform the confirmatory test for another cation that seems likely, until you know what your cation is.

If your sample does not dissolve in water, try 6 M nitric acid, HNO_3, as a solvent, using the same procedure as with water (about one milliliter of the solvent and fifty milligrams of solid). If the sample goes into solution, either cold or after being heated in the boiling-water bath, add about five milliliters of water, stir, and proceed to identify the cation present. With HNO_3 as the solvent, you may observe some bubbling when you test the solubility of your solid. This indicates the anion is either carbonate, CO_3^{2-}, or sulfite, SO_3^{2-}. If it is carbonate then odorless carbon dioxide, CO_2, gas will be given off, while if it is sulfite then gaseous sulfur dioxide, SO_2, with its characteristic sharp volcanic smell (like the smell of burnt matches), will be released. Nearly all the colored salts in the set are soluble in either hot water or hot HNO_3. If you by chance have one that will not dissolve in either of those, carry out the solubility test using 6 M hydrochloric acid, $HCl(aq)$, as the solvent. That should work if the first two solvents have not. In this part of the experiment, you should not need to use solvents D, E, or F.

B. Analyzing for the anion in a colored ionic solid

When analyzing the anion in your unknown, it is helpful to know that some anions are much more common than others in the chemicals that are generally available from chemical suppliers. It is easiest for chemists to work with chemicals that are soluble in water, so that is what chemical suppliers tend to sell. Nitrates, chlorides, sulfates, and acetates are often soluble, and are the most common salts found on laboratory shelves. Thiocyanates and sulfites are sometimes soluble but are usually found as sodium or potassium salts, so they are less likely to be salts you encounter in this experiment. Carbonates and phosphates are insoluble in water but are occasionally available. Hydroxides and oxides are relatively easy to make, but they are generally insoluble and will not be one of your unknowns.

Given this information and the spot tests for common anions, determine which anion is in your solid. For each test, use about one or two milliliters of the solution you prepared. You do not need to do all the spot tests, only those that seem reasonable in light of the previous paragraph. If you used water as your solvent, any positive test for an anion should be conclusive. If you used nitric acid, HNO_3, a test for nitrate ion in that solution will always come out positive because of the nitrate put into the solution by the nitric acid. Similarly, you cannot carry out a meaningful test for chloride ion in an HCl solution.

Some anions may be difficult to determine by the usual spot test, because the cation present may interfere. That is the case with acetate and nitrate, where the test solution is basic and will often cause the cation to precipitate. With acetates, a simple test is to add a milliliter of 6 M sulfuric acid, H_2SO_4, to about fifty milligrams of the solid, as if you were testing for solubility. Heat the solution, or solid-liquid mixture, in the boiling-water bath. If the sample contains acetate ion, that ion will turn into acetic acid when acidified—some of which will enter the gas phase and can be detected by its characteristic vinegar smell. Put your stirring rod in the warm liquid, remove it, and carefully smell the rod. If there are no interfering anions (and in your pure salt there cannot be), the odor of vinegar is an excellent test for acetate. With nitrates, begin by adding about one milliliter of 6 M sodium hydroxide, NaOH, to a small sample of solid unknown. If you get a hydroxide precipitate, you can try proceeding with the spot test. If the cation forms a hydroxide, OH^-, complex, you can try the spot test for nitrate, and it may work. In many cases, the aluminum metal will <u>reduce</u> (add electrons to) the cation to (form) metal, and you will get a black solid. With excess Al, you may still be able to detect the smell of ammonia gas, NH_3, and establish the presence of nitrate. The test for nitrate is perhaps the most difficult, and if worse comes to worst, you can treat your acidic solution of unknown with 1.0 M sodium carbonate, Na_2CO_3, until bubbling stops and the solution is definitely basic; then add about half a milliliter more. Centrifuge out the solid carbonate, which will contain your unknown's cation and can be discarded. After pouring it off into a new tube, acidify the liquid with 6 M sulfuric acid, H_2SO_4, drop by drop until bubbles of carbon dioxide, CO_2, stop forming and the solution is acidic. Carefully boil this solution down to about two milliliters and use it for the nitrate test and any other anion tests that seem indicated.

Once you have determined the anion and cation present in your colored unknown, you know what salt it must be composed of. On the report page, record the name and formula of the salt you believe is your colored unknown, as well as its ID number.

C. Analyzing a white or colorless ionic solid

The second unknown you will be given to analyze will be a white or colorless solid, which may have very limited solubility in water. It may also be insoluble in some of the acids and bases in the solvent list at the start of Part A.

The procedure—at least initially—is the same as that used with the colored sample. Check the solubility of the solid in water, both at room temperature and in the boiling-water bath. Some of the solids do dissolve. With these, and the rest of the unknowns, the solubility in the other solvents is likely to furnish clues about which cation is present. So, carry out solubility tests with all six of the solvents in the Part A solvent list (again, use about one milliliter of solvent added to about fifty milligrams of solid). Before deciding on the solubility in a given solvent, make sure you give the solid time to dissolve if it is going to, with stirring, particularly in the boiling-water bath. If the solid does not dissolve, add a milliliter of water and stir; that may help. Solids containing cations that form complexes with Cl^- are likely to dissolve in 6 M hydrochloric acid, HCl(aq), while those that form OH^- complexes are likely to dissolve in 6 M sodium hydroxide, NaOH. Those with cations that form NH_3 complexes will tend to dissolve in 6 M aqueous ammonia, NH_3(aq). The solubility depends on the relative solubility of the solid and stability of the complex, so a complexing solvent may dissolve one solid and not another with the same cation. Nitrate ion and sulfate ion are not very good ligands (complexing agents), so 6 M HNO_3 and 6 M H_2SO_4 will not dissolve solids by complex ion formation. If they dissolve a solid, it will be by reacting with an anion in the solid to form a weak acid. There are a few salts in the set of possible unknowns that will not dissolve in any of the six solvents, or may dissolve in only one of them. Those salts are all discussed in Appendix B2.

When you have completed the solubility tests on your solid, compare your results with those you obtained in your Advance Study Assignment using information from Appendix B2. You should be able to narrow down the list of possible cations to only a few, and in some cases you will be able to decide not only which cation is present but also which anion. If you can dissolve the solid in water or in acid, you can proceed to carry out the qualitative analysis scheme confirmatory test for any likely cations. With the white or colorless unknowns, the anion is even more likely to be one commonly found on laboratory shelves. If the sample will dissolve in water, or in HNO_3 or HCl, you can carry out spot tests for any anions that are not present in the solvent. For some solids, you will need to identify the anion purely on the basis of the solubility (or insolubility!) of the salt.

On the report page, note all your observations and the name and formula of the salt you believe is your white or colorless unknown, as well as its ID number.

Disposal of reaction products: When you are finished with this experiment, discard all the reaction products in your 250-mL beaker into a waste container, unless your instructor directs otherwise.

Experiment 39

Observations and Analysis: Identifying a Pure Ionic Solid

A. Identifying the cation in a colored ionic solid

Color of first unknown solid _____

Solubility properties of first, colored unknown:

_____

_____

_____

Cation(s) most likely to be present in colored unknown _____ _____

Confirmatory test(s) performed to identify cation in colored unknown:

Procedures used

_____

_____

_____

Results observed

_____

_____

_____

Conclusion: _____ is present

B. Analyzing for the anion in a colored ionic solid

Spot tests performed to identify anion in colored unknown:

Procedures used

_____

_____

_____

Results observed

_____

_____

_____

Conclusion: _____ is present

Name and formula of colored unknown: _____

ID number of colored unknown: _____

(continued on following page)

C. Analyzing a white or colorless ionic solid

Solubility of white or colorless unknown in:

Water _____   6 M HNO_3 _____

6 M HCl _____   6 M NaOH _____

6 M NH_3 _____   6 M H_2SO_4 _____

Reasoning and results of confirmatory tests:

_____

_____

_____

_____

_____

_____

_____

_____

_____

_____

_____

_____

Name and formula of white or colorless unknown: _____

ID number of white or colorless unknown: _____

Experiment 39

Advance Study Assignment: Identifying a Pure Ionic Solid

Use the information in Appendix B1 and Appendix B2 to find answers to these questions.

1. What are the formulas and colors of the colored cations used in this experiment?

   _____

   _____

   _____

2. Which of the *colorless* cations, at moderate (not high, but in excess) ligand concentrations where applicable,

   a. will form complex ions with NH_3? _____

   b. will form complex ions with OH^-? _____, _____, _____, _____

   c. will form complex ions with Cl^-? _____, _____, _____

   d. have brightly colored, rather than dark, sulfides? _____, _____
      Note: Sulfides are generally black; if a sulfide color is not specified, you may assume it is black.

   e. have one or more water-soluble salts
      (salts that dissolve in water alone)? _____, _____, _____, _____

3. Give the formula of a white or colorless salt formed from the ions involved in this experiment that

   a. dissolves in 6 M NH_3 _____

   b. turns black when NH_3 is added _____

   c. is soluble in 6 M NaOH but not in 6 M H_2SO_4 _____

   d. is appreciably soluble in hot water, but not in cold water _____

   e. is not soluble in any of the solvents used in this experiment _____

4. A white or colorless ionic solid is insoluble in water, and in 6 M NH_3, but will dissolve in 6 M HCl and 6 M NaOH. Identify two salts that have these properties.

   _____ and _____

Experiment 40

The Ten Test Tube Mystery

The aim of <u>qualitative analysis</u> is to determine which substances are present in an unknown (rather than in what amount).

In this experiment you will apply what you have learned about qualitative analysis to a related but somewhat different kind of problem. You will be given a set of ten numbered test tubes, each tube containing a solution of a single <u>solute</u> (dissolved substance). You will be provided, one week in advance, a list of formulas and concentrations of each of the ten solutes that will be in your set. Your challenge in the laboratory will be to find out which solution is in which test tube—that is, to assign a test tube number to each of the solution compositions. You will do this by combining small volumes of the solutions in the test tubes. You will not be able to use any external <u>reagents</u> (chemicals) or acid-base indicators such as litmus. You will be permitted, however, to use the odor and color of the different solutions, as well as the consequences of their reactions with each other, in attempting to identify them. Be sure to note whether any reactions change the temperature, as well as to use heat to drive reactions that do not noticeably occur at room temperature. Each student may have a different set of (test tube number)–(solution composition) pairings, and there may be several different sets of compositions as well.

Of the ten solutions, four will be common laboratory reagents. These will be 6 M hydrochloric acid, $HCl(aq)$; 3 M sulfuric acid, H_2SO_4; 6 M aqueous ammonia, $NH_3(aq)$; and 6 M sodium hydroxide, $NaOH$. The other solutions will usually be 0.10 M nitrate, chloride, or sulfate solutions of the <u>cations</u> (positive ions) you have studied in previous qualitative analysis experiments.

To determine which solution is in each test tube, you will need to know in advance what happens when the various solutions are mixed with one another. In some cases, nothing happens that you can observe. This will often be the case when a solution containing one of the cations is mixed with a solution of another cation. When one of the common laboratory reagents is mixed with a cation solution, you may get a precipitate—white, colorless, or colored—and that precipitate may dissolve in excess reagent by complex ion formation. In a few cases a gas may be released. When one laboratory reagent is mixed with another, you may find that the resulting solution gets very hot and/or that a visible cloud is produced in the air above it.

There is no way that you will be able to solve your particular test tube mystery without doing some preliminary work. You will need to know what to expect when any two of your ten solutions are mixed. You can find this out—for any pair of solutions—by studying Appendix B1, your chemistry text, and references on qualitative analysis.* A convenient way to tabulate the information you obtain is to set up a matrix (table) with ten columns and ten rows, one for each solution. The key information about a mixture of two solutions is put in the space where the row for one solution and the column for the other intersect [as is done in Appendix B1 for various cations and <u>anions</u> (negative ions)]. In that space you might put "NR", to indicate no apparent reaction on mixing of the two solutions. If a precipitate forms, write in "P", followed by an arrow pointing to "D" if the precipitate dissolves in excess reagent. If the precipitate or final solution is colored, indicate the color in parentheses. If the mixture heats up, write "H"; if bubbles or a cloud are formed, write in "B" or "C". Because mixing solution X with solution Y is the same as mixing Y with X, not all 100 spaces in the 10×10 matrix need to be filled in. Actually, there are only 45 relevant pairs, because combining X with X is not very informative either.

---

* For example, the following references may be useful:

[1]*Qualitative Analysis and the Properties of Ions in Aqueous Solutions.* 2nd ed., by E. J. Slowinski and W. L. Masterton, Saunders College Publishing, 1990.

[2]*Handbook of Chemistry and Physics.* Chemical Rubber Publishing, 2017.

[3]https://chem.libretexts.org/Bookshelves/Analytical_Chemistry/Supplemental_Modules_(Analytical_Chemistry)/Qualitative_Analysis

Because you are allowed to use the odor and/or color of a solution to identify it, the problem is somewhat simpler than it might first appear. In each set of ten solutions you will probably be able to identify at least two solutions simply by odor and/or color observations. Knowing the identities of those solutions, you can make mixtures with the other solutions in which one of the components is known. From the results obtained with those mixtures, and the information in the matrix, you can identify other solutions. These can be used to identify still others, until finally the entire set of ten is identified with certainty.

Let us now go through the various steps that would be involved in the solution to a somewhat simpler problem than the one you will solve. There are several steps: constructing the reaction matrix, identifying solutions by observing them in isolation, and devising efficient mixing tests. Let us assume we have to identify the following six solutions, in numbered test tubes:

0.10 M NiSO$_4$	0.10 M BiCl$_3$(in 3 M HCl)
0.10 M BaCl$_2$	6 M NaOH
0.10 M Al(NO$_3$)$_3$	3 M H$_2$SO$_4$

A. Constructing a reaction matrix

In the reaction matrix for these six solutions there will be six rows and six columns, one of each for each solution. The matrix should be set up as shown in Table 40.1.

Each solution contains a cation and an anion, both of which need to be considered. We need to include every possible pair of solutions: for six solutions there are 15 unique pairs. So we need 15 pieces of information regarding what happens when a pair of solutions is mixed. We begin on the top row of Table 40.1. The first pair of solutions we might consider is Al(NO$_3$)$_3$ and BaCl$_2$, containing Al^{3+}, NO$_3^-$, Ba^{2+}, and Cl$^-$ ions. Consulting Appendix B1, we see that in the Al^{3+}, Cl$^-$ space there is an "S", meaning that AlCl$_3$ is soluble and will not precipitate when aluminum and chloride ions are mixed. Similarly, Ba(NO$_3$)$_3$ is soluble (almost all nitrates are soluble). This means that there will be no reaction when the solutions of Al(NO$_3$)$_3$ and BaCl$_2$ are mixed. So, in the space for Al(NO$_3$)$_3$–BaCl$_2$, we have written "NR". The same is true if we mix solutions of Al(NO$_3$)$_3$ and NiSO$_4$, so we put "NR" in that space as well. Because BiCl$_3$ will not react with Al(NO$_3$)$_3$ solutions, there is also an "NR" in that space. However, when Al(NO$_3$)$_3$ is mixed with NaOH, one of the possible products is Al(OH)$_3$, which we see in Appendix B1 is insoluble in water but dissolves in (even weak) acid as well as in excess OH$^-$ ion ("A$^-$, C", with OH$^-$ listed as a complexing agent in the very last, rightmost column). This means that on adding NaOH to Al(NO$_3$)$_3$, we would initially get a precipitate of Al(OH)$_3$, "P", but that it would dissolve in excess NaOH, "D". Looking up the properties of Al(OH)$_3$ and the aluminum–hydroxide complex Al(OH)$_4^-$, we find the precipitate is white and the complex ion is colorless, so we have just "P" leading to "D", written as "P → D", with no color indications, in the space for that mixture.

Moving to the BaCl$_2$ row, we do not need to consider the first two columns, because they would give no new information. If BaCl$_2$ and NiSO$_4$ were mixed, BaSO$_4$ is a possible product. In Appendix B1 we see that BaSO$_4$ is insoluble in all common solvents ("I"). Hence, on mixing those solutions, we would get a precipitate of BaSO$_4$, "P". Using the *Handbook of Chemistry and Physics,* or some other source, we find that BaSO$_4$ is white, so no color indication goes into that space. With BiCl$_3$ there would be no reaction, thus "NR". Combining BaCl$_2$ with NaOH, however, may result in a slight precipitate ["S$^-$" for Ba(OH)$_2$ in Appendix B1], so we have written "P?" in the appropriate space. With H$_2$SO$_4$, we would again expect to get a white precipitate of white BaSO$_4$, so just "P".

In the row for NiSO$_4$ we note that the color of the solution is green, because in Appendix B1 we see that the Ni^{2+}(aq) ion is green. The first non-redundant column is that for BiCl$_3$; no precipitate forms on mixing solutions of NiSO$_4$ and BiCl$_3$, so "NR" is in that space. With NaOH, Ni(OH)$_2$ would precipitate, because the entry in Appendix B1 at the intersection of Ni^{2+} with OH$^-$ is not "S". From the *Handbook of Chemistry and Physics* or other sources we find that Ni(OH)$_2$ is green, so in the space we have written "P (green)". In Appendix B1 the Ni(OH)$_2$ entry is "A$^-$, C", which, along with the very last (rightmost) column on complexes, tells us that the precipitate will dissolve in 6 M NH$_3$ or weak acid, but not in 6 M NaOH as Al^{3+} does. There would be no reaction between solutions of NiSO$_4$ and H$_2$SO$_4$, so that spot in the grid gets "NR".

Table 40.1

Reaction Matrix for Six Solutions						
	$Al(NO_3)_3$	$BaCl_2$	$NiSO_4$	$BiCl_3$	NaOH	H_2SO_4
$Al(NO_3)_3$		NR	NR	NR	P → D	NR
$BaCl_2$			P	NR	P?	P
$NiSO_4$ (green)				NR	P (green)	NR
$BiCl_3$					H → P	NR
NaOH						H
H_2SO_4						

With $BiCl_3$ in 3 M HCl we have both a salt and an acid in the solution. Adding 6 M NaOH should produce a white precipitate of $Bi(OH)_3$, "P". There would also be a substantial exothermic acid–base reaction between H^+ ions in the HCl and OH^- ions in the NaOH, so there would be a very noticeable rise in temperature on mixing, which we denote with "H". The neutralization reaction happens more quickly than does the precipitation (this is usually the case), so the tube will first heat up, then the precipitate will form as more NaOH is added. We indicate this by "H → P", heat followed by precipitate. There would be no reaction with H_2SO_4.

When 6 M NaOH is mixed with 3 M H_2SO_4, there should be a large heat effect because of the acid–base reaction that occurs (but no precipitate would form), so in the space for NaOH–H_2SO_4 we have "H" (only).

The matrix in Table 40.1 summarizes the reaction information that would be needed to identify the solutions in the six test tubes of this example. Your reaction matrix can be constructed by similar reasoning.

B. Identifying solutions by observing them in isolation

When we made the matrix for the six-test-tube system, we noted that the $NiSO_4$ solution would be green, because the Ni^{2+}(aq) ion is green. So, as soon as we see the six text tubes, we can pick out the 0.10 M $NiSO_4$. Let us assume that it is in Test Tube 4. We have now identified one solution out of the six.

C. Selecting efficient mixing tests

Knowing that Test Tube 4 contains $NiSO_4$, we mix the solution in Test Tube 4 with all of the other solutions. For each mixture, we record what we observe. The results might be as follows:

4 + 1	no obvious reaction
4 + 2	no obvious reaction
4 + 3	green precipitate forms
4 + 5	no obvious reaction
4 + 6	white precipitate forms

Referring to the matrix, we would expect precipitates for mixtures of $NiSO_4$ with NaOH, and for $NiSO_4$ with $BaCl_2$. The former precipitate would be green, the latter white. So the solution in Test Tube 3 must be 6 M NaOH, and the solution in Test Tube 6 must be 0.10 M $BaCl_2$. At this point three solutions have been identified.

Because 6 M NaOH reacts with several of the solutions, and because we know now that it is in Test Tube 3, we mix the solution in Test Tube 3 with those that remain unidentified, with the following results:

3 + 1	solution becomes hot
3 + 2	solution gets hot, then a white precipitate forms when more of Solution 2 is added
3 + 5	no apparent reaction

Again, referring to the matrix, we would expect an increase in temperature upon mixing NaOH with H_2SO_4, as well as when NaOH reacts with the HCl in the $BiCl_3$. In addition, we would expect to get a white precipitate with $BiCl_3$. This indicates that Solution 1 is 3 M H_2SO_4 and that Solution 2 must be 0.10 M $BiCl_3$ in 3 M HCl. By elimination, Solution 5 must be $Al(NO_3)_3$. We repeat the 3 + 5 mixture, because there should have been an initial precipitate that dissolved in excess NaOH. This time, using 1 drop of Solution 3 added to about one milliliter of Solution 5, we get a precipitate. In an excess of Solution 3 the precipitate dissolves as the aluminum–hydroxide complex ion is formed.

The key to successfully carrying out this experiment is to consider *all* the possible explanations for an observation, and then make informed choices about additional observations you can make that will allow you to conclusively identify each unknown, usually one unknown at a time. Like any good forensic scientist, you must avoid the temptation to draw conclusions beyond what the data will support, while making inferences (guesses) that allow you to collect useful data in an efficient way. [Mixing every unknown with every other unknown is not only time-consuming, but leaves you with a huge pile of data to sift through—which is often more challenging than starting with an important clue (such as a colored unknown, or one with a smell) and following it where it leads.]

Experimental Procedure

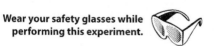

Wear your safety glasses while performing this experiment.

Using the approach outlined in our example, make observations and carry out tests as necessary to identify the solutions in your set. After you have tentatively assigned each solution to its test tube, carry out confirmatory mixing reactions until you have at least two observations that independently support your conclusion(s) about each tube. If it has one, be sure to record the ID code of your unknown set on the report page.

Disposal of reaction products: Dispose of all reaction products in a waste container, unless your instructor directs otherwise.

Take it further (optional): This experiment is an exercise in detective work! It might seem to be a challenge for the sleuth (the student), but not for the mystery-setter (the instructor). That is actually not the case, as setting up a good mystery has many pitfalls! Give it a try: generate your own set of unknowns for someone else to identify. Construct the reaction matrix of results they should get upon mixing each pair of unknowns, and ensure that those results would allow them to conclusively identify each tube.

Experiment 40

Observations and Analysis: The Ten Test Tube Mystery

A. Constructing a reaction matrix

Construct your reaction matrix on the back of the Advance Study Assignment page, following the instructions on the front of the Advance Study Assignment.

B. Identifying solutions by observing them in isolation

Solutions identified by observation in isolation

Test tube #	Observations	Conclusion(s)	Solution identity

C. Selecting efficient mixing tests

Mixing tests: first series

Test tubes combined	Observations	Conclusions

(continued on following page)

Mixing tests: second series

Test tubes combined	Observations	Conclusions

Confirmatory tests (combine at least two strategically chosen pairs of tubes not previously combined, to double-check your identifications)

Test tubes combined	Prediction(s)	Observations	Conclusion(s) supported

Final identifications: # 1 _____ # 2 _____ # 3 _____

4 _____ # 5 _____ # 6 _____ # 7 _____

8 _____ # 9 _____ # 10 _____

Which unknown set did you have (if applicable)? _____

Experiment 40

Advance Study Assignment: The Ten Test Tube Mystery

1. Construct the reaction matrix for your set of ten solutions. This should go on the reverse side of this page. The matrix can be made using information in Appendices B1 and B2, or from other references; or, if your instructor allows it, you can go into the lab prior to the scheduled session and work with known solutions to obtain the needed information.

2. Which solutions should you be able to identify by observing them in isolation, without mixing any solutions?

3. Outline the procedure you will follow in identifying the solutions that will require mixing tests. Be as specific as you can about what you will look for and what conclusions you will be able to reach from your observations.

Experiment 41

Making Aspirin

One example of an organic chemical reaction is the formation of an ester (a molecule containing a —COO— unit) by combining a carboxylic acid (a molecule with a carboxyl group, a —COOH unit, at one end) with an alcohol or phenol (molecules with a hydroxyl group, an —OH unit, connected to a carbon atom):

$$
\underset{\substack{\text{a carboxylic}\\\text{acid}}}{R-\overset{\overset{\textstyle O}{\|}}{C}-OH} + \underset{\substack{\text{an alcohol}\\\text{or a phenol}}}{HO-R'} \rightleftharpoons \underset{\text{an ester}}{R-\overset{\overset{\textstyle O}{\|}}{C}-O-R'} + H_2O \tag{1}
$$

In Reaction 1, R and R′ represent the "rest of the molecule"—organic molecular fragments made up of mostly hydrogen and carbon. There are many esters, because there are many carboxylic acids and alcohols/phenols, but they all can be formed, in principle at least, by Reaction 1. The driving force for Reaction 1 is, in general, not very great, so a mixture of ester, water, acid, and alcohol is obtained when Reaction 1 reaches equilibrium.

Some esters are solids—because of their high molecular weight or other properties. Most of these esters are not very soluble in water, so they can be separated from aqueous solutions by crystallization. This experiment deals with an ester of this kind: the substance commonly called aspirin.

Aspirin has been used since the beginning of the twentieth century to treat pain and fever. It remains one of the most widely used medicines in the world, with a daily production of about one hundred tons, or about thirteen regular-strength pills per year for each person on Earth. Aspirin actually works in two ways: as an anti-inflammatory, which explains its pain-killing properties, and as an anticoagulant, which is important for its use in preventing heart attacks and strokes. It works by inhibiting an enzyme that catalyzes the formation of precursors to prostaglandins—rather large organic molecules that regulate tissue inflammation—and of thromboxanes, which stimulate platelet aggregation: the first step in blood clotting.

Aspirin can be made (technical term: synthesized) by reacting the hydroxyl (—OH) group in salicylic acid, HOC_6H_4COOH, with the carboxyl (—COOH) group in acetic acid, CH_3COOH:

acetic acid   salicylic acid   aspirin (salicylic acid acetate)

A better preparative method, which you will use in this experiment and is shown in Reaction 3, replaces acetic acid with acetic anhydride, $(CH_3CO)_2O$. The anhydride can be considered to be the product of a reaction in which two acetic acid molecules combine, with the elimination of a molecule of water. The anhydride reacts with the water produced in Reaction 1, which helps drive the reaction to the right. An acid, normally sulfuric or phosphoric acid, is often added to act as a catalyst: speeding up the reaction without being itself changed or consumed in the process.

$$\underset{\text{acetic anhydride}}{\begin{array}{c}\text{O}\\\text{CH}_3-\text{C}\\\text{O}\\\text{CH}_3-\text{C}\\\text{O}\end{array}} + \underset{\text{salicylic acid}}{\begin{array}{c}\text{OH}\\\text{O}=\text{C}\quad\text{H}\\\text{C}=\text{C}\\\text{HO}-\text{C}\quad\text{C}-\text{H}\\\text{C}-\text{C}\\\text{H}\quad\text{H}\end{array}} \underset{\text{H}_3\text{PO}_4}{\rightleftharpoons} \underset{\text{aspirin}}{\begin{array}{c}\text{OH}\\\text{O}=\text{C}\quad\text{H}\\\text{O}\quad\quad\text{C}=\text{C}\\\text{CH}_3-\text{C}-\text{O}-\text{C}\quad\text{C}-\text{H}\\\text{C}-\text{C}\\\text{H}\quad\text{H}\end{array}} + \underset{\text{acetic acid}}{\begin{array}{c}\text{O}\\\text{CH}_3-\text{C}-\text{OH}\end{array}} \quad (3)$$

The aspirin you will <u>synthesize</u> (make) in this experiment is relatively impure; it should certainly not be taken internally, even if the experiment gives you a headache!

There are several ways in which the purity of your aspirin can be estimated. Probably the simplest way is to measure its melting point. If the aspirin is pure, it will melt sharply at the literature value of aspirin's melting point. If it is impure, the melting point will be lower than the literature value by an amount that is roughly proportional to the amount of impurity present, and it will melt over a wider temperature range.

A more numerical (technical term: <u>quantitative</u>) measure of the purity of your aspirin sample can be obtained by determining the mass percent of salicylic acid it contains. Salicylic acid is the most likely impurity in the sample because, unlike acetic acid but like aspirin, it is a solid that is not very soluble in water. Salicylic acid forms a strongly colored magenta complex with the iron(III) ion, Fe^{3+}. By measuring the absorption of light by a solution containing a known amount of your (impure) aspirin in excess Fe^{3+} ion, you will be able to determine the concentration of salicylic acid present in it.

Wear your safety glasses while performing this experiment.

Experimental Procedure

A. Synthesizing aspirin

Start warming up a water bath, which you will use later. Fill a 250-mL beaker with about one hundred and seventy milliliters of deionized water. If using a hotplate, set the surface temperature to 175 °C (about 20% power) and place the beaker directly on top. If using a laboratory burner, support the beaker with a wire gauze on an iron ring above a 3-cm-tall soft blue flame. Suspend a thermometer in the water and continue with the steps in the next paragraph while the bath warms up.

Add 2.0 g of salicylic acid, HOC_6H_4COOH, to a *dry* 50-mL Erlenmeyer flask. Measure out 5.0 mL of acetic anhydride, $(CH_3CO)_2O$, with a *dry* graduated cylinder, and pour it into the flask so that it washes any crystals of salicylic acid on the walls down to the bottom. Add about five drops of $85\%_{mass}$ phosphoric acid, H_3PO_4, to serve as a catalyst. **CAUTION:** **Both acetic anhydride and concentrated phosphoric acid are reactive chemicals that can give you a bad chemical burn, so use caution when handling them. If you get either chemical on your hands or clothes, wash thoroughly with soap and water. Acetic anhydride is a powerful lachrymator (a chemical that causes tears, like cutting onions does), so keep its fumes away from your eyes!**

Clamp the flask in your hot water bath, which should be at 75 ± 10 °C. (If it is not, make adjustments to get it there.) Stir the liquid in the small flask occasionally with a stirring rod, which you should leave in the flask. Maintain the heating bath temperature for about fifteen minutes, by which time Reaction 3 should reach equilibrium. **Cautiously**, add 2 mL of deionized water to the flask to decompose the excess acetic anhydride. There will be some hot acetic acid gas, $CH_3COOH(g)$, released as a result of this decomposition, so **keep your eyes and nose away**.

Without delay, remove the flask from the hot-water bath and add an additional 20 mL of deionized water. Let the flask cool slowly for a few minutes in air, during which time crystals of aspirin should begin to form. Put the flask in an ice bath to bring about more complete crystallization and increase the yield of aspirin. If crystals are slow to appear, it may help to scratch the inside of the flask with a stirring rod. Leave the flask in the ice bath for at least 5 minutes, stirring or swirling the flask periodically to help it cool completely.

Weigh a circle of dry filter paper to fit a Büchner funnel. Collect the aspirin by filtering the cold liquid in the flask through the Büchner funnel, under suction. (See Appendix D for more information.) Turn off the vacuum line, break the suction by twisting the vacuum hose as you pull it off the suction flask, and pour about five milliliters of ice-cold deionized water over the crystals; after about fifteen seconds, apply suction again to remove this wash liquid, along with most of the impurities in your product. Repeat the washing process with another five-milliliter sample of ice-cold water. Then pull air through the funnel for a few minutes to help dry the crystals; weigh a clean and dry watch glass while you wait. When you are ready, transfer the contents of the cup of the Büchner funnel onto the watch glass and determine the mass of the watch glass, product crystals, and filter paper to ± 0.1 g.

B. Analyzing synthesized aspirin

1. Solubility properties

Test the solubility properties of your aspirin by transferring as many crystals as will fit on the tip of a spatula into separate 1.0-mL samples of each of the following solvents, and stirring:

1. toluene, $C_6H_5CH_3$ (nonpolar <u>aromatic</u>: contains a ring with chemical resonance)
2. heptane, C_7H_{16} (nonpolar <u>aliphatic</u>: does not contain a ring with chemical resonance)
3. ethyl acetate, $C_2H_5OCOCH_3$ (aliphatic ester)
4. ethyl alcohol, C_2H_5OH (polar aliphatic, hydrogen-bonding)
5. acetone, CH_3COCH_3 (polar aliphatic, non-hydrogen-bonding)
6. water, H_2O (highly polar, hydrogen-bonding)

2. Melting point

To determine the melting point of your aspirin, add a small amount of your dried crystals to a melting point tube, as directed by your instructor. (The melting point tube will be made from 5-mm-diameter glass tubing

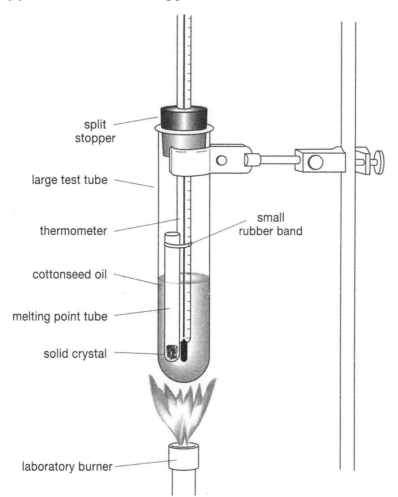

Figure 41.1 The apparatus shown here may be used to determine the melting point of your aspirin. **It is critical that the split stopper holding the thermometer not seal the large test tube completely.**

if the apparatus shown in Figure 41.1 is used; with most commercial melting point devices it will be much smaller.) Bring the solid down to the bottom of the melting point tube by tapping the tube on a hard surface, using enough solid to give you a depth of about five millimeters, or the amount called for by your instructor. If your lab has a commercial melting point apparatus, your instructor will show you how to use it. Otherwise, set up the apparatus shown in Figure 41.1. Fasten the melting point tube to the thermometer with a small rubber band, which should be above the surface of the oil. The thermometer bulb and sample should be about two centimeters above the bottom of the large test tube. Heat the oil bath *gently,* especially after the temperature gets above 100 °C. As the melting point is approached, the crystals of aspirin in the melting point tube will begin to soften and flow. Report the melting point as the temperature at which the last solid crystals disappear. You want the oil bath temperature to be rising *slowly* as this happens!

3. Impurity content

To analyze your aspirin for its salicylic acid impurity content, weigh out 0.10 ± 0.01 g of your crystals into a 100-mL beaker. Dissolve them in 5 mL of $95\%_{mass}$ ethanol, CH_3CH_2OH. Add 5 mL of 0.025 M iron(III) nitrate, $Fe(NO_3)_3$, in 0.5 M hydrochloric acid, HCl(aq), as well as 40. mL of deionized water. Make all volume measurements with a graduated cylinder. Stir with a stirring rod to mix the contents of the beaker well.

Blank a spectrophotometer cell at 525 nm using deionized water. (See Appendix D for more information.) Rinse out the cell with a few milliliters of the solution in the 100-mL beaker and then fill the cell with that solution. Measure the absorbance of the solution at 525 nm. From the calibration curve or equation provided, calculate the percent by mass of salicylic acid in your aspirin.

4. Observation

Carefully observe and record the appearance and smell of your dried product. **Do not taste it.**

> **Disposal of reaction products:** In most jurisdictions the contents of the suction flask may be poured down the drain but the toluene and heptane from the solubility tests should be put in a waste container. However, follow your instructor's directions regarding all waste materials.

Name ________________________ Section ________________________

Experiment 41

Data and Calculations: Making Aspirin

A. Synthesizing aspirin

Mass of salicylic acid used ________ g

Volume of acetic anhydride used ________ mL

Calculated mass of acetic anhydride used (density = 1.08 g/mL)

________ g

Mass of filter paper ________ g

Mass of empty watch glass ________ g

Mass of watch glass + aspirin + filter paper ________ g

Mass of aspirin obtained

________ g

Theoretical yield of aspirin

________ g

Percent yield of aspirin (by mass)

________ $\%_{mass}$

(continued on following page)

B. Analyzing synthesized aspirin

1. Solubility properties

Observed solubility properties of aspirin:

Toluene   _____   Ethyl alcohol   _____

Heptane   _____   Acetone   _____

Ethyl acetate   _____   Water   _____

S = soluble   I = insoluble   SS = slightly soluble

Circle the characteristics in the following list that seem to be important in a good solvent for aspirin.

organic   aliphatic   polar   hydrogen-bonding

inorganic   aromatic   nonpolar   non-hydrogen-bonding

2. Melting point

Melting point of synthesized aspirin   _____ °C

3. Impurity content

Absorbance of synthesized aspirin solution at 525 nm   _____

Mass percent of salicylic acid impurity in synthesized aspirin   _____ $\%_{mass}$

4. Observation

Observations on your product crystals (synthesized aspirin):
(What color are the crystals, if they have a color at all? Do they have a regular shape? How large are they? Do they smell like anything? *Note*: **Do not *taste* your product!**)

Experiment 41

Advance Study Assignment: Making Aspirin

1. Calculate the theoretical yield of aspirin from Reaction 3, if the starting materials are 2.1 g of salicylic acid and 5.0 mL of acetic anhydride (density = 1.08 g/mL), and the reaction is assumed to go to completion (to create as much aspirin as the limiting reagent would allow, if it were completely consumed).

_____ g

2. If 1.9 g of aspirin were obtained in this experiment, what would the percent yield by mass be?

_____ %$_{mass}$

3. The name "acetic anhydride" implies that the compound with that name will react with water to form acetic acid. Write the balanced equation for the reaction by which this occurs.

4. Identify R and R′ in Reaction 1 when the ester, aspirin, is made from salicylic acid and acetic acid.

5. In the presence of water, esters such as aspirin can decompose by the reverse of Reaction 1. Write the balanced equation for the reaction by which aspirin decomposes in a humid (wet) environment.

Experiment 42

The Kinetics of Aspirin Decomposition

Once aspirin has been prepared, it is relatively stable: in tablet form, stored in an airtight container, it resists decomposition for years. But if it is exposed to moisture, especially at elevated temperatures, it will slowly break down by the following reaction:

$$\text{aspirin} + \text{water} \rightleftharpoons \text{salicylic acid} + \text{acetic acid} \tag{1}$$

This reaction is one of many good arguments against storing medications in a bathroom medicine cabinet! Cool, dry places are far kinder to most medicines.

The study of the rates of chemical reactions is called <u>chemical kinetics</u>. The rate at which a chemical reaction occurs depends on several factors, including the nature of the reaction, the concentrations of the reactants, the temperature, and the presence of <u>catalysts</u> (substances that increase the rate of a chemical reaction without being chemically altered themselves). Each of these factors can have a big influence on the observed rate of a chemical reaction.

Some reactions at a given temperature are very slow: the reaction of oxygen with hydrogen gas, or with wood, does not proceed to a visible extent even after 100 years at room temperature. Other reactions happen as quickly as their reactants come in contact: the precipitation of silver chloride when solutions containing silver ions and chloride ions are mixed, and the formation of water when acidic and basic solutions are mixed, are examples of extremely fast reactions. In this experiment you will study a reaction that, at room temperature, proceeds at a rate you will be able to measure: not too fast, not too slow.

For a given reaction, the rate typically increases with an increase in the concentration of any reactant. Consider the balanced chemical reaction

$$a\,\text{A} + b\,\text{B} \rightarrow c\,\text{C}$$

in which A, B, and C are chemical substances and a, b, and c are stoichiometric coefficients. The relationship between reaction rate and concentrations, called the *rate law*, can usually be expressed by the equation

$$\text{rate} = k[\text{A}]^m[\text{B}]^n \tag{2}$$

in which m and n are generally, but not always, counting numbers: specifically 0, 1, 2, or possibly 3. [A] and [B] are the concentrations of A and B (ordinarily in moles per liter) and k is a constant, called the <u>rate constant</u> of the reaction. The numbers m and n are called the <u>orders of the reaction</u> with respect to A and B. If m is 1, the reaction is said to be <u>first order</u> with respect to the reactant A. If n is 2, the reaction is <u>second order</u> with respect to reactant B. The <u>overall reaction order</u> is the sum of m and n. In this example, the reaction would be <u>third order</u> overall. Note that the orders of the reaction do not have a reliable relationship to the stoichiometric coefficients (a, b, and c) of the overall balanced reaction.

The rate of a reaction also strongly depends on the temperature at which the reaction occurs, which is reflected in a strong temperature dependence of the rate constant, k. An increase in temperature increases the rate; a general rule that a $10\,\text{C}°$ rise in temperature will double the rate often works pretty well, though it is only a general approximation. Nevertheless, it is clear that an increase in temperature on the order of $100\,\text{C}°$ can really change the rate of a reaction!

A more precise equation relating temperature to reaction rate is known, but it is more mathematically complicated. It is based on the idea that in order to react, reactants must collide with a certain minimum amount of energy, typically provided by the kinetic (motion) energy of the reactants—which is indicated by their temperature. This required amount of energy is called the <u>activation energy</u> for the reaction. The equation relating the rate constant, k, to the absolute temperature, T, and the activation energy, E_a, is called the <u>Arrhenius equation</u>, and is most often written in this form:

$$\ln k = \frac{-E_a}{RT} + \text{constant} \qquad (3)$$

In Equation 3, $\ln k$ is the natural logarithm of the rate constant, k, and R is the gas constant [for E_a in joules per mole, 8.3145 joules/(mole·K)]. By measuring k at different temperatures it is possible to determine the activation energy for a reaction.

In this experiment you will study the rate at which aspirin decomposes in aqueous solution at 80°C. Like many organic reactions, this one is slow at room temperature. Using this elevated temperature will allow you to carry out the reaction in a few minutes rather than several hours. To analyze the reaction mixture, you will determine the amount of salicylic acid present in it after a measured amount of time. Iron(III) forms an intensely colored violet complex with salicylic acid, so after adding a source of iron(III) to the reaction mixture you can use the amount of light the mixture absorbs to find out how far Reaction 1 has progressed. This will allow you to determine the rate of Reaction 1 at different concentrations. You will also study Reaction 1 at 70°C, which will take longer but allow you to investigate the temperature dependence of the rate constant. Your goal will be to use the observed rates to establish the order of Reaction 1, its rate constant (at two different temperatures), and its activation energy.

Under the conditions of this experiment (with the water "concentration" constant, because it takes place in aqueous solution), the rate expression for the decomposition of aspirin in hot water can be simplified to just

$$\text{rate} = k \, [\text{aspirin}]^n \qquad (4)$$

in which n is the order of the reaction in aspirin.

Experimental Procedure

Wear your safety glasses while performing this experiment.

Obtain a thermometer and, if so directed, a 10-mL graduated measuring pipet or autopipet.

A. Measuring the rate of reaction

Step 1. Setting the stage Set up a hot-water bath, using a 600-mL beaker about three-quarters full of water. As a heat source, use an electric hotplate with a magnetic stirrer if these are available. Otherwise, use a laboratory burner, supporting the beaker on a wire gauze on an iron ring. If you use a burner, you should put a larger iron ring around the beaker, near the top, to prevent it from tipping over. Clamp this larger ring to the ring stand as well. If using a hotplate, set the surface temperature to 185°C (about 20% power); if using a laboratory burner, use a 3-cm-tall soft blue flame. Suspend a thermometer in the water and leave the bath to warm up. Prepare an ice bath by filling a 400-mL beaker with crushed ice, then adding water until the level is at the top of the ice. Fill three medium (18 × 150 mm) test tubes with deionized water and let them sit for at least a minute before pouring them out. Repeat this at least one more time: the goal is to ensure that no traces of acids or other possible catalysts for Reaction 1 remain in the tubes. Let the tubes drain, upside down in a test tube rack, for at least a minute. Now rinse the three tubes with the same few milliliters of acetone: pour the acetone into one of the tubes, shake it, then pour this same acetone into the next tube and repeat; pour the acetone into the last test tube and shake a final time. Pour out the acetone and dry the tubes upside down in the test tube rack or with a gentle stream of compressed air. Number the test tubes and put them in the test tube rack.

Step 2. Weighing the sample While the hot-water bath is heating, weigh out about sixty milligrams of aspirin, to ± 0.0001 g, into the first test tube. This is most easily done by putting the test tube into a beaker on the balance, so that it is held vertically, and taring it out. Then use a small spatula to add a small amount of aspirin. Add more aspirin, a very small amount at a time, until you have 60 ± 5 mg (0.060 ± 0.005 g) in total. Record the added mass on the report page.

Step 3. Preparing the reaction mixture With a twice-rinsed 10-mL graduated cylinder or pipet, add 10.0 mL of deionized water to the aspirin in Tube 1. The aspirin will not dissolve in the water; some will stay at the bottom of the tube, and some will float on the water. Rinse your stirring rod well (to ensure it has no catalysts on it) and then put it in the test tube.

Step 4. Carrying out the reaction By now the water in the 600-mL beaker should be getting warm. Adjust your heat source as necessary to hold the bath steady at $80 \pm 1\,°C$ for at least 5 minutes. Once you have managed that, you are ready to take your first kinetic measurement. Record the initial temperature of the water bath, under Item 8 on the back of the report page. Pick up the tube with the reaction mixture in it and, noting the time to the nearest second or starting a stopwatch, put the tube in the hot-water bath. Stir the mixture continuously, to help dissolve the aspirin and warm the reaction mixture up quickly. Once the aspirin has dissolved completely, which should happen within a minute, keep the stirring rod in the tube, stirring occasionally. Measure and record the temperature of the hot-water bath at 1-minute intervals as the reaction proceeds; it will fall at first (because of the cold test tube placed in the bath), but should come back to the original temperature fairly quickly. Leave the tube in the bath for 6.0 minutes. Then remove the tube and put it in the ice-water bath to quickly stop the reaction, stirring the reaction mixture to cool it as quickly as possible. Within a minute the solution should be cool, and you can put the tube back in the test tube rack.

Step 5. Adding the indicator Use a graduated cylinder or pipet to measure out 5.0 mL of 0.025 M iron(III) nitrate, $Fe(NO_3)_3$, in 0.5 M hydrochloric acid, HCl, into the reaction mixture in the cooled test tube. The solution should turn violet as the iron–salicylic acid complex forms. Stir the mixture well.

Step 6. Determining the salicylic acid concentration Measure the absorbance of the violet solution in the test tube at 525 nm on a spectrophotometer. (If this is your first time measuring absorbance, the procedure is described in more detail in Appendix D.) After blanking the spectrophotometer cell, rinse it twice with small amounts of the solution from Step 5. Then add a few milliliters of that solution to the cell and record the absorbance. Find and record the concentration of salicylic acid in the reaction mixture, using the graph or equation you are given relating absorbance and concentration. The graph or equation will provide the salicylic acid concentration, [Sal], in the reaction mixture *before* the indicator was added. Reaction 1 indicates this concentration will equal the change in the aspirin concentration, $\Delta[Asp]$, because you can assume that the aspirin starts out pure—that is, containing no salicylic acid—and that all of the salicylic acid present came from Reaction 1.

Step 7. Correcting for warm-up time Once the reaction mixture warms up to $80.\,°C$, the rate of Reaction 1 will obey Equation 2; but during the first minute, when the aspirin is dissolving and the mixture is warming up, the reaction is much slower. To correct for this, you might run the same size sample at $80.\,°C$ for 1.0 minute, then subtract the absorbance for that sample from the total absorbance, using 5.0 minutes as the time over which the reaction occurred. In developing this experiment we actually tried this, but we found that taking account of the warm-up period by using the total absorbance and 5.5 minutes for the time of reaction gave results that were essentially the same as with the more accurate—but more complicated—procedure. So, when calculating the rate, you should use this simpler approach: use 5.5 minutes as the reaction time, Δt, along with the concentration change, $\Delta[Asp]$, indicated by your work in Step 6.

B. Determining the order of the reaction and the rate constant

Reset your water bath to the temperature you used in Part A, and prepare a fresh ice bath.

Repeat Steps 1 through 7, again at $80 \pm 1\,°C$, but using about forty milligrams of sample, weighed to ± 0.0001 g, in Test Tube 2. If you used it to measure out the acidic iron solution, be sure to rinse out your graduated cylinder or pipet with deionized water before measuring out the 10.0 mL of water. [Any acidity carried over from the acidic iron(III) solution you measured out in Step 5 would catalyze Reaction 1, causing it to happen much faster than it should.] The results you obtain with this sample, along with those from Part A, will allow you to find the order of Reaction 1 and its rate constant at 80.°C. On the report page carry out the calculations for the order and the rate constant of the reaction. Calculate the average (mean) value of the rate constant and the standard deviation. (See Appendix H.) Then proceed to Part C.

C. Determining the activation energy

The last piece of information to be obtained in this experiment is the activation energy for Reaction 1, which is the energy that must be present for the slowest step in the process by which the reaction proceeds. It is found by measuring the rate constant's dependence on temperature.

This can be determined using nearly the same procedure as in Part A, Steps 1 through 7, except that you will carry out the reaction in Test Tube 3 at 70.°C instead of 80.°C. The sample size should be about sixty milligrams, weighed to ± 0.0001 g. Bring the water bath to $70 \pm 1\,°C$ and hold it there for 5 minutes. Make a fresh ice bath, and again—if you used it to measure out the acidic iron solution—make sure you rinse your pipet or graduated cylinder before measuring out the water.

At this lower temperature, the reaction will go much more slowly than in Part A. In order to get easily measured absorbance values, run the reaction for 12 minutes rather than 6 minutes. The aspirin will be slower to dissolve, so the warm-up period will be 2 minutes rather than 1 minute. It will require a considerably longer time to dissolve the aspirin, but, with continuous stirring, it should go into solution in 2 minutes. Record the temperature at 2-minute intervals. Record the absorbance that you obtain. To correct for warm-up time, calculate the reaction rate using the measured absorbance but assume the effective time, Δt, to be *11 minutes*.

With the data you have obtained, you can now calculate the activation energy, E_a, for Reaction 1. Do this using the rate constants at 80.°C and 70.°C and the Arrhenius equation, Equation 3.

Disposal of reaction products: All of the reaction products from this experiment may be poured down the sink, unless your instructor directs otherwise.

Take it further (optional): Bring in a sample of commercial aspirin and determine the extent to which, if any, it has decomposed to salicylic acid and vinegar. (Sniffing the container when you first open it will also give you a pretty good indication!) Note that medication with a water-resistant coating is unlikely to have decomposed at all, since the coating keeps water away from the aspirin molecules inside.

Experiment 42

Data and Calculations: The Kinetics of Aspirin Decomposition

Let Asp = aspirin (MM_{Asp} = 180.2 g/mol) and Sal = salicylic acid

	Part A Tube 1	Part B Tube 2	Part C Tube 3
1. Mass of aspirin in tube {grams}	═══════	═══════	═══════
Moles of aspirin in tube {mol}	_____	_____	_____
(Initial) aspirin concentration, [Asp] (solution volume = 10.0 mL) {mol/L} (Record temperature measurements on the back of this page.)	_____	_____	_____
2. Absorbance of analyzed solution, Abs	═══════	═══════	═══════
Salicylic acid concentration in reaction mixture, [Sal] {mol/L}	═══════	═══════	═══════
Change in [Asp], Δ[Asp] = −[Sal] {mol/L} (*Note*: Δ[Asp] << [Asp], so [Asp] can be approximated as constant.)	_____	_____	_____

3. Rate of Reaction 1 = $\dfrac{-\Delta[\text{Asp}]}{\Delta t}$ = $-k[\text{Asp}]^{n}$

                                                                        (4a)

    Rate of Reaction 1, based on Δt = 5.5 min {mol/(L·min)}   _____ _____

    Rate of Reaction 1, based on Δt = 11 min {mol/(L·min)}   _____

4. $\dfrac{\text{Rate in tube 1}}{\text{Rate in tube 2}}$ =   _____

(*continued on following page*)

	Part A	Part B	Part C
	Tube 1	Tube 2	Tube 3

5. $\dfrac{[\text{Asp}] \text{ in Tube 1}}{[\text{Asp}] \text{ in Tube 2}}$ _____

Reaction 1 is _____ order (Use Equation 4a for Tube 1 divided by Equation 4a for Tube 2; round off your result to the nearest counting number, and show your work or explain your reasoning here.)

6. Value of rate constant, k {min^{-1}}   _____ _____ _____

Average (mean) value of k at 80.°C _____   Standard deviation in $k_{80°C}$ _____

7. $\dfrac{k_{\text{mean}} \text{ at } 80.°C}{k \text{ at } 70.°C} =$ _____

8. Temperature measurements, t, taken during reactions:

Tubes 1 and 2 at
time {min}

	0	1	2	3	4	5	6	Average
								(mean)
Tube 3 at time {min}	0	2	4	6	8	10	12	temperature

t in Tube 1 {°C}   ===== ===== ===== ===== ===== ===== =====   _____

t in Tube 2 {°C}   ===== ===== ===== ===== ===== ===== =====   _____

t in Tube 3 {°C}   ===== ===== ===== ===== ===== ===== =====   _____

9. $\ln k = -E_a/RT + \text{a constant}$ (Requires T in K, not °C!)   **(3)**

Solve for E_a:

The activation energy, $E_a =$ _____ J = _____ kJ

Experiment 42

Advance Study Assignment: The Kinetics of Aspirin Decomposition

1. A student studies the rate at which aspirin decomposes by Reaction 1,

$$\text{aspirin} + \text{water} \rightleftharpoons \text{salicylic acid} + \text{acetic acid} \qquad \qquad (1)$$

She weighs out 59.4 mg of aspirin ("Asp") and dissolves it, making a 10.0-mL solution in water. She heats the solution for 5.0 minutes at 90.°C (which means 90 ± 1 °C), and finds that 10.% ($10\% \pm 1\%$) of the aspirin is converted to salicylic acid and acetic acid.

a. How many moles of aspirin were in the initial solution, $n_{\text{Asp, initial}}$? What was the initial molarity of the aspirin, $[\text{Asp}]_{\text{initial}}$? The molar mass of aspirin is 180.2 g/mol.

_____ moles _____ M

b. What is the rate of reaction? (See Equation 4a on the report page; treat [Asp] as constant.)

_____ moles/(L · min)

c. In a similar experiment with a smaller amount of aspirin, and thus a lower [Asp], she finds that once again 10.% of the aspirin decomposes in 5.0 minutes at 90.°C. What is the order of Reaction 1? How do you know?

d. What is the rate constant for Reaction 1, based on this data? Be sure to include its units! Show your work here.

$k = \text{_____}$

(continued on following page)

e. In another experiment she determines that the activation energy, E_a, for Reaction 1 is 95 kJ/mol. How long would it take for 10.% of an aspirin sample to decompose at body temperature, 37 °C? Show your work in detail. (*Note*: In reality, because Reaction 1 is catalyzed by acids and your stomach is very acidic, aspirin decomposes much more quickly in your stomach; but what you calculate here is the approximate rate at which aspirin would decompose in your bloodstream, at least without any enzymes acting on it.)

_____ minutes

Experiment 43

Analysis for Vitamin C

Vitamin C, known chemically as ascorbic acid, is an important component of a healthy diet. In the mid-eighteenth century the British Royal Navy found that adding citrus fruit to sailors' diets prevented the illness called scurvy. We humans are one of the few members of the animal kingdom unable to make (technical term: synthesize) vitamin C ourselves, meaning we need to have sources of it in our diets in order to remain healthy. The National Academy of Sciences has established a threshold of 75 mg/day for (nonpregnant) female adults and 90 mg/day for male adults as the Reference Daily Intake (RDI). Linus Pauling, a famous chemist, recommended an intake of 500 mg/day to help ward off the common cold. He also suggested that large doses of vitamin C are helpful in preventing cancer. Although Pauling's claims were based on experimental data, he turned out to be a better chemist than he was a pharmacologist; his judgements about the importance of vitamin C were clouded by his own personal obsession with it.

Vitamin C does play an important role in the body as a reducing agent (a chemical that can easily and permanently give up electrons in a chemical reaction). Because vitamin C is oxidized more easily than most substances in our bodies, it acts as an antioxidant, meaning it can react with and thereby neutralize potentially damaging oxidizing agents (chemicals that permanently take away electrons) in our bodies, preventing them from reacting with the body itself.

The vitamin C content of foods can be determined by oxidizing (taking electrons away from) ascorbic acid, $C_6H_8O_6$, to (form) dehydro-L-ascorbic acid, $C_6H_6O_6$:

$$C_6H_8O_6 \quad \rightarrow \quad C_6H_6O_6 \quad + 2\,H^+ + 2\,e^- \tag{1}$$

Written using Lewis structures, the same electrochemical half-reaction looks like this:

$$+ 2\,H^+ + 2\,e^- \tag{1}$$

This reaction is very slow for truly dry ascorbic acid, but contact with moisture speeds it up. An oxidizing (electron-removing) agent that is particularly good for oxidizing vitamin C is an aqueous solution of iodine, I_2. Because I_2 is not very soluble in water, it is dissolved in a solution of potassium iodide, KI, in which the I_2 exists mainly as I_3^-, a (much more soluble) complex ion. The reaction with ascorbic acid involves aqueous elemental iodine, $I_2(aq)$, which is reduced (gains an electron) to (form) iodide ion, $I^-(aq)$, in this electrochemical half-reaction:

$$2\,e^- + I_2(aq) \rightarrow 2\,I^-(aq) \tag{2}$$

yellow   colorless
to red

In the overall reaction, one mole of ascorbic acid requires one mole of I_2 for complete oxidation—because the two electrons generated in Reaction 1 are consumed in Reaction 2 when the reactions are added together to give the overall reaction. (A balanced overall reaction can not create or destroy electrons!)

A convenient method for determining the amounts required for complete reaction with a sample is called titration. A solution of known concentration (the titrant) is added in measured amounts to the unknown sample until the reaction stops occurring.

In this experiment, you will titrate samples containing vitamin C using aqueous iodine, I_2(aq), as the titrant. When the colored I_2 titrant is added to the colorless ascorbic acid solution, the characteristic yellow to reddish brown (depending on concentration) iodine color of the titrant disappears because of Reaction 2. Although the first permanent survival of the yellow color of dilute iodine in the titration flask could be used to mark the endpoint of the titration, better results are obtained when starch is added as an indicator. Starch reacts with I_2 to form an intensely colored blue complex. In the titration I_2 reacts quickly with ascorbic acid, so its concentration remains very low until the ascorbic acid is completely oxidized. At that point, the I_2 concentration begins to go up, and the reaction with the starch indicator starts to occur:

$$I_2(aq) + starch \rightarrow starch\text{--}I_2\,complex(aq) \tag{3}$$
yellow   intense blue

Because an I_2 solution cannot be prepared accurately by direct weighing, it is necessary to <u>standardize</u> the I_2 titrant (determine its concentration both accurately and precisely) against a reference substance of known purity (a <u>primary standard</u>). You will use pure ascorbic acid as your primary standard. Once the iodine titrant is standardized, you can use it for the direct determination of vitamin C in any kind of unknown sample.

Experimental Procedure

Wear your safety glasses while performing this experiment.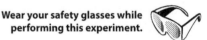

A. Standardizing the iodine titrant

Obtain a buret and an unknown vitamin C sample. Weigh out three pure ascorbic acid samples (not your unknown!) of approximately one-fifth of a gram, to ±0.0001 g, into clean 250-mL Erlenmeyer flasks. Dissolve each sample in approximately one hundred milliliters of water.

Clean your buret thoroughly. Rinse the buret with a few milliliters of the I_2 titrant, draining through the stopcock, three times. Then fill the buret with the I_2 titrant. (See Appendix D for more details.)

After taking and recording an initial buret reading (you may find looking toward a light source will make it easier to see the bottom of the meniscus in this dark titrant), add about one milliliter (about twenty drops) of starch indicator to the first ascorbic acid sample and titrate it with the iodine solution. Note the disappearance of the I_2 titrant's color as it mixes into the contents of the Erlenmeyer flask. Swirl the flask gently and continuously during the titration. Continue adding the iodine titrant, using progressively smaller volume increments, until the sample solution turns (and remains) a distinct blue. After reading and refilling the buret, titrate the other two sample solutions using the same technique—being sure to add the starch indicator first and to read your buret at the start and end of each titration.

B. Analyzing an unknown containing vitamin C

Use a similar procedure to determine the vitamin C content of your unknown sample. You will need to select a sample size; keep in mind that your unknown will not be pure ascorbic acid, so you will need more than a fifth of a gram of it, and that you may need to carry out some initial treatment of the sample. In particular, if you are working with fruit juice or another sample containing insoluble impurities it may be helpful to first filter the sample through filter material of some kind, and then rinse the filter with water to bring through any remaining vitamin C. In choosing your sample size, note that it is difficult to obtain accurate results if the titrant (I_2 solution) volume used in a titration is less than 15 mL. Carrying out a rough initial titration using a small sample of your unknown can provide you with an indication of how much to scale up for your "real" titrations.

If your unknown is a solid, report your results as the percent of vitamin C in the solid, by mass. For liquid unknowns, report how many milligrams of vitamin C are present in 100 mL of the liquid. In each case, calculate the average (mean) and standard deviation of the measured vitamin C concentrations in your unknown, as well as the amount of unknown required to provide the nonpregnant female RDI of vitamin C.

Disposal of reaction products: All reaction products from this experiment may be poured down the drain, unless your instructor directs otherwise.

Experiment 43

Data and Calculations: Analysis for Vitamin C

A. Standardizing the iodine titrant

Titration #	1	2	3
Mass of pure ascorbic acid used {g}	_____	_____	_____
Moles of ascorbic acid (MM_{Asc} = 176.14 g/mol) {mol}	_____	_____	_____
Initial buret reading {mL}	_____	_____	_____
Final buret reading {mL}	_____	_____	_____
Volume of I_2 titrant added {mL}	_____	_____	_____
Moles of I_2 consumed {mol}	_____	_____	_____
Concentration of I_2 titrant {moles/L}	_____	_____	_____

Average (mean) concentration of I_2 titrant: _____ M

Standard deviation in I_2 titrant concentration: _____ M

(continued on following page)

B. Analyzing an unknown containing vitamin C

Titration #	1	2	3
Mass or volume of unknown {g or mL}	═══	═══	═══
Initial I_2 titrant buret reading {mL}	═══	═══	═══
Final I_2 titrant buret reading {mL}	═══	═══	═══
Volume of I_2 titrant added {mL}	───	───	───
Moles of iodine added {mol}	───	───	───
Moles of vitamin C in sample {mol}	───	───	───
Mass of vitamin C in sample (MM_{Asc} = 176.14 g/mol) {g}	───	───	───

Concentration of vitamin C:
$\%_{mass}$ in solid sample or {mg/100 mL} in liquid sample _____   _____   _____

Average (mean) vitamin C concentration in unknown _____ $\%_{mass}$ or {mg/100 mL}

Standard deviation in vitamin C concentration in unknown _____ $\%_{mass}$ or {mg/100 mL}

Calculate how much of your unknown is required to provide the nonpregnant female RDI amount of vitamin C. (Be sure to specify the units of your answer!)

Experiment 43

Advance Study Assignment: Analysis for Vitamin C

1. Write a balanced equation for the reaction between I_2 and ascorbic acid ("Asc"). Identify the oxidizing agent (the substance that gives up electrons) and the reducing agent (the substance that gains electrons).

2. A solution of I_2 was standardized with ascorbic acid. Using a 0.2026-g sample of pure ascorbic acid, 23.43 mL of I_2 titrant was required to reach the starch endpoint.

   a. How many moles of ascorbic acid were titrated? (The molar mass of Asc is 176.14 g/mol)

   _____ mol

   b. How many moles of iodine, I_2, would be required to completely oxidize that much vitamin C?

   _____ mol

   c. How many liters of titrant were used in the titration?

   _____ L

   d. What is the concentration of the iodine titrant, in moles per liter?

   _____ M

(continued on following page)

3. A sample of fresh grapefruit juice was pureed, filtered, and titrated with the I_2 solution from Problem 2. A 500.-mL (the decimal point here indicates 500. has three significant figures: it indicates 500 ± 1 mL) sample of the juice required 21.86 mL of the iodine titrant to reach the starch endpoint.

a. How many moles of I_2 were required for complete titration?

_____ mol

b. How many moles of vitamin C does that indicate were present in the juice?

_____ mol

c. How many milligrams of vitamin C is that?

_____ mg

d. What is the concentration of vitamin C in the juice in (mg vitamin C)/(100 mL of juice)?

_____ mg/100 mL

e. What quantity of this juice will provide the male RDI amount of vitamin C?

_____ mL

Experiment 44

Fundamentals of Quantum Mechanics

For many years, scientists relied on their first-hand experience with the physical world to guide their scientific interpretation of it: if an idea did not make physical sense in light of their own experience, they discarded it. It was often a scientist's interactions with the physical world (such as Newton seeing a falling apple) that guided and inspired the creation of scientific theories. However, in 1900 the physicist Max Planck, as part of an effort to make electric light bulbs more efficient, was attempting to mathematically describe blackbody radiation (the light emitted from something glowing-hot, such as an incandescent light bulb filament), and he just could not get it to work. In desperation, he made a mathematical assumption that at the time he thought was physically ridiculous: he assumed that light energy could not come in just any amount, but rather that it had to be a multiple of a minimum quantity—he assumed light was quantized. Remarkably, this made the mathematics work out, and with this assumption he was able to mathematically describe blackbody radiation perfectly. The idea of quantization eventually became the foundation of a whole new way of looking at the world, called quantum mechanics. It is a strange view, one that at first does not seem to connect with our direct experience of the physical world at all. But because it underlies our understanding of a lot of modern technology, from lasers to computer chips, as well as today's understanding of chemical bonding, it is worth trying to gain a better feel for.

As you will learn in this experiment, there are some situations where quantization *is* something we can experience first-hand. You may also get a pretty good workout doing this experiment!

We now understand that *everything* is quantized: from the speed at which a bicycle or a baseball can travel to the color of light that can be emitted by the relaxation of an excited atom. However, for macroscopic objects (objects large enough to see with the naked eye), the spacing between allowed levels is so small that we can never pick up on the quantization. That is why the concept of quantized states seems so foreign to us. Consider the mass of water in a bucket: in a way, it is quantized—in that you can only add or take away a discrete number of water molecules. You cannot add one-third of an H_2O molecule, or take half of one away. But the mass of a single water molecule is so incredibly small that we do not observe this quantization on the human scale. In our experience, the mass of water in a bucket can be varied continuously—set to any amount we wish. The quantization only becomes noticeable when working with an incredibly small mass of water—that is, on a submicroscopic (or "quantum") scale. However, there are a few everyday things that exhibit quantization we *can* observe first hand, and almost all of these things have some relationship to musical instruments. In this experiment you will study one of these: the oscillation (movement in a regular pattern) of a vibrating (oscillating about an equilibrium point) string.

Consider a rubber band stretched between two fixed points. If you flick this rubber band, it will vibrate back and forth for a while before coming to rest again. If you listen carefully, you will even hear it make a tone as it vibrates. What is interesting is that this tone has the same pitch whether you flick the band gently or with a lot of force. (The sound may be louder or softer depending on how hard you flick, but the pitch—the note it plays—is the same.) The pitch remains (essentially) constant throughout the time the rubber band vibrates. This is because the vibration is quantized. So too are the vibrations of a guitar string, a piano string, the reed in a woodwind instrument such as a clarinet, and the air in the cavity of a brass instrument such as a trombone. That is why these are musical instruments: they produce a reliable, fixed tone, whether played loudly or softly.

As with most musical instruments, we can tune a rubber band. Its pitch depends on how long it is, the mass of the material it is made of, and how much tension it is under. Generally, the vibrations of a rubber band are too rapid for us to count easily (which is why we can hear them), so we are going to use a different, more slowly moving model to learn some of the general rules that apply to quantized systems. Our model system will consist of spinning a series of jump ropes (or something similar), sometimes in ways that would not work so well for the actual game of jump rope.

Experimental Procedure

You will be given a 4-meter-long section of standard "rope" marked at the 3 m point—this "rope" may actually be lab tubing or something similar. You will study the oscillation of this rope—how it moves back and forth, up and down, or around and around, over and over again. (In this experiment the oscillation will be around and around: turning in a circle as in a game of jump rope.) You will investigate the <u>frequency</u> of oscillation—how often, per unit time, the repeated pattern of motion returns to the same point—under different conditions. You should find that the rope has a series of favored frequencies, called <u>resonant frequencies</u>, under a given set of conditions. Some conditions, such as the length and weight of the rope, matter, while others, such as the color of the rope or whether you are left- or right-handed, should have no effect.

A. The effect of quantum state on resonant frequency

Take one end of your rope and give the other end to your partner. Hold the very end of the rope: this ensures that the length of the oscillating rope is clearly defined. Also, hold the rope consistently at waist level, just in front of you. Have your partner move away from you until the rope is stretched out in a straight line between you, then move together until the rope is suspended just barely above the ground between you. The tension on the rope matters: you want to keep it constant, so in this part always hold your rope such that when it is stationary it just barely touches the ground. Try to stay in that position as you do the experiment.

Now, with your partner holding their hand still, rotate your end of the rope so that the entire rope starts to spin, as it would in a game of jump rope. You may have to make exaggerated loops with your arm to get started, but once you have, try to move the hand spinning the rope only a little. Spin the rope fast enough that its shape is consistent all the way around the loop—that is, do not go too slowly. Once you have the rope going in a regular, stable pattern, have your partner time (with his or her free hand, so that their end of the rope remains stationary) how long it takes for the rope to go around 50 times. Stop the rope and record this time on the report page.

Set up again, but this time when you spin the rope, get it going a good deal faster. If you do this right, you will end up having two segments of rope spinning between you and your partner, with a stationary point in the middle. This stationary point is called a <u>node</u>, and it indicates that the oscillation of the rope is now in a higher <u>quantum state</u> than it was when you were spinning it more slowly. Specifically, you are now in the <u>first excited state</u>, while before (with no nodes) you were in the <u>ground state</u>. You will find it takes a lot more energy to maintain the oscillation in the excited state—you will get tired doing this! So without delay, your partner should start timing and again determine how long it takes for the rope to go around 50 times. Stop the rope and record this time on the report page.

Finally, set up as before but this time spin the rope still faster, so you get *two* nodes. (This is called the <u>second excited state</u>.) Again, time how long it takes the rope to go around 50 times and record the result.

Each quantum state is assigned a number, called the <u>principal quantum number</u>, n; it is equal to one more than the number of nodes observed in that quantum state. This means the lowest possible n is 1, and in the first excited state, with one node, $n = 2$. In the second excited state, with two nodes, $n = 2 + 1 = 3$.

Optional   Try to spin the rope so fast you get *even more* nodes. (These are called higher-order excited states: the third, the fourth, and so on.) This will be hard to do! You do not have to record anything here, but it will give you an even clearer sense of the relationships between nodes, excited states, and energy. ▪

B. The effect of length on resonant frequencies

Repeat the previous steps (time the ground state, first excited state, and, if possible, the second excited state) with one of you holding the rope at the 3-meter mark, again controlling the tension by holding the rope so that it barely touches the ground when it is at rest. You or your partner may want to coil up the extra rope and hold it, so that it does not get tangled in the spinning section of the rope. Record your results on the report page.

C. The effect of tension on resonant frequencies

For this part, take your standard rope and prepare to do the experiment again, holding it at the end as you did in Part A. However, once you have the rope in the position you used previously, put one of your feet immediately behind the other (heel to toe) and step back that far (about ten inches). There should now be more tension in the rope, and it should be hanging a little above the ground. Repeat the steps in Part A, taking note of whether the frequency of oscillation in each quantum state is higher or lower than it was in Part A.

Optional **D. The effect of mass on resonant frequencies**

For this last part, repeat Part A with an alternate rope (made of a different material) to determine whether a change in mass (with approximately the same length and tension) causes an increase or decrease in resonant frequency. (If it is not obvious, your instructor may tell you whether the standard or the alternate rope has a larger mass per unit length.)

Experiment 44

Data and Calculations: Fundamentals of Quantum Mechanics

A. The effect of quantum state on resonant frequency

Time required for standard rope to complete 50 cycles in ground
(lowest energy, zero node, $n = 1$) state: _____ sec

Time required for standard rope to complete 50 cycles in first excited
(one node, $n = 2$) state: _____ sec

Time required for standard rope to complete 50 cycles in second excited
(two node, $n = 3$) state: _____ sec

Resonant frequency of standard rope in ground (lowest energy, zero node, $n = 1$)
state: _____ Hz

Resonant frequency of standard rope in first excited (one node, $n = 2$) state: _____ Hz

Resonant frequency of standard rope in second excited (two node, $n = 3$) state: _____ Hz

Does the resonant frequency tend to increase or decrease when going to higher
excited states (states with more nodes and higher principal quantum numbers)?

increase / decrease
(circle one)

B. The effect of length on resonant frequencies

Time required for standard rope held at 3-m mark to complete 50 cycles in ground
(lowest energy, zero node, $n = 1$) state: _____ sec

Time required for standard rope held at 3-m mark to complete 50 cycles in first
excited (one node, $n = 2$) state: _____ sec

Time required for standard rope held at 3-m mark to complete 50 cycles in second
excited (two node, $n = 3$) state: _____ sec

Resonant frequency of standard rope held at 3-m mark in ground (lowest energy,
zero node, $n = 1$) state: _____ Hz

Resonant frequency of standard rope held at 3-m mark in first excited
(one node, $n = 2$) state: _____ Hz

Resonant frequency of standard rope held at 3-m mark in second excited
(two node, $n = 3$) state: _____ Hz

In the same quantum state, are these resonant frequencies higher or
lower than those obtained with the longer segment of the same rope?

higher / lower
(circle one)

(continued on following page)

C. The effect of tension on resonant frequencies

Time required for elevated standard rope to complete 50 cycles in ground
(lowest energy, zero node, $n = 1$) state: _____ sec

Time required for elevated standard rope to complete 50 cycles in first excited
(one node, $n = 2$) state: _____ sec

Time required for elevated standard rope to complete 50 cycles in second excited
(two node, $n = 3$) state: _____ sec

Resonant frequency of elevated standard rope in ground (lowest energy,
zero node, $n = 1$) state: _____ Hz

Resonant frequency of elevated standard rope in first excited (one node, $n = 2$) state: _____ Hz

Resonant frequency of elevated standard rope in second excited (two node, $n = 3$)
state: _____ Hz

In the same quantum state, are these resonant frequencies higher or lower
than those obtained with less tension on the same rope?

higher / lower
(circle one)

Optional   ## D. The effect of mass on resonant frequencies

Time required for alternate rope to complete 50 cycles in ground
(lowest energy, zero node, $n = 1$) state: _____ sec

Time required for alternate rope to complete 50 cycles in first excited
(one node, $n = 2$) state: _____ sec

Time required for alternate rope to complete 50 cycles in second excited
(two node, $n = 3$) state: _____ sec

Resonant frequency of alternate rope in ground (lowest energy, zero node, $n = 1$)
state: _____ Hz

Resonant frequency of alternate rope in first excited (one node, $n = 2$) state: _____ Hz

Resonant frequency of alternate rope in second excited (two node, $n = 3$) state: _____ Hz

Are resonant frequencies in the same quantum state higher for a lighter
or a heavier rope, other things being equal?

lighter / heavier
(circle one)

Experiment 44

Advance Study Assignment: Fundamentals of Quantum Mechanics

1. This experiment investigates the phenomenon of resonance—specifically, in the oscillations of a length of rope—as an example of the quantization that underlies quantum mechanics. What three physical characteristics of the rope are intentionally varied in the course of this experiment (including optional Part D)?

2. When spun quickly enough, a rope will have one or more stationary points along its length: points that do not move or oscillate. What is the technical term for these stationary points?

3. The lowest-energy oscillation in a given system is called the <u>ground state</u>. This is the state obtained when the rope is spun just quickly enough to keep it rotating consistently. If the rope is spun quite a bit faster, at the correct speed, it can enter a higher-energy state with one or more of the stationary points from Question 2 along its length. What is the technical term for such a higher-energy state?

4. Frequency is measured in units of hertz, abbreviated Hz. One hertz is one oscillation per second. If it takes 67 seconds for a rope to go around 50 times, what is the frequency of its oscillation, in hertz? Show your work here.

_____ Hz

Appendix A

The Vapor Pressure and Density of Liquid Water

Temperature {°C}	Vapor pressure {mm Hg}	Liquid density {g/mL}	Temperature {°C}	Vapor pressure {mm Hg}	Liquid density {g/mL}
0.0	4.59	0.99979	25.0	23.8	0.99700
0.5	4.75	0.99982	25.5	24.5	0.99687
1.0	4.93	0.99985	26.0	25.2	0.99674
2.0	5.30	0.99989	26.5	26.0	0.99661
3.0	5.69	0.99992	27.0	26.8	0.99647
4.0	6.10	0.99993	27.5	27.6	0.99633
5.0	6.54	0.99992	28.0	28.4	0.99619
6.0	7.02	0.99989	28.5	29.2	0.99605
7.0	7.52	0.99986	29.0	30.1	0.99590
8.0	8.05	0.99980	29.5	31.0	0.99576
9.0	8.61	0.99974	30.0	31.9	0.99561
10.0	9.21	0.99965	32.0	35.7	0.99499
12.0	10.5	0.99945	34.0	39.9	0.99433
14.0	12.0	0.99920	36.0	44.6	0.99364
15.0	12.8	0.99906	38.0	49.8	0.99292
15.5	13.2	0.99898	40.0	55.4	0.99218
16.0	13.6	0.99890	45.0	72.0	0.99017
16.5	14.1	0.99882	50.0	92.6	0.98800
17.0	14.5	0.99873	55.0	118.2	0.98566
17.5	15.0	0.99864	60.0	149.6	0.98316
18.0	15.5	0.99855	65.0	187.8	0.98052
18.5	16.0	0.99846	70.0	234.0	0.97773
19.0	16.5	0.99836	75.0	289.5	0.97481
19.5	17.0	0.99826	80.0	355.6	0.97177
20.0	17.5	0.99816	85.0	434.0	0.96859
20.5	18.1	0.99806	90.0	526.4	0.96530
21.0	18.7	0.99795	92.0	567.7	0.96394
21.5	19.2	0.99784	94.0	611.6	0.96257
22.0	19.8	0.99773	96.0	658.3	0.96118
22.5	20.5	0.99761	98.0	708.0	0.95978
23.0	21.1	0.99750	99.0	734.0	0.95906
23.5	21.7	0.99738	99.5	747.2	0.95871
24.0	22.4	0.99725	100.0	760.7	0.95835
24.5	23.1	0.99713			

W. Wagner and A. Pruss, *J Phys Chem Ref Data*, 31–2: 387–535, 2002, in *NIST Chemistry WebBook, NIST Standard Reference Database Number 69*, P. J. Linstrom and W. G. Mallard, eds., National Institute of Standards and Technology, Gaithersburg MD, 20899, https://doi.org/10.18434/T4D303 (retrieved June 28, 2010 and December 13, 2018).

Appendix B1

A Summary of the Solubility Properties of Ions and Solids

Cation (color in aqueous solution)	Cl^-, Br^- I^-, SCN^-	SO_4^{2-}	CrO_4^{2-}	PO_4^{3-}	$C_2O_4^{2-}$	CO_3^{2-}	SO_3^{2-}
Na^+, K^+, NH_4^+	S	S	S	S	S	S	S
Ba^{2+}	S	I	A	A^-	A	A^-	A
Ca^{2+}	S	S^-	S	A^-	A	A^-	A^-
Mg^{2+}	S	S	S	A^-	A^-	A^-	S
Fe^{3+} (yellow)	*S	S	A^-	A	S	D: A^-	D: S
Cr^{3+} (violet‡)	S	S	A^-	A	S	A^-	S
Al^{3+}	S	S	A^-, C	A, C	A^-, C	D: A^-, C	A^-, C
Ni^{2+} (green)	S	S	S	A^-, C	A, C	A^-, C	A^-
Co^{2+} (pink)	S	S	A^-	A^-	A^-, C	A^-	A^-
Zn^{2+}	S	S	A^-, C	A^-, C	A^-, C	A^-, C	S
Mn^{2+} (pale pink)	S	S	S	A^-	A^-	A^-	S
Cu^{2+} (blue)	**S	S	A^-, C	A^-, C	A, C	A^-, C	A^-, C
Cd^{2+}	S	S	A^-, C	A^-, C	A, C	A^-, C	A^-, C
Bi^{3+}	A	A^-	A	A	A	A^-	A
Hg^{2+}	†S	S	A	A^-	A	A^-	D: O
Sn^{2+}, Sn^{4+}	A, C	A, C	A, C	A, C	A, C	A, C	A, C
Sb^{3+}	A, C	A, C	A, C	A, C	A^-, C	A, C	A, C
Ag^+	††C	S^-	A, C	A, C	A, C	A^-, C	A, C
Pb^{2+}	C, HW	C	C	A, C	A, C	A^-, C	A, C
Hg_2^{2+}	O^+	A	A	A	O	A	D: O

Key: S, soluble in water; no precipitate on mixing cation, 0.10 M, with anion, 1.0 M

    S^-, slightly soluble; tends to precipitate on mixing cation, 0.10 M, with anion, 1.0 M, but precipitation may be slow and/or small in quantity

    HW, soluble in hot water

    A^-, soluble in 1.0 M CH_3COOH (acetic acid)

    A, soluble in acid (6 M HCl or some other nonprecipitating, nonoxidizing acid)

    A^+, soluble in 12 M HCl

    O, soluble in hot 6 M HNO_3 (a strong, oxidizing acid)

    O^+, soluble in aqua regia, a (dangerous!) aqueous oxidizing acid and complexing agent combination

    CAUTION: Acidic solutions are corrosive, more so when hot, and even more so if they are oxidizing. Use such solutions with great care.

    C, soluble in solutions containing a good complexing ligand (see the rightmost column in this table, on the next page)

    D, unstable; decomposes to a product with solubility as indicated

    I, insoluble in any common solvent or solution

*FeI_3 is unstable; it decomposes to FeI_2 and I_2.

**CuI_2 is unstable; it decomposes to CuI and I_2.

†HgI_2 is insoluble, but dissolves in excess I^-.

††AgCl and AgSCN dissolve in 6 M NH_3, but AgBr and AgI do not.

‡Blue-gray in natural light; other colors are often observed—particularly green—in the presence of chloride, sulphate, and most other anions that can substitute in as ligands. Ligand substitution is often slow at room temperature.

Nomenclature: SCN^- = thiocyanate ion; CrO_4^{2-} = chromate ion; $C_2O_4^{2-}$ = oxalate ion; SO_3^{2-} = sulfite ion; S^{2-} = sulfide ion; O^{2-} = oxide ion; ClO_3^- = chlorate ion; CH_3COO^- = acetate ion; NO_2^- = nitrite ion; $S_2O_3^{2-}$ = thiosulfate ion

Cation (color in aqueous solution)	S^{2-}	OH⁻, *O^{2-}	NO_3^-, ClO_3^-, CH_3COO^-, NO_2^-	Complexing agents**
Na^+, K^+, NH_4^+	S	S	S	—
Ba^{2+}	S	S⁻	S	NH_3
Ca^{2+}	D: A⁻	S⁻	S	—
Mg^{2+}	D: A⁻	A⁻	S	$C_2O_4^{2-}$
Fe^{3+} (yellow)	D: A⁻	A⁻	S	$C_2O_4^{2-}$, SCN^-
Cr^{3+} (violet‡)	D: A⁻	A⁻	S	OH^-, $C_2O_4^{2-}$, (NH_3), (CH_3OO^-)
Al^{3+}	D: A⁻, C	A⁻, C	S	OH^-, $C_2O_4^{2-}$, (SCN^-)
Ni^{2+} (green)	O	A⁻, C	S	NH_3, $C_2O_4^{2-}$
Co^{2+} (pink)	O	A⁻	S	$C_2O_4^{2-}$, NH_3, (Cl^-)
Zn^{2+}	A⁻, C	A⁻, C	S	OH^-, NH_3, $C_2O_4^{2-}$
Mn^{2+} (pale pink)	A⁻	A⁻	S	CH_3COO^-, OH^-, $C_2O_4^{2-}$
Cu^{2+} (blue)	O	A⁻, C	S	(OH^-)†, NH_3, (I^-)†, $C_2O_4^{2-}$
Cd^{2+}	A	A⁻, C	S	OH^-, NH_3, $C_2O_4^{2-}$, I^-
Bi^{3+}	A⁺, O	A⁻	A⁻	I^-, Br^-, Cl^-
Hg^{2+}	O⁺	A⁻	S	all except CH_3COO^-
Sn^{2+}, Sn^{4+}	A, C	A, C	A, C	$C_2O_4^{2-}$, OH^-, Cl^-
Sb^{3+}	A, C	A, C	A, C	OH^-, Cl^-
Ag^+	O	A⁻, C	S	$C_2O_4^{2-}$, I^-, $S_2O_3^{2-}$, SCN^-, Br^-, NH_3, (Cl^-)
Pb^{2+}	O	A⁻, C	S	OH^-, CH_3COO^-, $C_2O_4^{2-}$, $S_2O_3^{2-}$, (I^-), (Cl^-)
Hg_2^{2+}	D: O⁺	D: O	S	—

Key: S, soluble in water; no precipitate on mixing cation, 0.10 M, with anion, 1.0 M

  S⁻, slightly soluble; tends to precipitate on mixing cation, 0.10 M, with anion, 1.0 M, but precipitation may be slow and/or small in quantity

  A⁻, soluble in 1.0 M CH_3COOH (acetic acid)

  A, soluble in acid (6 M HCl or some other nonprecipitating, nonoxidizing acid)

  A⁺, soluble in 12 M HCl

  O, soluble in hot 6 M HNO_3 (a strong, oxidizing acid)

  O⁺, soluble in aqua regia, a (dangerous!) aqueous oxidizing acid and complexing agent combination

  CAUTION: **Acidic solutions are corrosive, more so when hot, and even more so if they are oxidizing. Use such solutions with great care.**

  C, soluble in solutions containing a good complexing ligand (see the rightmost column in this table)

  D, unstable; decomposes to a product with solubility as indicated

*Oxides behave like hydroxides, but may be slow to dissolve.

**Complexing agents capable of dissolving solids containing the indicated anion are listed in order of decreasing complexing strength (best complexing agents first); complexing agents listed in parentheses require high concentrations in order to be effective (to form a significant concentration of complex ions with the indicated cation).

†OH^- coordinates strongly to Cu^{2+}, but $Cu(OH)_2$ is quite stable and will only dissolve to a complex in very concentrated hydroxide solutions; I^- complexes Cu^{2+} but also reduces it to Cu^+, precipitating CuI.

‡Blue-gray in natural light; other colors are often observed—particularly green—in the presence of chloride, sulphate, and most other anions that can substitute in as ligands. Ligand substitution is often slow at room temperature.

Nomenclature: SCN^- = thiocyanate ion; CrO_4^{2-} = chromate ion; $C_2O_4^{2-}$ = oxalate ion; SO_3^{2-} = sulfite ion; S^{2-} = sulfide ion; O^{2-} = oxide ion; ClO_3^- = chlorate ion; CH_3COO^- = acetate ion; NO_2^- = nitrite ion; $S_2O_3^{2-}$ = thiosulfate ion

Appendix B2

Some Properties of the Cations in Groups I, II, and III

In qualitative analysis, the chemical and physical properties of the cations in Groups I, II, and III are used to separate these ions from one another. This appendix summarizes many of the properties that are typically used to identify and isolate these ions.

Lead(II) Ion, Pb^{2+}

Most lead compounds are insoluble in water; the only exceptions are lead(II) nitrate, $Pb(NO_3)_2$, and lead(II) acetate, $Pb(CH_3COO)_2$. Lead(II) chloride, $PbCl_2$, is slightly soluble in water at room temperature, while it is much more soluble in hot water and in solutions in which the Cl^- concentration is high. The lead(II) ion forms stable complexes with OH^-, CH_3COO^-, and, to a lesser degree, with Cl^-. Lead(II) sulfate, $PbSO_4$, an acid-insoluble sulfate, is white and will dissolve in 6 M NaOH to form the $Pb(OH)_4{}^{2-}$(aq) complex ion or in a concentrated solution of acetate ions to form $Pb(CH_3COO)_2$(aq), a neutral but soluble substance that does not dissociate appreciably in solution. Lead(II) chromate, $PbCrO_4$, is a bright yellow insoluble solid, often used as a confirmatory test for lead. It dissolves in 6 M NaOH.

Silver Ion, Ag^+

Most silver compounds are insoluble; the nitrate, $AgNO_3$, is just about the only exception. Silver acetate, $AgCH_3COO$, is slightly soluble. Silver chloride, $AgCl$, is white and insoluble in water, but it is soluble in 6 M ammonia, NH_3, due to the formation of the $Ag(NH_3)_2{}^+$ complex ion. These properties of AgCl are usually used in the confirmatory test for silver ion. They may also be used to test for the presence of Cl^- ion. In NaOH solution, Ag^+ precipitates as brown silver oxide, Ag_2O; this oxide is soluble in nitric acid, HNO_3(aq), and in concentrated ammonia solutions.

Mercury(I) Ion, $Hg_2{}^{2+}$

No salts of the Hg(I) "mercurous" ion are soluble in water. Solutions of mercury(I) nitrate, $Hg_2(NO_3)_2$, always contain fairly high concentrations of nitric acid, HNO_3. Calomel, Hg_2Cl_2 (containing the stable $Hg_2{}^{2+}$ cation, which is the dimer of the unstable odd-electron Hg^+ ion), precipitates upon adding hydrochloric acid, HCl, to any solution of Hg(I) ion. Calomel is white and essentially insoluble in all common solvents and solutions. If treated with 6 M ammonia, NH_3, it turns gray-black due to a reaction forming Hg metal (black) and white, insoluble, mercury(I) amidochloride, $HgNH_2Cl$. This is the definitive test for the Hg(I) ion.

Bismuth(III) Ion, Bi^{3+}

There are no water-soluble bismuth salts. The usual solution is highly acidic and contains bismuth(III) nitrate, $Bi(NO_3)_3$, or bismuth(III) chloride, $BiCl_3$. The bismuth(III) ion forms complexes with Cl^- and the other halides, but not with OH^- or NH_3. Bismuth(III) hydroxide, $Bi(OH)_3$, is white and insoluble in water, but soluble in strong acids. If you add a few drops of an acidic $BiCl_3$ solution to water, a cloudy white precipitate of bismuth(III) oxychloride, BiClO, forms. This is the usual test for the Bi^{3+} ion, but care must be taken if there is a risk that antimony(III), Sb^{3+}, which behaves similarly, is also present. (See the entry on antimony(III) for more information.) Another confirmatory test is to add a 0.10 M solution of tin(II) chloride, $SnCl_2$, to $Bi(OH)_3$ in a strongly basic environment, in which Bi(III) will be reduced to black Bi metal.

Tin(II) or Tin(IV) Ion, Sn^{2+} or Sn^{4+}

No common tin compounds are water-soluble. The usual 0.10 M tin solution contains tin(II) chloride, $SnCl_2$, in 0.10 M hydrochloric acid, HCl. Tin exists in either the +2, called "stannous", or +4, called "stannic", state. The Sn(II) ion, particularly in basic solution, is a good reducing agent, and can reduce Bi(III) and Hg(II) to the metals. Tin in either oxidation state forms many complex ions, and in solution is usually present as a complex ion. In basic solution Sn(II) exists as $Sn(OH)_4^{2-}$. In HCl the complex ion has the formula $SnCl_4^{2-}$. Tin compounds are white, for the most part. SnS is dark brown, and its formation at pH 0.5 can suggest the presence of Sn(II), though Bi_2S_3 is also dark brown. The classic confirmatory test is to add $SnCl_2$ in HCl to a solution of mercury(II) chloride, $HgCl_2$. The tin(II) ion reduces the mercury(II) to form a precipitate of white calomel, Hg_2Cl_2 (which may darken to gray if the reduction proceeds all the way to metallic mercury). A less toxic and more environmentally friendly confirmatory test for Sn(II) uses the organic compound Janus Green (JG), which is easily reduced by Sn^{2+}. A color change from blue to either violet-red (JG^-) or colorless (JG^{2-}) confirms the presence of Sn(II). (The color change observed depends upon the relative amounts of Sn^{2+} and JG that are present.)

Antimony(III) Ion, Sb^{3+}

There are no common antimony salts that are soluble in water. Antimony salts are not easily dissolved. The usual solution of Sb(III) contains antimony(III) chloide, $SbCl_3$, in 3 M hydrochloric acid, HCl. In this solution the antimony exists as the $SbCl_6^{3-}$ complex ion. Antimony also forms a stable complex ion with hydroxide ion, $Sb(OH)_6^{3-}$. If a few drops of $SbCl_3$ in acidic solution are added to water, a white precipitate of antimony(III) oxychloride, SbClO, forms—very similar to what is observed with bismuth(III) chloride, $BiCl_3$. The confirmatory test for antimony is the formation of the characteristic bright red-orange sulfide precipitate antimony(III) sulfide, Sb_2S_3, upon adding thioacetamide, CH_3CSNH_2, to antimony-containing solutions at pH 0.5. If the cation is Bi(III), the precipitate will instead be dark brown bismuth(III) sulfide, Bi_2S_3.

Copper(II) Ion, Cu^{2+}

The copper(II) ion, Cu^{2+}, is colored in most of its compounds and solutions. The hydrated ion, $Cu(H_2O)_6^{2+}$, is cyan or aqua (greenish blue), but in the presence of other complexing agents the Cu(II) ion may be green or dark blue. The sulfate, chloride, nitrate, and acetate all dissolve in water. If 6 M ammonia, NH_3, is added to a solution of Cu(II) ion, light blue copper(II) hydroxide, $Cu(OH)_2$, initially precipitates, but in excess ammonia the hydroxide easily dissolves due to the formation of the very characteristic dark blue copper(II)–ammonia complex ion $Cu(NH_3)_4^{2+}$. This complex ion serves as an excellent confirmatory test for the presence of copper. Copper ion is easily reduced to the metal by more active metals, such as zinc.

Nickel(II) Ion, Ni^{2+}

Nickel is another of the colored aqueous cations. The chloride, nitrate, sulfate, and acetate are water-soluble and form green solutions containing the $Ni(H_2O)_6^{2+}$ complex ion. Nickel forms other complex ions, but the one most commonly observed is the pale blue $Ni(NH_3)_6^{2+}$ ion, formed upon adding an excess of 6 M ammonia, NH_3, to Ni(II) solutions. The blue color is much less intense than that of the copper–ammonia complex ion. Ni^{2+} precipitates as $Ni(OH)_2$ and does not dissolve in excess hydroxide. The usual test for the presence of nickel is the formation of a rose-red precipitate of nickel dimethylglyoxime, $Ni(C_4H_7N_2O_2)_2$, upon adding a solution of dimethylglyoxime, $C_4H_8N_2O_2$, to a slightly basic solution containing Ni^{2+}.

Iron(III) Ion, Fe^{3+}

Iron has two common oxidation states, +2 and +3. Iron(II), called "ferrous" ion, is less commonly observed because it is easily oxidized by oxygen in the air to the +3 state. The rest of this discussion will be limited to iron(III), or "ferric", salts. Iron(III) salts are often colored, usually yellow in aqueous solution. The color is due to hydrolysis (chemical attack by water), because the $Fe(H_2O)_6^{3+}$ ion is itself almost colorless, having a very pale violet color. The nitrate, chloride, and sulfate dissolve in water but tend to hydrolyze to form basic salts that may require a slightly acidic solution to dissolve completely. Iron(III) hydroxide, $Fe(OH)_3$, is very insoluble and is the dark orange color of rust. Iron forms several complex ions, but the one most commonly encountered in qualitative analysis is the iron(III)–thiocyanate complex ion, $FeSCN^{2+}$, which is red-orange (blood red at high concentrations) and often used as a confirmatory test for Fe(III). Iron does not form complex ions with NH_3 or OH^-.

Chromium(III) Ion, Cr^{3+}

Chromium(III), as its name implies, is colored ("chroma" is the Greek word for color): depending on the lighting and complexing ions present it may be violet, green, or gray-blue in aqueous solution. The chromium(III) ion usually exists as a complex that—unlike many complex ions—is often slow to undergo ligand exchange. Chromium's two most common oxidation states are +3 and +6, with the +3 state being the most stable. In the +6 state chromium(VI) is found as the yellow "chromate" ion, CrO_4^{2-}, or, if the solution is acidic, as the orange "dichromate" ion, $Cr_2O_7^{2-}$. The chloride, nitrate, sulfate, and acetate of Cr^{3+} dissolve in water, but the dissolution process may be more rapid in acidic solution. If excess 6 M sodium hydroxide, NaOH, is added to a solution of Cr(III), the initial jelly-like gray-green precipitate of chromium(III) hydroxide, $Cr(OH)_3$, may dissolve, and a dark green but clear solution of the $Cr(OH)_4^-$ complex ion may result. With 6 M ammonia, NH_3, the same insoluble gray-green hydroxide is initially formed, which in a large excess of ammonia may slowly dissolve to yield a pink to violet complex ion. Boiling a solution of either complex causes insoluble $Cr(OH)_3$ to re-form. In most qualitative analysis schemes chromium is oxidized to the +6 state, which under acidic conditions forms a characteristic—but short-lived—deep blue solution upon adding hydrogen peroxide, H_2O_2. When chromium(III) is the only cation in a sample, the formation of the dark green hydroxide complex in excess 6 M NaOH is definitive. If this test is inconclusive, oxidizing Cr(III) to Cr(VI) and adding H_2O_2—as just described—is the best approach.

Aluminum Ion, Al^{3+}

Aluminum salts are typically white. The nitrate, chloride, sulfate, and acetate dissolve in water but have a tendency to form basic salts if no acid is added. Aluminum forms a very stable hydroxide complex ion, $Al(OH)_4^-$, so it is difficult to precipitate otherwise insoluble aluminum hydroxide, $Al(OH)_3$, by adding NaOH to Al(III) solutions. $Al(OH)_3$ *will* precipitate upon adding 6 M ammonia, NH_3, to a solution containing Al^{3+} buffered with NH_4^+ ion. The precipitate is usually light (it does not sink), white to almost transparent, and jelly-like. A good confirmatory test for aluminum involves first precipitating $Al(OH)_3$ from a solution of the sample, then dissolving the solid in very dilute acetic acid, CH_3COOH. Add a few drops of catechol violet, HCV. If aluminum is present, the solution will turn blue due to the formation of the aluminum complex ion $AlCV^{2+}$.

Appendix C

Standard Atomic Weights of the Elements*
(Scaled Relative to Carbon-12 := <u>12</u> g/mol)

	Symbol	Atomic number	Atomic molar mass {g/mol}		Symbol	Atomic number	Atomic molar mass {g/mol}
Aluminum	Al	13	26.98	Molybdenum	Mo	42	95.95
Antimony	Sb	51	121.76	Neodymium	Nd	60	144.24
Argon	Ar	18	39.95	Neon	Ne	10	20.18
Arsenic	As	33	74.92	Nickel	Ni	28	58.69
Barium	Ba	56	137.33	Niobium	Nb	41	92.91
Beryllium	Be	4	9.012	Nitrogen	N	7	14.01
Bismuth	Bi	83	208.98	Osmium	Os	76	190.23
Boron	B	5	10.81	Oxygen	O	8	16.00
Bromine	Br	35	79.90	Palladium	Pd	46	106.42
Cadmium	Cd	48	112.41	Phosphorus	P	15	30.97
Calcium	Ca	20	40.08	Platinum	Pt	78	195.08
Carbon	C	6	12.01	Potassium	K	19	39.10
Cerium	Ce	58	140.12	Praseodymium	Pr	59	140.91
Cesium	Cs	55	132.91	Protactinium	Pa	91	231.04
Chlorine	Cl	17	35.45	Rhenium	Re	75	186.21
Chromium	Cr	24	52.00	Rhodium	Rh	45	102.91
Cobalt	Co	27	58.93	Rubidium	Rb	37	85.47
Copper	Cu	29	63.55	Ruthenium	Ru	44	101.07
Dysprosium	Dy	66	162.50	Samarium	Sm	62	150.36
Erbium	Er	68	167.26	Scandium	Sc	21	44.96
Europium	Eu	63	151.96	Selenium	Se	34	78.97
Fluorine	F	9	19.00	Silicon	Si	14	28.09
Gadolinium	Gd	64	157.25	Silver	Ag	47	107.87
Gallium	Ga	31	69.72	Sodium	Na	11	22.99
Germanium	Ge	32	72.63	Strontium	Sr	38	87.62
Gold	Au	79	196.97	Sulfur	S	16	32.06
Hafnium	Hf	72	178.49	Tantalum	Ta	73	180.95
Helium	He	2	4.003	Tellurium	Te	52	127.60
Holmium	Ho	67	164.93	Terbium	Tb	65	158.93
Hydrogen	H	1	1.008	Thallium	Tl	81	204.38
Indium	In	49	114.82	Thorium	Th	90	232.04
Iodine	I	53	126.90	Thulium	Tm	69	168.93
Iridium	Ir	77	192.22	Tin	Sn	50	118.71
Iron	Fe	26	55.85	Titanium	Ti	22	47.87
Krypton	Kr	36	83.80	Tungsten	W	74	183.84
Lanthanum	La	57	138.91	Uranium	U	92	238.03
Lead	Pb	82	207.2	Vanadium	V	23	50.94
Lithium	Li	3	6.94	Xenon	Xe	54	131.29
Lutetium	Lu	71	174.97	Ytterbium	Yb	70	173.05
Magnesium	Mg	12	24.31	Yttrium	Y	39	88.91
Manganese	Mn	25	54.94	Zinc	Zn	30	65.38
Mercury	Hg	80	200.59	Zirconium	Zr	40	91.22

*Standard atomic weights describe the typical average molar masses of elements as found on Earth. Only elements with isotopes stable enough that a characteristic isotopic abundance can be defined have standard atomic weights.

Appendix D

Laboratory Techniques and Taking Measurements

For centuries, people who made measurements set up systems of <u>units</u>—scales of measurement—to describe length, area, volume, and mass. These units often had descriptive names and were created for a specific purpose. Some examples include:

acre	barrel	bolt	bushel
carat	chain	cord	cubit
dram	ell	em	fathom
furlong	gill	grain	hand
league	noggin	pace	perch
quire	rod	stone	ream

These days many of these units are still used, but usually in connection with only one application, such as horse racing (furlong), a gemstone's weight (carat), or commercial printing (em).

Scientists realized early on that it would be helpful to have one unified system of measurement used by everyone, and several systems were suggested, starting in the 1800s. These were mostly based on the <u>metric</u> approach, with different sizes related to one another by factors of 10. For example, there is the meter, equal to <u>100</u> centimeters, equal to <u>1000</u> millimeters. In 1960, a group of scientists set up the <u>International System of Units, or "SI"</u>: a coherent system of seven base units that can be used to describe most measurements when used with metric prefixes and/or combined to form <u>derived units</u>, which are ratios or products of the base units. The six of these base units that we will use, and their abbreviations, are:

meter (m)   kilogram (kg)   second (s)   Kelvin (K)   mole (mol)   ampere (A)

Although in principle we might expect to use these units directly in reporting data, some of them turn out to have awkward magnitudes in many situations. We use SI base units when it is convenient, but for the first two in our list we more often work with related metric units, such as the centimeter and millimeter. We will discuss our approach in the following sections. (Perhaps not surprisingly, none of the units in our very first list are SI units, or metrically related to SI units.)

Good scientific data has three essential components: a numerical value, a unit, and an uncertainty. A measurement lacking any one of these three components is of little use, if it can be called data at all! That this is true of the numerical value rarely raises debate, but many need convincing that it is true of both units and uncertainty. After all, often, in common speech—as when a child reports "I'm seven!"—only a numerical value is given, yet (almost) everyone understands that the child is reporting that they have celebrated their seventh birthday, but not yet their eighth. The unit and the uncertainty are implicitly understood, without being explicitly stated. However, consider what you might make of someone reporting that a building is "six tall." Six what? Feet? Meters? Inches? (It might be a toy house!) If you knew enough about the building, you might make an educated guess, but you cannot be sure. Even if the reporter clarified that the building is "six meters tall," you might well need more information. Suppose you were trying to find a ladder that would (safely) get you onto the roof of this building: the question of how reliable the measurement was could be very important. Is the building *exactly* six meters tall, or is that value only an estimate... or was it rounded off to the nearest meter? These types of concerns arise even more often in science than they do in everyday life, so it is essential that every scientific measurement you record or report be complete: that it consist of a numerical value with *both* its units *and* its uncertainty. Uncertainty and how it is reported are the subject of Appendix E. For now, let us return to units!

Mass

Some of the most important measurements you will make in lab involve determining the <u>mass</u> of a sample: the amount of substance present based on how much it resists changes in velocity (technical term: inertia). The unit you will almost always use is the <u>gram</u>, rather than the kilogram, because a kilogram (about two pounds) is just too large for bench-scale chemistry. The gram, abbreviated g, is not a base unit in SI (despite its non-prefixed name) but is easily converted to the kilogram when necessary. A small metal paperclip and a raisin each have a mass of about one gram.

Mass is almost always actually measured by determining the force exerted on a sample by gravity: a process called <u>weighing</u>. The device that is used is called a <u>balance</u>, because the mass of the sample is established by balancing it against a standard mass or force. Two different electronic balances are shown in Figure D.1; they differ in their <u>precision</u>—that is, in how small a difference in mass they can reliably detect. When attempting to detect very small mass differences even the flow of air through a lab can be important, which is why the balance with the greater precision (ability to detect smaller mass differences) has a glass box (which can be opened and closed) enclosing the weighing area.

To use a balance, you should first zero the balance such that it displays a steady value of zero (all zeroes out to the last significant digit the balance is capable of, usually 0.00 g or 0.0000 g) with nothing on the balance pan. There is a button for this purpose (usually marked "zero"), but to get a stable reading you must ensure that all the doors on the balance are *closed* (often including one on top, which can be easy to miss), if it has any, and that the surface on which the balance sits is not moving (which for a precise balance can be caused by something as seemingly minor as someone leaning or writing on the surface!). To measure the mass of an object, open a balance door (if the balance has them), place the object on the balance pan, close the door, and read the mass when the balance gives a steady reading.

To find the mass of a sample you wish to add to a container, first place the empty container on the balance and <u>tare</u> it out (usually with a button marked "tare") so that the balance reads zero with the container on it. Then when you add your sample to the container, the mass of just the added sample will appear directly on the digital readout. Where the size, mass, or something else about the container your sample will be transferred into prevents putting it on the balance, another option is <u>weighing by difference</u>, in which you calculate the mass of sample *removed* from a balance-compatible container by weighing it before and after transferring sample into your desired container. An example of the calculations associated with weighing by difference appears in Subsection D of Appendix E.

There are many models of balances, each slightly different, so your instructor will describe the details of the operation of those you have access to before you use them. Here are some guidelines for successful balance operation that apply to weighing a sample on any balance:

1. Ensure the balance is level; usually this is indicated by a small float level containing an air bubble that should be within a marked circular region.
2. Be certain the balance has been "zeroed" (reads 0 grams) before you place anything on the balance pan.
3. Never weigh chemicals directly on a balance pan. Always use a suitable container or weighing paper.
4. Be certain that air currents are not disturbing the balance pan. If the balance you are using has them, always shut the balance case doors when taking measurements you are recording.
5. When measuring mass with great precision (± 0.001 g or better), the surface on which the balance sits must not be subject to vibrations—even the small ones caused by leaning on or writing on the surface!
6. Do not attempt to weigh hot—or even warm—objects. The elevated temperature will cause convection currents (the air near the object will heat up and rise), and the balance will yield inaccurate measurements.
7. Record your measurement, to the proper number of significant figures, directly on the report page or in your notebook.
8. After finishing your measurements, brush out any chemicals you may have spilled, close the balance doors, and be certain that the balance registers zero again.

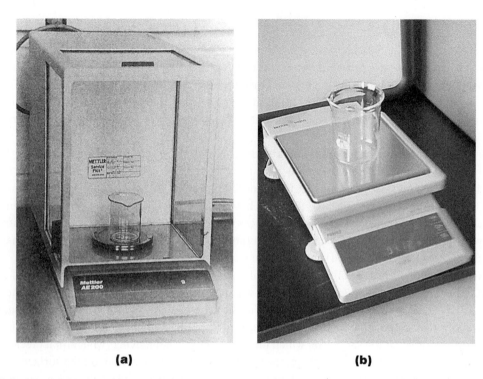

Figure D.1 (a) An analytical balance, used to determine masses with great precision. Its maximum load is 120 g, and it is reliable to ± 0.0001 g. The box around the sample prevents ambient airflow from making the reading unstable. (b) A top-loading balance with a maximum load of 3.6 kg and reliable to ± 0.01 g. Newer balances of each of these types differ functionally only in the extent to which they automatically recalibrate themselves to account for changes in the ambient temperature. *(Mary Lou Wolsey)*

Volume

Volume, the amount of substance present based on how much space it occupies, is another quantity commonly measured in chemistry laboratories. The base SI unit of volume is the cubic meter, m^3. Again, this is a much larger scale than you ordinarily encounter in a chemistry lab. A large bathtub might contain a single cubic meter of water, which is roughly 250 gallons. The units of volume ordinarily used in lab settings are the liter, L, and more often the milliliter, mL. A liter equals $\underline{0.001}$ m^3, and a mL equals $\underline{1 \times 10^{-6}}$ m^3, so both are derived from the SI base unit.

Another name for the milliliter is the cubic centimeter, cm^3, sometimes called the "cc" (pronounced "see see")—a term most often used in medicine.

$$\underline{1} \text{ mL} = \underline{1} \text{ cc} = \underline{1} \text{ cm}^3$$

If lab instructions direct you to add "about one-hundred and fifty milliliters of water" during an experiment, approximate volumes are fine. You might use a beaker or flask on which there are some volume markings called graduations [as shown in Figure D.2(a)], the best of which are reliable to about five percent of the volume of the beaker (but look to the markings for an indication of how reliable they actually are: some are ±20%, others claim only to be "approximate"). The graduations on a beaker might also be good enough if the instructions tell you to add "150 mL of water," which means between 140 and 160 mL (see Appendix E); but this approach definitely would not be exact enough if the instructions direct you to add "150. mL of water," which means ±1 mL (again, see Appendix E for information about how significant figures in lab instructions guide you as to how exacting you need to be), far less than 5% of 150 mL.

Somewhat more precise volumes can be obtained with a graduated cylinder, which can yield measurements within about one percent of a needed volume. Especially for smaller cylinders, however, using one properly requires understanding the role of the meniscus—the curvature of the surface of the contained liquid—and of the position of your eye. To properly read any cylindrical glass volume measuring device, you should do so at the center of the meniscus, where it is furthest from the glass. With water, the meniscus dips down as it moves further from the

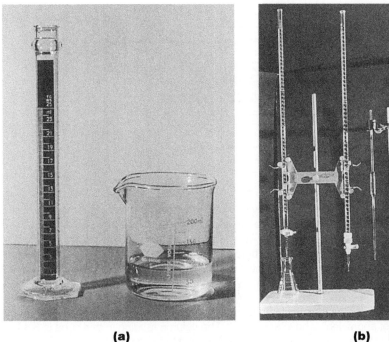
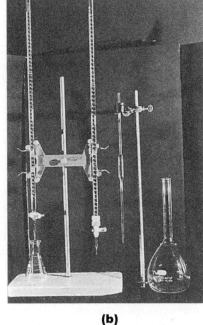

(a)   (b)

Figure D.2 (a) A 25-mL graduated cylinder, reliable to ± 0.2 mL, and a 250-mL beaker with graduations (volume markings), reliable to ± 20 mL. (b) A pair of 50-mL burets, a 10-mL volumetric transfer pipet, and a 1000-mL volumetric flask. The Class A versions of these are reliable to ± 0.05 mL for the burets, ± 0.02 mL for the glass pipet, and ± 0.3 mL for the volumetric flask. The uncertainties are twice as large for Class B volumetric glassware. *(Sherman Schultz)*

glass, so you should read from the bottom of the meniscus, as shown in Figure D.3. It is also essential that your eye be physically positioned at the about same level as the meniscus, not well above or below it, to avoid introducing something called "parallax error" into your measurement.

A. Pipets

Still more precise volume measurements are often required; these are possible with <u>pipets</u> (also spelled pipettes, but in this manual we will use the shorter spelling), which are available in many different volumes. There are several different kinds, but the simplest is a glass pipet, shown in Figure D.2(b). Glass pipets are calibrated to deliver a specific volume when the meniscus of the measured liquid is at the horizontal line etched around the upper pipet stem. Here are some important things to know about using a glass pipet:

1. To work properly, a glass pipet must be clean. When it drains, no droplets should remain on the inner walls of the pipet. This can be challenging to attain and maintain—but when it is not the case, the volumes measured with a glass pipet will not be nearly as exact as specified by the pipet's tolerance.
2. **Always use a pipet bulb, never your mouth, when pulling liquid into a glass pipet!** There are specialized pipet pumps and bulbs for this purpose: use them. Never run the risk of getting a liquid or vapor into your mouth.
3. To use a glass pipet, place a pipet bulb or pump, emptied of air, on the upper end of the pipet. Immerse the tip of the pipet into the liquid you wish to measure and pull up enough of it to get some into the main body of the pipet. Remove the bulb or pump and immediately push a finger tightly on the top of the pipet. Hold the pipet in a horizontal position and swirl the liquid around inside, rinsing the upper stem and body. Drain the liquid into a beaker; repeat this rinsing process twice with small volumes of the liquid to be measured. This ensures that the contents of the pipet will be the liquid you wish to work with, not diluted by water or contaminated with a previously used chemical. Now fill the pipet almost completely, pulling in liquid a centimeter or two above the calibration mark on the upper stem. Remove the bulb or pump and quickly push the tip of your index finger onto the top of the pipet, to prevent the liquid from draining out. Place the tip of the pipet against a vertical glass surface, such as the wall of the beaker you drained into previously, and release the pressure exerted by your finger slightly: enough to allow liquid to slowly drain out of the pipet. Allow liquid to drain

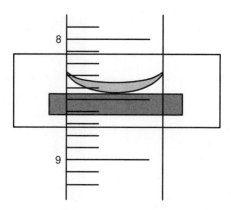

Figure D.3 To properly read a liquid level in any cylindrical glass volume measurement device, such as this buret, do so at its center: at the extreme of the meniscus (the curved surface of the liquid). With aqueous solutions in glass the meniscus curves downward, and readings should be taken from the bottom of the meniscus. A dark line on a light background, held behind the meniscus, can help make it easier to see; but if this is done, it is critical that it be done consistently—for all measurements that will be compared to one another. When reading any graduated device (one with ruler-like markings on it), keep in mind the direction of the markings (increasing or decreasing) when reading or <u>interpolating</u> between markings (estimating values between them) on the scale. It is also essential that your eye be physically positioned at essentially the same level as the meniscus, not appreciably above or below it. The volume reading in this buret is 8.44 mL, as the bottom of the meniscus is located just slightly above the midpoint between the 8.4 mL (above) and 8.5 mL (below, longer) markings, making the interpolated value slightly closer to 8.4 than to 8.5 mL.

until the center of the meniscus is at the calibration mark when viewed with your eye positioned at the level of the mark and with the pipet tip still touching a vertical glass surface. This is most easily accomplished if your index finger is not wet! Holding your finger tightly on the top of the pipet, transfer the pipet tip to the receiving container, with the tip touching the wall. Lift your index finger to release the liquid and let it flow into the container. Hold the tip against the wall for about ten seconds (longer for a thick liquid) after it appears that the liquid has stopped flowing: this is required in order for the correct amount of liquid to drain out! Almost all glass pipets are designed "to deliver" and will be marked with the code "TD". These will accurately deliver the indicated amount of an *aqueous* liquid (one composed of mostly water) when a small amount (the amount that remains naturally when the pipet tip is held against a clean vertical glass surface) remains in the tip—that last bit should *not* be blown out with a pipet bulb or pump!

More mechanized versions of glass pipets, called autopipets and repipets and shown in Figure D.4, can make it easier to deliver precise volumes. These speed up the pipetting procedure enormously, and are easier to learn how to use. <u>Repipets</u> provide very reproducible volumes provided they are pumped smoothly—without stopping—and at a moderate speed, on both the upward and downward strokes. They can be tricky to set up to deliver an exact volume, but that is something that will be done for you. <u>Autopipets</u> require proper technique as well, and are calibrated to work with <u>aqueous</u> liquids (liquids containing mostly water). They look simpler to use than they really are! Autopipets are generally made "to contain" (abbreviated "TC"), so every last drop in their tips must be expelled in order to obtain precise volumes. Their plungers have a two-position stroke to make this possible. Here are some critical things to know about measuring exact volumes with an autopipet:

1. Autopipets use disposable plastic tips with a water-repellent coating. This coating is important, and it does wear off with use—so while you can use the same disposable tip for numerous measurements of the same liquid, tips should be replaced often.
2. To install a tip, push it on firmly with a slight twisting motion. A leak-free seal is essential! The twisting motion is important: a tip pushed straight on, even with great force, is far more likely to leak.
3. The piston on an autopipet has two "stops," one just before the other, at the end of the piston's travel. You can feel these as you press the plunger all the way down: the first as a slight hesitation in the smooth travel of the piston just before it hits bottom, and the second as a hard stop.
4. To provide exact volume measurements, the autopipet tip must first be wet with (and the air in the autopipet saturated with the vapor of) the solution to be measured. Accomplish this by pushing the plunger down to

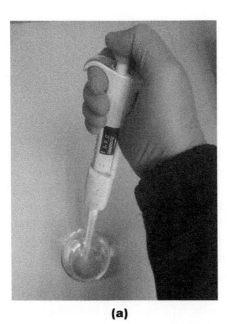

(a)   **(b)**

Figure D.4   (a) An autopipette is a handheld, mechanized version of a glass pipet that is easier to learn to use and employs disposable tips to avoid the need for rinsing. (b) A repipet (also called a bottle-top dispenser) is like a "pipet station" that can be set up to quickly and easily dispense the same precisely measured volume of the same liquid, over and over. (*Robert Rossi*)

the first stop, immersing the tip into the liquid to be measured, and allowing the plunger to smoothly rise all the way back up with the tip still in the liquid. Smooth operation is important here: if the plunger moves suddenly, small droplets of the liquid can fly everywhere inside the tip, causing problems later. Expel this rinsing into a waste container (or, if the liquid is precious, the original container) by touching the tip to the side wall of the container and pushing the plunger smoothly all the way down, to the second stop. No liquid droplets should remain in or on the tip: at the very end, inside, or on the outside of it.

5. To make the actual volume measurement, push the plunger down to the first stop again, with the tip in the air. Lower the tip into the liquid to be measured, submerging it about 3 mm below the surface. Holding the tip at that depth, allow the plunger to smoothly rise all the way up. Wait a second, then smoothly lift the tip out of the source container and move it into contact with the wall of the receiving container. Keeping it there, smoothly push the plunger all the way down, to the second stop, such that all of the liquid leaves the tip. Keep the tip there for a second before removing it.

6. When using an autopipet, keep it near vertical, especially when it has liquid in it. This is not only important for exact measurements, but to protect its inner workings from damage. Only the autopipet *tip* is designed to come into contact with liquids; its other parts must remain dry.

7. Tip immersion depth matters. If the tip is not deep enough, a vortex will form as liquid is pulled into the tip, and droplets will fly around even if air is not actually sucked in with the liquid. If the tip is too deep, the hydrostatic pressure of the liquid changes the amount pulled into the autopipet (by increasing the pressure of—and compressing—the air in the pipet body).

It takes skill and practice to make good use of *any* kind of pipet. (Autopipets and even repipets are NOT foolproof exceptions to this statement!) You are *strongly* encouraged to *practice* with any measuring device that is new to you before relying on it for exacting measurements. With volume, using a balance to confirm the mass of pure water transferred in a practice measurement is an easy way to do so, because the density of pure water is known (see Appendix A).

B. Burets

A buret (also spelled burette, but in this manual we will use the shorter spelling) is a long, calibrated tube fitted with a stopcock (a valve) to control the release of liquids contained inside it [see Figure D.2(b)]. Burets are used in a procedure called titration, and are sometimes used in pairs. Titration allows the addition of precisely measured volumes of a liquid until a so-called <u>equivalence point</u> is reached: at which the amount of reactant in the added liquid is equal to the amount needed to completely react with another substance present in the container the liquid is being dispensed into. The liquid whose volume is measured with the buret is called the <u>titrant</u>.

As with a pipet, working with a buret requires skill and practice. The following procedure is recommended:

1. Determine whether the buret is clean by filling it with deionized water and allowing it to drain out. A clean buret will leave an unbroken film of water on the interior walls, with no beading or droplets. If necessary, clean the inside of your buret with soapy water and a long buret brush. Rinse several times with tap water, and finally with deionized water, confirming the water now drains out without leaving any droplets behind.

2. Rinse the buret three times with a few milliliters of the titrant. Tip the buret to wash the walls with the titrant, then drain through the tip into a waste beaker. With the stopcock closed, fill the buret with titrant to above the top graduation (volume marking) and briefly open the stopcock completely, several times, to ensure all air bubbles are removed from the tip by the rapid flow of titrant. If necessary, top up the buret so that the titrant level is slightly above the top graduation again. Open the stopcock carefully and let the titrant level drop to just below the top (zero) graduation. Read the initial buret volume to the nearest 0.01 mL. For clear and colorless titrants it can be helpful to place a white card with a sharp black rectangle on it (a piece of black tape is ideal) behind the buret, so that the reflection of the bottom of the meniscus is just above the upper black line, and at eye level (see Figure D.3). If this is done, it is essential it be done in the same way with *all* buret readings that are to be compared—because it systematically changes what is read from the buret.

3. Add titrant to the <u>titration vessel</u> (where the material to be titrated is) until you reach the <u>endpoint</u> of the titration, which is usually established by a color change in a small added amount of a strongly colored chemical called an <u>indicator</u>. To hit the endpoint accurately, you need to add titrant in very small quantities as you near the endpoint, though you can add it more rapidly and in larger quantities at the beginning of a titration. Quickly twisting a stopcock a half turn, through the open position, is often a more foolproof way to add very small volumes of titrant than trying to deliver individual drops slowly, by barely opening the stopcock; but both techniques are widely used. Swirl the titration vessel to ensure its contents are well-mixed. Often you get a clue from the indicator that you are near the endpoint. If you go past the endpoint (technical term: <u>overshoot</u> it), you can use your first titration as a guide for a second or third trial. Read the final liquid level in the buret as before, to ± 0.01 mL.

4. A good titration requires all of the chemicals in the titration vessel be well-mixed as the endpoint is approached. Frequently swirling the titration vessel is essential, as is ensuring that any chemicals stuck to its walls are mixed into the bulk of the solution (and if solid, fully dissolved). Those new to titration often focus so much on endpoint color that they neglect this critical fact!

5. The more visible the endpoint color change, the easier titration becomes—but adding more indicator is rarely a good way to achieve this. Instead, carry out titrations against a white background, or in a white titration vessel under bright light (except where the titration chemistry makes this ineffective, as is sometimes the case). If the color change associated with a given indicator is not distinct to your eyes, an alternative indicator can sometimes be found that provides a color change you can more easily see.

C. Volumetric flasks

A volumetric flask is one that has been exactly calibrated to hold a specified volume: 10.00 mL, 25.00 mL, or some other amount. It has a special shape, shown in Figure D.2(b), that allows for very reliable volume measurements. These flasks are used when very accurate dilutions are required in analytical experiments. Place a measured quantity of the substance to dissolve (technical term: the <u>solute</u>) or solution to dilute into a previously cleaned, but not necessarily dry, volumetric flask. (If the flask starts out wet with whatever solvent will be used, that is not a problem because you will be adding more solvent to the flask anyway.) Fill the flask about half full of solvent and swirl the contents until any solid has dissolved. Then add solvent to bring the meniscus up to the mark on the flask: once the liquid volume reaches the narrow neck of the flask, add the solvent with a wash bottle or Pasteur pipet, with the last volume increments added one drop at a time until the meniscus is at the mark on the flask. Stopper the flask and mix thoroughly by inverting at least 13 times, allowing the bubble of trapped air to move all the way from the top of the neck to the bottom of the flask, and back, each time. (Shaking the flask or inverting it only a few times will not result in the contents of the flask being thoroughly mixed!)

Temperature

The SI unit of temperature is the Kelvin, which is equal in size to the Centigrade degree, C°. At 0 °C, the Kelvin temperature is 273.15 K. The degree sign, °, is not used when reporting Kelvin temperatures.

Temperatures are usually measured with a <u>thermometer</u>. Your lab drawer probably contains a standard laboratory thermometer suitable for temperature readings between about −10 °C and about +100 °C, with graduations at 1 °C intervals. You will use this thermometer in experiments when exact readings are not required.

A typical glass thermometer has a hollow, thin-walled bulb at one end, connected to a fine <u>capillary</u> tube (a tube with an internal diameter similar to the diameter of a human hair). It contains a liquid, usually alcohol or mercury, which fills the bulb and part of the capillary. As the temperature goes up, the liquid expands and the level rises in the capillary. To make the graduations on a thermometer, in principle you could find the level at 0 °C and the level at 100 °C, then divide the distance between them into 100 equal parts. With mercury, that works quite well. With most other liquids, however, the thermometer must also be calibrated at several temperatures in between in order to provide reliable measurements. The scale on a standard glass thermometer spanning a 110 C° range can be read to ±0.2 °C, but the actual error is likely to be greater than that.

In some cases more exact temperature readings than those possible with a standard liquid-filled thermometer are required. A digital thermometer that operates over a wide temperature range and reads directly to ±0.1 °C is often the best way to do so. The sensing tip on such a thermometer contains a <u>thermistor</u>—a temperature-sensitive resistor—whose electrical resistance changes reliably with temperature.

Glass thermometers are fragile, so be careful when using them. If you slip a split rubber stopper around the thermometer and support it with a clamp you will minimize breakage. **Should you ever break a mercury thermometer, contact your lab instructor immediately**. Liquid mercury has a low vapor pressure, but that vapor is very toxic—it is very important to clean up mercury spills *completely*, which is not easy. Do not heat a liquid-filled thermometer above its maximum readable temperature, because that can ruin the thermometer or even cause it to explode.

Glass thermometers are marked with an <u>immersion line</u>: a line well below the lowest temperature graduation. For reliable temperature readings, the thermometer must be at the same temperature—the temperature you wish to measure—at all points below the immersion line. With mercury this is not as important, because mercury is a metal and therefore an excellent thermal conductor—but mercury thermometers are now rare. For glass thermometers filled with other liquids, the immersion line matters: if only the bulb is exposed to the temperature of interest, they will read temperatures slightly closer to room temperature than they should.

When taking temperature readings, allow enough time for a thermometer reading to become steady before noting the final temperature. This usually takes less than a minute—if you check the reading over that period of time, you should be able to detect when it has reached a stable and reliable value. Most digital thermometers will only provide an exact liquid temperature reading when the tip of the probe, where the thermistor is located, is moving relative to the liquid. This can be accomplished by stirring the liquid rather than moving the probe itself. Digital thermometers read the temperature in a very localized location (at the tip of the probe), so be aware that a probe tip resting on the bottom of a beaker, touching the glass, will not read the temperature of the bulk of the liquid if the beaker is being heated or cooled, especially if the liquid is not being actively stirred.

Pressure

In studying the behavior of gases it is important to be able to measure gas pressure. One way to accomplish this is by comparing the pressure of interest with that exerted by the atmosphere, using a device called a <u>manometer</u>. A manometer is a glass U-tube partially filled with a liquid, usually water or mercury. A manometer indicates the gas pressure difference between the two ends of the tube, measured as the height difference in the liquid used to fill the manometer tube (most often as mm Hg or mm H_2O), as shown in Figure D.5.

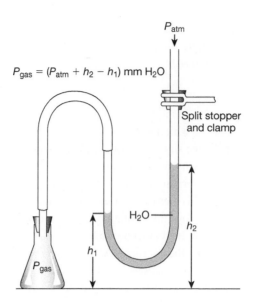

P_{atm}

$P_{gas} = (P_{atm} + h_2 - h_1)$ mm H_2O

Split stopper
and clamp

H_2O

h_2

h_1

P_{gas}

Figure D.5 Measuring the pressure of a gas with a water-filled manometer. The pressure difference between the pressure in the sample container, P_{gas}, and the pressure exerted by the atmosphere, P_{atm}, is equal to the difference in the two liquid heights, $h_2 - h_1$.

The atmospheric pressure itself can be measured with a mercury barometer, but its electronic equivalent is typically used. A mercury barometer consists of a straight glass tube about eighty centimeters long, which is initially completely filled with mercury. The tube is then inverted while the open end is immersed in a pool of mercury. The mercury level in the tube falls. The observed height of the mercury column above the pool, as read from a scale behind the tube, is equal to the pressure exerted by the atmosphere, called the atmospheric pressure. Electronic barometers use a microelectromechanical device (a microscopic, pressure-sensitive moving part) to measure pressure. These days almost all smartphones contain such a barometer, which assists with GPS location finding. With the right app, the atmospheric pressure can be retrieved: but heed the warnings later in this section about "barometric pressure"!

While many different units are used for pressure, we will default to using mm Hg or mm H_2O since these are what can be read directly from a manometer. These units are related to each other by the relative density of mercury to that of water, such that 1 mm Hg = 13.57 mm H_2O. Another common pressure unit is the atmosphere, abbreviated atm; 1 atm = 760.00 mm Hg = 10313 mm H_2O, which is approximately the pressure exerted on us by the atmosphere every day. The (derived) SI unit for pressure is the pascal, abbreviated Pa, which is equal to 1 kg/(m·s²). By definition, 1 atm := 101325 Pa; because the pascal is a small unit, the bar, equal to 10^5 Pa, is often used: 1 bar is approximately one atmosphere (more precisely, 1 atm := 1.01325 bar).

Pressure measurements are very useful in weather forecasting—a fact that has shaped the reporting of such measurements and even the terms used to describe them. You cannot rely on a reading marked "atmospheric pressure" or "barometric pressure" to be the value you need for chemistry! Air pressure decreases with increasing elevation, but weather forecasting is aided by maps showing how air pressure varies at *sea level*, corrected for elevation differences. Because sea level is below ground in most locations, the air pressure there cannot be measured directly—it must be calculated based on the systematic way in which air pressure changes with altitude. If you go to a weather forecasting site, such as weather.gov, the "barometer" reading shown there will *not* be the same value that would be measured with a mercury barometer (and that we want to use as chemists). Rather, it shows the "mean sea level pressure", which is not the absolute pressure (what a mercury barometer would read, and what we want to use) but has been *corrected for elevation*. Unless the weather station is at or below sea level, the reported "barometer" reading will be lower than the true, absolute pressure, with the difference growing larger the higher the altitude of the reporting weather station. At 1000 feet of elevation the correction is 27 mm Hg, which is substantial. It is therefore important to understand this correction if using atmospheric pressure data reported by any electronic device or remote source. Weather forecasters call the absolute pressure the "station pressure", and you can find calculators online that convert between it and mean sea level pressure. Many electronic barometers will report the station pressure if they have their altitude set to zero; however, some determine altitude (for example, by GPS) rather than ask you for it, so it is often challenging to know for sure. The best way to check is to get local weather data online (the barometer and altitude values), correct the barometer reading to the station pressure using an online calculator, and ensure that your local device reads the same value, within the required tolerance.

Having found the atmospheric pressure, P_{atm}, a manometer can be used to measure the gas pressure in a flask, as shown Figure D.5. The pressure inside the flask is equal to the atmospheric pressure plus the pressure exerted by a water column of effective height $h_2 - h_1$, the difference between the water levels in the right and left arms of the U-tube. With these all in the same unit, the gas pressure in the flask, P_{gas}, is

$$P_{gas} = (P_{atm} + h_2 - h_1) \tag{1}$$

When reading the water levels in a water-filled manometer, take the height measurements at the bottom of the meniscus in each of the two arms, which can be read against any common (horizontal) zero, though usually the elevation above the lab bench is most convenient.

Light Absorption

In many chemical experiments the interaction of light with a sample provides useful information. Of particular importance is determining the concentration of a substance in a solution by measuring the amount of light at a given wavelength (of a given color) that is absorbed by the sample. An instrument called a spectrophotometer is made for this purpose. A spectrophotometer often contains a device called a monochromator (usually incorporating a prism or diffraction grating), which allows only a specific wavelength of light to pass through. The absorbance, A, of the sample at that wavelength, defined as the negative of the base 10 logarithm of the fraction of light that makes it through the sample, can be read directly from the instrument. An approximate mathematical relationship called Beer's law holds that the absorbance is proportional to the concentration, c, of the sample, which is usually expressed as a molarity:

$$A = K \times c \tag{2}$$

In Equation 2, K is a constant that depends on the sample, the container, and the wavelength of the light. Most samples obey Beer's law reasonably well. To find the proportionality constant, K, for a given a substance at a specific wavelength and in a given sample cell (technical term: cuvet), the absorbance of a solution (or, better and more often, a series of solutions) containing a known concentration of the substance is measured. From that absorbance, or several absorbances obtained at known concentrations, a calibrating graph or equation relating concentration to absorbance can be prepared and then used to analyze unknown solutions containing the same substance.

Figure D.6 A spectrophotometer provides a numerical measure of how much light is absorbed by a sample at a particular wavelength. (*Martin Shields/Alamy Stock Photo*)

Spectrophotometers come in a wide range of shapes and sizes (an example is shown in Figure D.6), but they are all functionally equivalent. They contain one or more light sources, a wavelength selection mechanism, a cell in which the sample is reproducibly positioned, an electronic light intensity detector, and a readout. There are two key controls: the wavelength of the light at which absorbance is measured and a means of setting the zero point for absorbance. Some instruments require a warm-up time before they can provide exact measurements.

Before any measurements are made, a spectrophotometer must be calibrated to zero absorbance by rinsing and then filling the cell three-quarters full with a solution that matches the standard and sample in other respects but does not contain the light-absorbing substance of interest. (Pure water can be used for less exacting work.) This is called the blank, and with it in place and the wavelength set to a value at which the substance of interest absorbs an appreciable amount of light, the absorbance reading is set to zero. The cell is then emptied, rinsed with the solution to be measured (a standard or a sample), and filled with that solution before the absorbance is read again at the same wavelength.

pH Measurements

The pH of a solution is mathematically related to the concentration of the H^+ ion, $[H^+]$, in that solution, and therefore its acidity. Acid-base indicators provide one way to obtain approximate pH values. Each indicator has a characteristic pH range over which its color change occurs, so it can be used to indicate whether the pH is higher, lower, or where it falls in that range. Using multiple indicators, the pH of a solution can be measured to within about half a pH unit. Acid-base indicators are also used to prepare pH test strips for various pH ranges. These accomplish the same chemistry conveniently; both approaches work well provided the solution of interest itself is not strongly colored.

More exact pH measurements can be made with a pH meter (see Figure D.7). A pH meter is a very high-resistance voltmeter that measures the difference in voltage between a reference electrode and a special glass electrode whose potential is a function of the H^+ ion concentration in the solution in which it is immersed. The meter indicates the pH directly. The two electrodes may be separate, but more commonly are together in one "combination electrode". The electrode(s) must be kept wet, in an appropriate storage solution, when not in use.

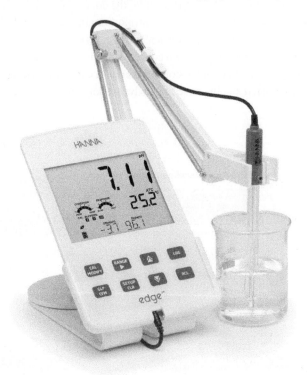

Figure D.7 A pH meter measures the concentration of H^+ ion in solution, using electrochemistry. The pH probe must be kept wet and the surface of the glass bulb at its tip never touched. [*Water Science School/U.S. Department of the Interior (USGS)*]

When you are asked to find the pH of a solution with a pH meter, you will first need to calibrate the pH meter if this has not already been done. Calibration requires at least one buffer solution with a well-defined pH to serve as a reference. Put about twenty-five milliliters of buffer in a small beaker. Remove the pH meter's electrode(s) from the storage solution, rinse with a stream of deionized water from your wash bottle, and then place the electrode(s) in the buffer. (pH electrodes are fragile and expensive, so treat them gently.) Allow the system to equilibrate until the pH reading becomes steady. Depending on the pH meter you are using and its specific operating instructions, the meter may ask for and/or automatically detect the pH of the reference solution when the reading has stabilized—but you will almost always have to press a button to confirm that a good calibration point has been obtained. If calibrating using more than one standard buffer, rinse the electrodes into a waste beaker with deionized water and then place them in the next buffer solution, taking whatever steps your meter requires to set an additional calibration point. Measuring the pH of a sample is similar, once calibration is complete.

Here are some important things to know about taking pH measurements with a pH meter:

1. Never touch the glass ball at the end of a pH electrode—with anything. Squirt water at it only. It is made of extremely thin glass, but the main reason for not touching it is that it relies on surface chemistry to operate. Keep it immersed in liquid when not in use.
2. Do not even wipe the outer shield of a pH probe with anything, except when putting it into long-term storage. Doing so can easily build up a static charge large enough to make the pH meter inaccurate until the charge dissipates.
3. Slow response or an inability to calibrate a pH meter usually results from a damaged, dirty, or clogged electrode. There is a tiny hole (a salt bridge) connecting the solution inside the electrode to the solution being measured: if this does not touch the solution, or if it has clogged (usually because the electrode dried out), the pH reading will never stabilize.
4. pH meters often show more digits than they can be trusted to measure reliably. The extra digits can be useful indicators of whether and to what extent the reading is drifting over time, but without temperature correction a pH meter should not be trusted to better than ± 0.1 pH unit. Even with temperature correction, great care is required to obtain a reading reliable to even ± 0.01 pH units.

Separating Solids from Liquids

A precipitate is a solid that appears in a solution as a result of a chemical or physical change. It is often initially composed of very small particles, but these can sometimes be made to grow larger, or to stick together (technical term: agglomerate), by boiling the solution, cooling it slowly, or letting it sit. A common lab objective in chemistry is the separation of a precipitate from the liquid it came out of (technical terms: the mother liquor, or supernatant). A number of different methods and tools have been developed to assist with this.

A. Centrifugation

A centrifuge (see Figure D.8) is used to aid in separating a solid from a liquid in a test tube. It spins the test tube at several hundred revolutions per minute, effectively increasing the force of gravity experienced by the tube to many times normal earth gravity. This speeds up separation based on density differences, and often overcomes the tendency for very small particles of solid to continue floating around in a liquid indefinitely (technical term: Brownian motion).

How to use a centrifuge:

1. Make sure the test tube is the correct size for the centrifuge.
2. Be certain that the test tube is not overly full—the liquid should be at least 3 cm below the top of the tube.
3. Make sure the test tube has no cracks, as these may cause the tube to break during centrifugation.

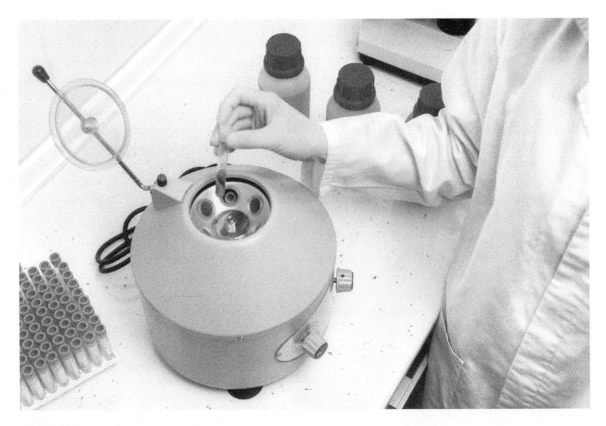

Figure D.8 A centrifuge spins samples at high speed, magnifying the effect of gravity and encouraging immiscible materials (substances that do not dissolve in or mix with one another) to more quickly and completely separate based on their density differences. (*Zoonar.com/DAVID HERRAEZ CALZADA/AGE Fotostock*)

4. Place the test tube containing the solid you wish to separate into one of the centrifuge slots.
5. Place a similar tube, containing the same amount of water as you have liquid in your sample tube, in the centrifuge slot opposite your sample—this serves as a counterweight and keeps the rotor of the centrifuge (the part that spins) balanced so that the centrifuge does not vibrate or jump around as it operates.
6. Turn on the centrifuge, allow it to spin for about a minute, and then turn it off.
7. Keep your hands away from the spinning rotor. After the rotor has stopped moving, remove the tube. If the centrifugation was successful, the solid (or denser liquid) will have all collected at the bottom of the test tube, and the liquid above it will be clear.

B. Decantation

Decantation is the technical term for removing (most of) the liquid from a solid by carefully pouring it off, leaving the solid behind, as shown in Figure D.9. Decantation requires that the solid first settle to the bottom of its container, or be made to do so (using a technique such as centrifugation). Although not essential, it can help to use a stirring rod when decanting. Place the stirring rod across the top of the container you wish to decant from, extending a centimeter or two beyond the edge you will be pouring out of and positioned in the spout, if the container has one. Hold the stirring rod in place with the index finger of the hand with which you hold the container, and pour out the liquid slowly and smoothly. The purpose of the stirring rod is to help direct the flow of the liquid, as well as to hold back solid particles that would otherwise escape as you near the end of the decantation process.

C. Filtration—the Büchner funnel

Filtration is used to recover a solid, in pure form, from a liquid. While with great patience this can be accomplished with gravity filtration, in general chemists use a Büchner funnel, as shown in Figure D.10, which contains a piece of filter paper of appropriate size covering the holes in the cup of the funnel and uses reduced pressure to pull liquid through much more quickly than would gravity alone. The Büchner funnel is connected by a rubber

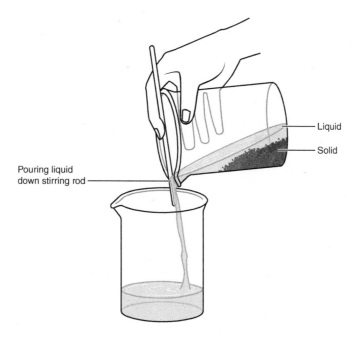

Liquid

Solid

Pouring liquid
down stirring rod

Figure D.9   Decantation is the technical term for removing (most of) the liquid from a solid by carefully pouring it off, leaving the solid behind. The purpose of the stirring rod is to help direct the flow of the liquid, as well as to hold back solid particles that would otherwise escape as you near the end of the decanting process.

stopper or other adapter to the top of a side-arm suction flask. The side-arm of the flask is connected through a series of rubber hoses and a safety trap to a water aspirator, a central vacuum system accessible through a jet on the lab bench, or another vacuum source (see Figure D.11). When the vacuum line is turned on, the reduced pressure created in the flask by the vacuum source will pull air or liquids through the funnel.

When you are ready to filter, moisten the filter paper completely, using a minimum amount of water, *then* turn on the vacuum line. With the vacuum source applying maximum suction, stir the combination of solution

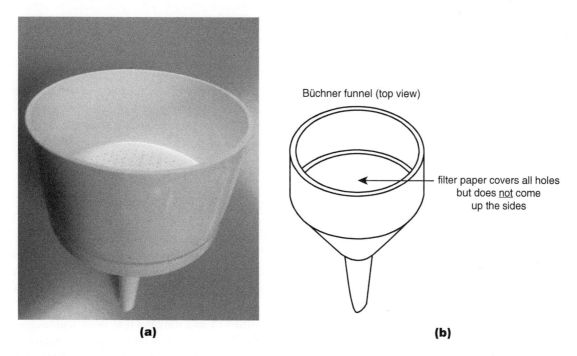

Büchner funnel (top view)

filter paper covers all holes
but does <u>not</u> come
up the sides

(a)   **(b)**

Figure D.10   (a) Some Büchner funnels are a single piece, usually made of porcelain, but others (such as the one shown here, made of plastic) come apart into a funnel portion (lower) and a cup portion (upper). (*Robert Rossi*) (b) A single disk of filter paper is placed into the cup of a Büchner funnel and wet before use. It must cover all of the holes at the bottom of the cup, but also lay flat on the bottom of the cup, ensuring no solid can bypass it.

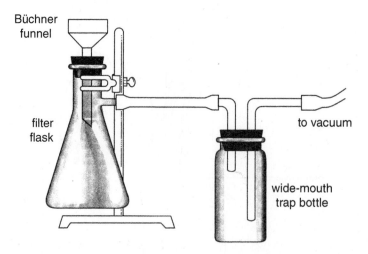

Figure D.11   To operate a Büchner funnel, put a piece of circular filter paper in the funnel. Thoroughly moisten the filter paper with deionized water from a wash bottle, but do not add more water than the filter itself will absorb. Just before pouring the sample into the filter, turn on the suction. Keep the suction on while filtering the sample.

and precipitate (technical term: <u>slurry</u>)—to suspend the solid—and pour the mixture into the cup of the Büchner funnel before the solid has a chance to settle. Usually you will want to wash out any remaining precipitate with a stream of liquid from your wash bottle. A procedure may call for rinsing the precipitate in the Büchner funnel, sometimes with an appropriate quick-to-evaporate liquid, before passing air through the filtered-out solid (technical term: filter cake) for some time to help dry the solid. Turn off the vacuum line, break the suction by twisting the vacuum hose as you pull it off the suction flask, and remove the funnel, or at least its cup (some plastic Büchner funnels are two-part systems in which the cup can be separated from the funnel portion; to free these cups, rotate them as you pull up).

Heating and Boiling

Heating is a common operation in most chemistry laboratories, and can be accomplished in a variety of ways. This section provides some general advice on how to safely and efficiently carry it out.

A. Laboratory burner

A <u>laboratory burner</u> is a common laboratory heat source: it mixes fuel and air in a reliable way, providing an easily controlled and stable flame whose temperature can be varied. The most common fuel used is methane (natural gas, CH_4), provided by a valved gas line on the lab bench, but some burners have their own gas supply. Laboratory burners come in a variety of sizes and shapes, some specialized for specific purposes, but a general laboratory burner, such as the one shown in Figure D.12(a), is multipurpose.

To use a laboratory burner:

1. Connect the burner's gas inlet to a natural gas line or other low-pressure gaseous fuel source.
2. Turn on the gas line valve gradually, until you hear gas coming through (but do not open it all the way). If gas does not flow, ensure that the fuel regulator on the burner itself is open, if it has one.
3. Attempt to ignite the burner by placing a flame or spark at the top of the vertical tube. If the flame does not light and stay lit, too much air is probably being mixed into the fuel. Either close the air control vent at the bottom of the tube or (partially) seal it with your fingers, to reduce the relative amount of air in the gas mixture emerging at the top of the tube, and try again.
4. Once the flame is lit, its size and temperature can be controlled by adjusting the flow of air and fuel into the burner. With only fuel (no air) entering the burner, the flame will be bright yellow and relatively cool, dancing around and rising higher above the tube. As more air is mixed in, the flame will become progressively shorter, hotter, louder, and bluer, until it eventually becomes unstable and goes out (which should be avoided). While a yellow flame is hot enough to do some things, its yellow color is the result of

(a) **(b)**

Figure D.12 (a) A laboratory burner mixes fuel and air in a reliable way, providing an easily controlled and stable flame whose temperature can be varied. (*Turtle Rock Scientific/Science Source*) (b) A hotplate is an electrical heating device and does not involve an open flame. In most modern hotplates, the temperature of the top surface is actively controlled. (*azrin_aziri/Shutterstock.com*)

incomplete combustion—there are actually tiny particles of solid carbon in the flame, glowing yellow— and black soot will form on anything placed in such a flame. Therefore yellow flames should not be used to heat things. Instead, use a blue flame, moving it further from the item you wish to heat and/or making the flame smaller (allowing in less fuel) to provide less intense heating. A "roaring" blue flame with a bright blue inner cone is at a nearly uniform temperature of 1500 °C at all points above the peak of its inner cone. A larger flame will heat larger objects more quickly and is recommended over a very small but very hot flame when heating a large object, so that the temperature of the heated object remains more uniform.

5. To safely heat a beaker over a laboratory burner, first clamp an iron ring to a ring stand and place a sheet of wire gauze on top of it. Ideally, the ring should have a diameter slightly larger than the base of the beaker you are heating because this provides maximum support for the beaker while also encouraging, rather than getting in the way of, heat transfer to the beaker. The ring should be positioned high enough that the beaker sits just above the top of the laboratory burner flame: this is about five centimeters above the top of the laboratory burner itself, in most situations. Place the beaker on the wire gauze and then ignite the burner. Adjust the flame for the degree of heating you need: a larger, louder blue flame for rapid heating or a smaller, quiet blue flame for gentle heating. Then move the burner under the iron ring, wire gauze, and beaker. If the top of the flame is not near the wire gauze, immediately remove the burner and beaker and make adjustments so that it is, before everything gets hot. Control the rate of heating by changing the fuel and air flow to the burner or by removing the burner from under the beaker periodically. If you are bringing a liquid to a boil, see also Subsection C, on boiling liquids.

6. When you are finished using a laboratory burner, turn the gas off completely, at the gas line valve.

B. Hotplate

Hotplates provide heat using electricity, without an open flame. An example is shown Figure D.12(b). Modern hotplates actively control the temperature of the top surface, switching the heating element on and off to maintain a target temperature. The maximum surface temperature is typically around 350 °C if the heated surface is made of aluminum, 540 °C if it is made of ceramic. Aluminum surfaces offer faster heating, better temperature uniformity, and superior thermal conductivity, at the expense of a lower maximum operating temperature. Hotplates often incorporate a magnetic stirrer: a rotating magnet under the heated surface that can cause another magnet, placed inside a container on top of the hotplate, to turn in unison with it. Some hotplates can be connected to an external probe and made to actively control the temperature at the probe (which is placed into what is being heated) rather than the temperature of the top surface. Always place a glass container on a hotplate **before** the hotplate surface has become very hot. Do not allow liquid to contact a hot hotplate surface, and do not heat a hotplate with a wet surface.

C. Boiling

A common objective in heating operations is to bring a liquid, most often water, to a boil. This means increasing its vapor pressure to the atmospheric pressure, such that bubbles of vapor (evaporating liquid—that is, the liquid in a gaseous state) constantly form within the liquid itself and the liquid is being constantly converted into gas. Understanding how to do this safely and efficiently is important:

1. Before heating any glass container, look it over carefully for cracks or chips, including cracks that can only be seen by looking through the glass. Glass is a poor thermal conductor and does not heat up evenly. As it heats, it expands slightly, making it more likely to break across temperature differences in the glass. Borosilicate (lab) glass is made to tolerate this, but if there is the slightest crack, that crack will tend to very quickly grow when heated, often shattering the glass. When using a laboratory burner, a wire gauze placed under a glass container evens out the temperature over the bottom surface of the glass, reducing the risk of thermal breakage.

2. The rate of thermal energy transfer from one object to another depends on the materials they are made of and temperature difference between them. Glass and ceramic are both poor thermal conductors, so large temperature differences are required to quickly transfer thermal energy through them. To bring a significant volume of liquid to a boil, use a large temperature difference—a hot laboratory burner flame placed close to the container, or a hotplate set to maximum—until the liquid just begins to boil. Then quickly reduce the temperature of the heat source by a significant fraction—to perhaps one third of what it was: enough to maintain boiling, but avoiding such vigorous boiling that the liquid splashes everywhere. With some hotplates (especially ceramic ones) it may be necessary to start lowering the temperature before boiling even starts, perhaps when condensation can first be seen forming on the upper (cooler) surfaces of the container.

3. Especially in very clean containers with no sharp edges or scratches in contact with the liquid, boiling can occur in a dangerous way: in vigorous bursts that can cause the liquid to jump out of its container (technical term: bumping). This happens because it is easier for a vapor bubble to grow larger than to form in the first place. Boiling can be made smoother, and more predictable, by providing a constant and reliable source of bubbles in the liquid. Chemists usually accomplish this using boiling chips, also called boiling stones, which are unreactive solids full of tiny holes. One or two of these should be added to a liquid being heated **before it approaches the boiling point**. Boiling stones are not very effective if reused, unless the pores inside them are thoroughly dried first.

Qualitative Analysis—A Few Suggestions

In the course of your laboratory program, you may do several qualitative analysis experiments, whose goal is to determine the presence or absence of a substance rather than its amount. Those experiments use small (13×100 mm) test tubes and small beakers as sample containers. Solids resulting from chemical reactions can be separated from the liquids in which they form by centrifugation.

In many procedures you will be told to add about one milliliter, or perhaps about half a milliliter, of liquid to a test tube. This is best accomplished by first finding out what 1.0 mL looks like in the test tube being used, as well as how many drops correspond to a milliliter with a given type of dropper or Pasteur pipet. To do so, count how many drops of deionized water it takes to reach the 1.0 mL mark in a small graduated cylinder, confirming this by counting how many additional drops it takes to get to the next two 1.0 mL graduations (markings) on the graduated cylinder. Then put 1.0 mL worth of drops into a test tube, noting the height reached by the liquid in the tube. From then on, use that height as a guide when you need to add about one milliliter; half that height indicates about half a milliliter. Counting drops is not a very reliable way to measure volumes across different liquids, but it usually works well enough. The height in the test tube is reliable for any liquid, and will work well across test tubes provided they have identical inner dimensions.

Many of the steps in qualitative analysis involve heating a mixture in a boiling-water bath. To make such a bath appropriate for small test tubes, fill a 250-mL beaker about two-thirds full of water. Support the beaker on a wire gauze resting on an iron ring and heat the water to the boiling point with a laboratory burner, adjusting the flame so as to keep the water just barely boiling. Or, place the beaker on a hotplate, adjusting the heat setting to accomplish the same result. Use this bath to heat a test tube when that is called for. You can remove test tubes from the boiling-water bath with your test tube clamp if they are too hot to handle with your fingers. Add water to your hot-water bath as needed to keep the level above half full.

In separating a solid from a liquid in qualitative analysis, centrifugation followed by decantation is typically used (see previous subsections for details of those techniques). The clarified liquid (technical term: the supernatant) is poured off into a test tube or discarded, depending on the procedure. The solid may then need to be washed free of any entrained (trapped) liquid. To do this, add the indicated wash liquid and stir with a glass stirring rod. Centrifuge again and discard the wash. Keep a set of stirring rods in a 400-mL beaker filled with deionized water, and return each rod to this beaker between uses in different liquids to prevent cross-contamination.

Pay attention to what you are doing, and do not just follow the directions as though you were baking a cake. Try to keep in mind what happens to the various chemicals in each step, so that you do not end up throwing away the material you want while keeping a waste product. When you need to put a fraction (a solid or liquid you isolate) aside while you are working on another part of the mixture, make sure you know where you put it by labeling the test tube or rack with the number of the step in which you will analyze that fraction.

Appendix E

Numbers in Science—
Significant Figures and Making Graphs

When taking a measurement or reporting a result in science, there are three *essential* things to include:

1. The numerical value
2. The *units*
3. An indication of how *reliable* the numerical value is: its <u>uncertainty</u>

Forgetting *any* of these three things is a scientific mistake: all three are *necessary*.

Consider the example of an environmental chemist measuring the arsenic level in a drinking water sample for a court case in which the legal limit is (exactly) <u>10</u> parts per billion (<u>ppb</u>). Their measuring device reports the concentration to be 0.01035 parts per million (<u>ppm</u>).

If they write down only the number that appears on the display, without its units, it will not be clear whether the arsenic level was above the legal limit in that sample. Is it in ppm or ppb? Or perhaps in moles per liter, or teaspoons per gallon? In science, units should *never* be implied or assumed; they must be *explicitly* recorded. This can be done in the heading of a table (as is often done in this manual), but the units of any reported value must be clear and unambiguous to a trained scientific eye.

Just as important, but more often overlooked, is the uncertainty in a scientific value. A device may read a value to four digits, but is it reliable to four digits? If the measurement is repeated, is the same value obtained? In a high-stakes case such as this, extensive statistical experimentation and analysis would have to go into establishing how reliable the measurement is. But in reporting the results, including that information is essential to the court case. Was the arsenic concentration above <u>10</u> ppb *beyond a reasonable doubt*? That is what would have to be proven, and even recording a result indicating a reliability greater than what is warranted can shatter a court case—by throwing doubt on the scientific reliability and competence of the chemist taking the measurements. The legal discovery process would go back to the chemist's original notebook, looking for evidence to use in the case...and any failure to properly handle either units or uncertainty could prove devastating in the hands of a competent lawyer, one who understands the nature of numbers in science.

The first part of this appendix deals with <u>significant figures</u>, which are one way of indicating how reliable a scientific measurement is. It discusses this "secret code" used to indicate numerical reliability, and how to determine the reliability of numbers calculated from other values.

The second part of this appendix deals with graphing, or plotting, numerical data, which is often essential in finding trends in scientific data as well as effectively and efficiently communicating it.

Significant Figures

Scientists make many kinds of measurements. The reliability of each measurement depends on the device used to make it. An analytical balance, when used properly, can reliably measure mass to ± 0.0001 g—so if the display on such a balance indicates a sample has a mass of 2.4965 g, there are *five* meaningful figures in the measurement. These figures are called <u>significant figures</u>. A given decimal place is considered significant if it is reliable to ± 1—that is, its uncertainty is not more than one. By recording a mass of 2.4965 g, a scientist is asserting that if the same measurement were taken again the result would be 2.4965 ± 0.0001 g: specifically, between 2.4964 g and 2.4966 g.

If the volume of a sample is measured using a graduated cylinder, the reliability of the volume reading depends on the cylinder used. If a volume of 6 mL is measured with a 100-mL cylinder, it is difficult to tell the difference between a volume reading of 6 mL and a reading of 7 mL; the volume reading would be ± 1 mL and should be reported as 6 mL, with *one* significant figure. With a 10-mL cylinder, the volume could be measured more precisely—the volume could be reported as 6.0 mL, with confidence that both figures in the result are meaningful and that it deserves *two* significant figures. Reporting the result as 6.0 mL amounts to a scientific claim that the measurement is ± 0.1 mL, or in other words that the true value lies somewhere between 5.9 mL and 6.1 mL.

In laboratory procedures and instructions, significant figures indicate how carefully an amount needs to be measured and often implicitly indicate which device should be used. For example, suppose there is an instruction to add 1.00 mL of water to a reaction vessel. This could not be accomplished to three significant figures with a beaker, or even with a graduated cylinder. A properly used pipet *can* reliably deliver 1.00 mL of water, however. In this case, it would be necessary to use a pipet, or another mechanism capable of providing a 1.00-mL volume measurement with three significant figures. On the other hand, when instructions call for adding 1.0 mL of water to a reaction vessel, only two significant figures are called for—using a graduated cylinder is not only appropriate, but the wise choice. Doing so allows the water to be measured out more quickly and efficiently than with a pipet, and the instructions indicate that the amount need not be more precise than that. (Counting drops from a dropper bulb or basing the volume on the liquid height in a small test tube would also suffice in this example.) If you were asked to add 1 mL of water to a reaction vessel, it would technically indicate that anything between 0 and 2 mL would work (which is rarely true, because 0 mL amounts to adding nothing, making the step unnecessary). We will instead use written-out words for such approximate quantities, calling for "about one milliliter of water" in the example at hand. In these cases there is no need to measure at all, once you have developed a sense of approximate amounts at the lab bench, and routinely measuring such quantities with far greater precision than needed, say with a pipet, may prevent you from completing lab work in a reasonable amount of time.

A. Exact numbers

There will be a whole number, such as 12, or 19, students in your lab section. There is no way there will be 14.3 students, unless there is something very disturbing going on. However, if you have not counted carefully, it is still possible that your count is off by an entire person or two, perhaps because someone was absent when you counted. A scientist reporting that there are 19 students in their lab section is implicitly saying there are 18, 19, or 20 (but not that there may be 19.2): this reported result still has uncertainty. Some numbers *are* exact and have no uncertainty at all: usually by definition. One way to indicate that numbers are exact is to underline them, and that is the approach we will use in this manual. Conversion factors, used to convert one set of units to another, often contain exact numbers: $\underline{1}$ meter := $\underline{100}$ cm. Both numbers are exact, with no uncertainty at all. With volumes, $\underline{1}$ liter := $\underline{1000}$ mL. Again, both numbers are exact. In converting from one measurement system to another, often only one of the numbers in the conversion factor is exact: $\underline{1}$ gallon = 3.785 liters, or $\underline{1}$ mole = 6.022×10^{23} molecules. In both of these cases the 1 is exact but the second number is not: it has four significant figures. The 1 is made exact so that the ratio of the two numbers (the conversion factor) is not limited to one significant figure (see Subsection C). The value 1 written with four or more significant figures, as 1.000 (or 1.0000, or 1.00000, and so on) would also work; what matters is that the number to the right of the equal sign indicate and control how reliable the conversion factor is. Note that a few conversions across different unit systems *are* exact, such as $\underline{1}$ inch := $\underline{2.54}$ cm. After the SI system came into wide use, the inch was actually redefined in terms of the centimeter, as *exactly* $\underline{2.54}$ cm.

B. Insignificant digits

It is sometimes helpful to be able to explicitly indicate that a digit is *not* reliable: that it is <u>insignificant</u>. In this manual we accomplish this by subscripting such digits. For example, $1._3$ mL indicates a volume that is only reliable to one significant figure, and is ± 1 mL, but that the best guess as to the tenths place value is a three. It indicates a volume of between 0.3 mL and 2.3 mL.

C. Identifying significant figures

If you are uncertain about the number of significant figures in a given number, there is a method for finding out that usually works. Write the number in scientific notation, expressing it as the product of a number between 1 and 10 and a power of ten:

$$26.042 = 2.6042 \times 10^1 \qquad 0.0091 = 9.1 \times 10^{-3} \qquad 605.20 = 6.0520 \times 10^2$$

The first number in the product indicates the number of significant figures. So, the first of the numbers in this set of examples has 5 significant figures, the second has 2 (leading zeroes are *not* significant), and the third has 5 (trailing zeroes *are* significant, having been written specifically to indicate that a value is reliable out to that digit.) If a number is recorded by a well-trained scientist, you should be able to rely on it being written with attention to significant figures. If you saw 2400, you would understand it to have 2 significant figures. If the scientist had intended three significant figures, she would have written 2.40×10^3 or 240_0, while if she wished to indicate four significant figures, she would have written 2.400×10^3 or 2400., with the explicit decimal point after the last zero making the two place-holding zeroes significant. She would have written 2.4000×10^3 or 2400.0 to indicate five significant figures.

D. Error propagation

Recording measurements with an appropriate number of significant figures and correctly interpreting significant figures in values reported by others are both important skills, but there is a third critical aspect to working with significant figures, called <u>error propagation</u>, which is perhaps the most challenging. The goal of error propagation is to track reliability through numerical calculations. Say, for example, you are weighing out a substance in a beaker by difference and get these results:

Mass of beaker before removing sample = 25.4329 g

Mass of beaker after removing sample   = 24.6263 g

To determine the mass of the sample you would subtract the mass of the (partly) emptied beaker from that of the full beaker: 25.4329 g − 24.6263 g = 0.8066 g. Even though the two measured masses each have 6 significant figures, the mass of the sample deserves only 4. What matters with addition or subtraction is the position of the left-most insignificant decimal place; that decimal place must also be insignificant in the sum or difference, because the sum or difference cannot be known with any greater precision (in this case ± 0.0001 g) than any of the values being added or subtracted. In another case—in which some components are mixed to make a solution, weighing each of the components separately on different balances—these results are obtained:

Mass of beaker	= 25.5329 g
Mass of water	= 14.0 g
Mass of salt	= 6.42 g
Total mass	= ?

Here each mass has a different left-most insignificant decimal place. The first mass is reliable to ± 0.0001 g, the second to ± 0.1 g, and the third to ± 0.01 g. Their sum is 45.9529 g, but a scientist should not report that value because it could be off, not by the ± 0.0001 g implied by writing it that way, but by ± 0.1 g, the possible error in the mass of the water. The total mass must be reported to ± 0.1 g. Because the sum is closer to 46.0 g than to 45.9 g, it should be reported as 46.0, with 3 significant figures. It might be even better to record and report 45.9_5 g, or 45.9_{53} g, as these also indicate a reliability of ± 0.1 g, but additionally convey best guess values for one or two insignificant digits. Carrying one or two insignificant digits through calculations helps to avoid <u>rounding error</u>, which is error introduced into a series of calculation steps as a result of rounding off at each step along the way. Generalizing, *when adding or subtracting numbers, the result should have the same left-most insignificant decimal place as the left-most insignificant decimal place in any of the things being added or subtracted.* The last significant digit can be rounded off based on the digits that follow it; but to avoid rounding error, it is better to round off at the first or second insignificant digit and record these insignificant digits in subscript.

When *multiplying* or *dividing* measured quantities, the rule is different: *the number of significant figures in the result is equal to the number of significant figures in the quantity with the fewest significant figures.* For example, suppose you want to determine the volume of a <u>pycnometer</u>, which is just a flask with a well-defined volume. You find that volume by weighing the empty pycnometer and then weighing the pycnometer again when it is full of water. You might obtain the following data:

Mass of pycnometer filled with water	60.8867 g
Mass of empty pycnometer	31.9342 g
Mass of water in pycnometer	28.9525 g

You can then use the density of water, which is 0.9973 g/mL at the temperature at which these measurements were taken, to calculate the volume of the pycnometer:

$$\text{density} = \frac{\text{mass}}{\text{volume}} \quad \text{so} \quad \text{volume} = \frac{\text{mass}}{\text{density}} = \frac{28.9525 \text{ g}}{0.9973 \text{ g/mL}} = 29.03_{0883} \text{ mL}$$

The mass has 6 significant figures, the density has 4, and the results from a calculator have 8 or more. Because the density has the fewest significant figures (4), the volume should be reported to 4 significant figures, as 29.03 mL. To avoid rounding error, some insignificant figures are often carried along through calculations, subscripted, as shown. (Though using just 29.03_1 mL would probably be enough to avoid rounding error, there is no harm in recording as many insignificant figures as you like, provided you subscript them.)

Significant figures are actually a shorthand and approximate way of dealing with uncertainty. They round off uncertainty to the next largest power of ten and use simplified rules for propagating uncertainty through calculations. In more exacting work, uncertainty is reported explicitly (for example, 25.4329 ± 0.0001 g and 24.6263 ± 0.0001 g) and propagated through independent calculations (the difference of these two masses would be 0.8066 ± 0.0002 g).

Graphing Techniques

In many chemical experiments we find that one measured quantity depends on another. If we change one, the other changes. In such an experiment we might collect data under several sets of conditions, with each data point associated with a pair of measured quantities. In such cases, we can represent the data on a two-dimensional graph which shows, in a continuous way, how the quantities are interrelated. The quantity actively controlled in an experiment is called the <u>controlled variable</u> and is plotted on the horizontal axis of a graph; the quantity you observe changing in response to the controlled variable is called the <u>responding variable</u>, or the <u>dependent variable</u>, and is plotted on the vertical axis.

A. Interpreting a graph

The graph in Figure E.1 describes how the solubilities of two compounds, KNO_3 and $CuSO_4 \cdot 5H_2O$, change with temperature. The temperature, which was the controlled variable when the data for the graph was collected, is shown along the horizontal axis (technical term: abscissa) over a range from 0 to 100 °C. The solubility, in grams per <u>100</u> grams water, was the measured quantity in the experiment that produced the data and is on the vertical axis (technical term: ordinate). The small symbols along each of the two curved lines in the graph are <u>data points</u>, indicating measured solubility values at specific temperatures.

On vertical lines in Figure E.1, the temperature is constant. At all points on the vertical line that passes through the 50 hatch mark on the horizontal (*x-*) axis the temperature is always 50 °C. On a vertical line you might draw in at 73 °C, about three-tenths of the distance between the 70 and 80 hatch marks, it is always 73 °C, and so on. Along horizontal lines solubility is constant. The horizontal line drawn through the 80 hatch mark on the vertical (*y-*) axis represents a solubility of 80 grams per <u>100</u> grams of water all along its length. When the data for the solubility of KNO_3 was collected, it was found that at 60 °C, 109 g of KNO_3 would dissolve in 100. g of water. That led to the data point, labelled "A", on the graph at 60 °C and 109 (g KNO_3/<u>100</u> g H_2O). The rest of the data points for the two compounds were placed in the same way. Smooth curves were then drawn through those data points.

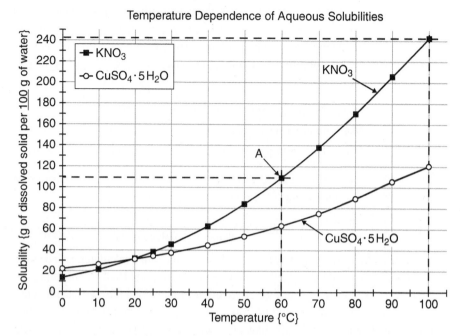

Figure E.1 The solubility of two different salts [potassium nitrate, KNO_3, and copper(II) sulphate pentahydrate, $CuSO_4 \cdot 5\,H_2O$] are plotted as a function of temperature. At 60 °C the solubility of KNO_3 is 10_9 g of KNO_3 per $\underline{100}$ grams of water, while at 100 °C it is 24_2 g / 100 g H_2O.

Given a graph such as Figure E.1, you can extract the data by essentially working backwards. To find the solubility of KNO_3 at 60 °C, draw a dashed line up from the 60 hatch mark, as shown. The solubility is given by the point (marked "A") at which that line intersects the KNO_3 graph line. Following the dashed horizontal line drawn from that intersection to the vertical axis allows you to read off the solubility of KNO_3, which is 109 (g / $\underline{100}$ g H_2O), though you might only be able to read it as 10_9 or 110 (g / $\underline{100}$ g H_2O), with two significant figures. If you wanted to find the amount of water you would need to dissolve 21 g of KNO_3 at 100 °C, you would draw a vertical line from the 100 hatch mark on the *x*-axis up to the KNO_3 line, as shown, then draw a horizontal line at the intersection with the KNO_3 curve. This indicates that at 100 °C, 24_2 g of KNO_3 will dissolve in $\underline{100}$ g of H_2O. (24_2 is 242 with two significant figures; the trailing 2 is subscripted to indicate that it is insignificant. Given the scale of Figure E.1, it is hard to be certain the correct value does not differ from 242 by more than one: that it is not 244, for example.) A conversion factor calculation then reveals that 21 g KNO_3 / 24_2 (g KNO_3 / $\underline{100}$ mL H_2O) = 8.6_{78} mL, or 8.7 mL, of water would be required to dissolve 21 g of KNO_3. There is a lot of information in Figure E.1, condensed into a compact visual form that makes it easy to spot trends.

B. Making a graph

Let us now construct a graph from some experimental data. Suppose the pressure of a fixed (constant) mass of air in a fixed volume was measured as a function of temperature, resulting in the following dataset:

Pressure {mm Hg}	674	741	783	824
Temperature {°C}	0.0	24.5	42.0	60.1

To construct a graph using this data, we first need to decide what should go where. Because the temperature was controlled and the pressure measured in response, we put the pressure on the vertical axis and the temperature on the horizontal axis, labeling those axes as shown in Figure E.2. The temperature data goes from 0 °C to 60 °C; we want the data we graph to fill the available space, not just be in one corner, so we select a temperature interval spacing that will cause the data to span as much of the available space as possible. Because we have about thirty vertical grid lines available, and need to cover a 60 °C interval, having each grid line correspond to 2 °C should work well. Rather than marking every 2 °C interval, however, we put 10 °C hatch marks at intervals of five grid lines. Because there are about twenty horizontal grid lines available, and

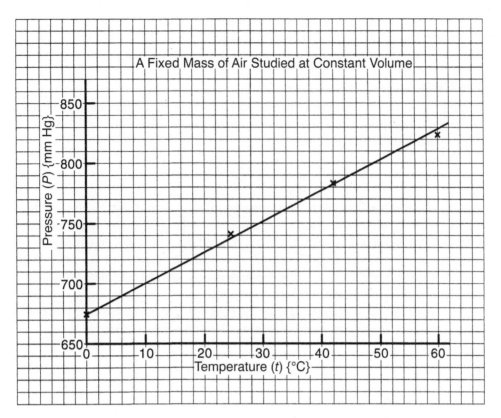

Figure E.2 A properly prepared graph has complete axis labels (including units!), makes good use of the available space, and has a title—focused on relevant information not available elsewhere in the graph.

we need to cover a pressure change of 150 mm Hg, we make the lowest pressure 650 mm Hg and the interval between grid lines equal to 10 mm Hg. We decide to put in hatch marks at 50-mm intervals.

Having laid out the axes, we then plot each of the data points—using ×s, dots, or some other (consistent!) data point marker. The data points fall on a nearly straight line, so we draw such a line, minimizing as best we can the sum of the distances from the data points to the line. If we wish, we can find the equation for that line, which, because it is straight, is of the form $y = mx + b$ in which m is the slope of the line and b the value of y when x equals zero (technical term: y-intercept). Plugging two (well-separated) points on the line,

$(x_1, y_1) = (0.0, 674)$ and $(x_2, y_2) = (60.0, 829)$, into the point-point formula for a line, $y - y_1 = \left(\dfrac{y_2 - y_1}{x_2 - x_1} \right)(x - x_1)$,

and simplifying, we find the equation of the line to be $y = 2.58_{33}x + 674$. This equation can be used to predict the pressure in this system (under the same conditions that were held constant) at any arbitrary temperature, perhaps even outside the bounds of the graph. But the further afield you go, the riskier that gets: note that the data points in Figure E.2 have a <u>systematic</u> (not random) downward curvature to them, and the straight line does not perfectly capture what is going on. It is a good estimate over the range of the plot, but is likely to become less reliable as the temperature rises further above 60 °C or drops further below 0 °C. Such <u>extrapolation</u> (assuming the same behavior or equation applies beyond the range of values actually studied) also assumes that nothing else important changes (such as the phase of the currently gaseous substance under study) as a result of the change in the controlled variable—which will not always be true!

In summary, the first key to successfully constructing graphs is to properly select the variables and assign them to the two axes. (The variable you control should be on the horizontal axis, and the quantity of interest that responds to changes in the variable you control should be on the vertical axis.) Label those axes. Then, given the range of the variables in your dataset, select grid intervals that will cause your data to (nearly) fill the area of the grid. Put in hatch marks for each variable at appropriate intervals and label them with the values of the variable. Carefully plot the data points on the grid. If the data is theoretically expected to fall on a line, draw a straight line through the data so as to minimize the sum of the distances from the data points to the line. Finally, place a title at the top of the graph, summarizing what it shows. The title should focus on information not available elsewhere in the graph, such as in the axis labels or a plot legend.

C. Linearization

Many relationships between variables in science are not linear. However, it is often possible to manipulate more complicated relationships into linear forms. For example, the vapor pressure of a liquid depends on temperature according to the following equation:

$$\ln VP = \frac{-\Delta H_{vap}}{RT} + C \tag{1}$$

In Equation 1, $\ln VP$ is the logarithm of the vapor pressure to the base e, T is the Kelvin temperature, and ΔH_{vap} and R are constants equal to the enthalpy of vaporization and the gas constant, respectively. A plot of VP vs T would certainly not be linear. However, if we let $x = \ln VP$ and $y = 1/T$, then Equation 1 does take on the linear form $y = mx + b$—the equation of a straight line—with a slope, m, equal to $-\Delta H_{vap}/R$. By plotting $(\ln VP)$ against $(1/T)$, we can find the slope of the resulting straight line and determine ΔH_{vap} for the liquid. It is very possible that Equation 1 was initially found by an imaginative scientist who measured vapor pressure as a function of temperature and tried plotting the data against different functions of the variables, seeking a straight line relationship. From such efforts many great discoveries have been made!

Appendix F

Suggested Locker Equipment

2 beakers, 30 or 50 mL
2 beakers, 100 mL
2 beakers, 150 mL
2 beakers, 250 mL
2 beakers, 400 mL
1 beaker, 600 mL
2 Erlenmeyer flasks, 25 or 50 mL
2 Erlenmeyer flasks, 125 mL
2 Erlenmeyer flasks, 250 mL
1 graduated cylinder, 10 or 25 mL
1 graduated cylinder, 50 or 100 mL
1 glass lab thermometer,
   ($-20\,°C$ or $-10\,°C$) to ($+110\,°C$ or $+150\,°C$)

2 watch glasses: 75, 90, or 100 mm
1 crucible and cover, size #0
1 evaporating dish, small
2 medicine droppers
5 medium test tubes, 18×150 mm
8 small test tubes, 13×100 mm
1 test tube brush
1 spatula
1 test tube clamp, Stoddard or equivalent
1 test tube rack
1 pair of lab tongs
1 plastic wash bottle
3 glass stirring rods

Appendix G1

An Introduction to Excel

Software called a "spreadsheet" has come into general use for processing scientific data obtained in the laboratory, as well as data involved in business activities. Essentially, a spreadsheet contains many cells in which data can be stored and manipulated. It is particularly useful when there are several, or many, experimental values for a property—such as the volume of a gas obtained at different temperatures under constant pressure—and the same calculation needs to be conducted on those values over and over. Carrying out such calculations by hand, with a calculator, is time consuming and prone to error. Setting up a spreadsheet to conduct the same calculations takes more time to learn, but a spreadsheet will not make calculation mistakes and will instantly recalculate everything in the event an error is found and you need to change a number. This often makes it very worthwhile to set up a spreadsheet, and knowing how to do so is a skill that will serve you well in nearly any line of work. In this appendix we offer a short introduction to a widely used spreadsheet package called Excel, in some of its simpler scientific applications. To get started, you will need to have a computer on which the software is available. There are many versions of Excel, but we try here to provide instructions for most of them. These instructions are specific to Excel but should be easily adaptable to other spreadsheet packages, with only minor changes. Appendix G2 offers specific instructions for Google Sheets, another common spreadsheet application.

Basic Layout

Click on the Excel icon on your computer. If there is none, your instructor will tell you how to call it up. If using one of the latest versions, double-click on [move over and press the primary (usually left) mouse button twice, in quick succession] **Blank workbook** in the upper right. You should see a screen on which there is a grid containing many rectangular boxes, called cells. Each cell has an associated label, with a letter (above it) indicating its column and a number (to the left of it) indicating its row; the letter + number combination referring to a given cell is called a cell reference. If you click on cell B5, it will be highlighted. If you double-click in that cell, type a number or a word, and press the **enter** key on the keyboard, what you type will appear in the cell. You can change the width of the cell by clicking and holding on the grid line in question where it intersects with the rectangle containing a letter (called the column heading) up at the top. Double-clicking there will automatically adjust the width of the column to accommodate the widest entry in that column.

Setting Up Columns of Data

Start by entering the column heading "Temperature {°C}" in cell A1. (*Note*: To get the ° symbol on a PC, hold down the **Alt** key, type "248" *on the numeric keypad*, then release **Alt**; to get it on a Mac, hold down the **shift** and **option** keys and type an "8". With a Chromebook, hold down the **ctrl** and **shift** keys and type "U", then release all keys, type "b0", and press **enter**. There are other ways to get symbols, but this is the fast way!) Change the cell width, if necessary, to make the column heading fit. In cell A3, type in "20" and press the **enter** key. The highlighted cell will automatically become A4. Enter the formula "=A3+20" in that cell, press **enter**, and 40 should appear in A4. Click cell A4 *once* with the mouse, and then move the cursor (what the mouse moves around) to the lower right-hand corner of that cell. You will find that the cursor will change from a white plus symbol to a solid black plus symbol or an open square. While the cursor is a black plus or a square, click and drag (press and hold the primary mouse button as you move the mouse) down to select all the cells into which you want to copy that formula; in this case, go down to cell A12. When you release the mouse button, the numbers 60, 80, 100, and so on, up to 200, will appear in column A. This method is very useful if you have a column of data related by a mathematical formula.

Now enter the heading "Temperature {K}" in cell B1, adjusting the width of that column as necessary.to make it fit. In cell B3, enter the formula "=A3+273", press enter, and 293 will appear in cell B3. Using the same approach used for column A, copy this formula down into cells B4 through B12 to produce the Kelvin temperatures in column B for each of the Celsius temperatures in column A.

In cell C1, type the heading "Volume {L}". You will use the ideal gas law, $PV = nRT$, to find the volume (in liters) of $\underline{1}$ mole of gas at $\underline{1}$ atmosphere pressure and the Celsius temperatures in column A. By the ideal gas law, $V = nRT/P$; in this case, $n = \underline{1}$ mol, $P = \underline{1}$ atm, $R = 0.08206$ L $\cdot$ atm / (mole $\cdot$ K), and T is the Kelvin temperature. So in column C, V equals $\underline{1}*0.08206*T/\underline{1}$, or just $0.08206*T$. [*Note*: Some mathematical symbols used for mathematical operations in a spreadsheet do not correspond to those often used elsewhere: $+$ and $-$ work for add and subtract, but $*$ must be used for multiply, / for divide, and ^ for raising something to a power. We write 2^3 as 2^3 in Excel. A number such as 3.45×10^5 should be entered and will be displayed as 3.45E5 in a spreadsheet. Some mathematical operations are (only) available through <u>functions</u>: for example, "SQRT(2)" returns the square root of 2 in Excel.]

To calculate the volumes in column C, first enter the gas constant into a cell and label it: type "$R =$" into cell A14, and "0.08206" in cell B14. You should specify the units for R, so in cell C14 enter "L atm /(mol K)". In cell C3, type the formula "=\$B\$14*B3" and press **enter**. (Here is a handy shortcut: rather than typing in B3 from the keyboard, you can click on that cell in the spreadsheet when you get to that point when typing in the formula. This can be very helpful when you are constructing your own formulas and see the cell you want to use; you can just click on it to get the right cell reference!) In cell C3 the number 24.055 should appear (possibly with more or fewer digits shown, as Excel knows nothing of significant figures and will just display as many digits as it has room to display). Highlight cell C3 by clicking on it once, position the cursor over the lower right-hand corner of the cell, and—when the black plus symbol or square appears—click and drag down to cell C12 and release. The volumes, in liters, of $\underline{1}$ mole of an ideal gas at the various temperatures should appear in column C.

You may wonder why you needed to use \$B\$14 instead of just B14 in the formula in cell C3. Try replacing \$B\$14 with B14 in the formula in cell C3, and see what happens! [A handy trick for doing this is to click on B14 in the formula and then press the **F4** key on the keyboard (you may also have to hold down **fn**) a few times.] At first, everything will look fine; but if you drag down the cursor to copy this cell, the program will increment not only B3 to B4, but also B14 to B15 in the copied formula, and zeroes will appear in column C because B15 and the cells below it are empty (and therefore treated as a value of zero in numerical formulas). In many formulas, you will find that you will wish to have all or part of a cell reference remain unchanged when the formula is copied. You can accomplish this by preceding the letter, the number, or both in a cell reference with a dollar sign (\$).

Some Comments on Excel

Excel assigns the contents of each cell to one of three categories: number, text, or formula. The number and text categories cause trouble when one is mistaken for the other. To avoid this, when entering numerical values you intend to use in calculations or a plot, be certain to enter *only* the number—do not, for example, add units to it in the same cell. (Specifying units is an excellent habit! But to do so in a spreadsheet, use a column heading or a next-door-neighbor cell—as demonstrated in the preceding steps.) Formulas most often cause trouble when Excel does not realize you are entering one. You must get into the habit of beginning every formula you enter into Excel with =, +, or −; otherwise Excel will interpret your entry as text. [Some spreadsheets assume a formula, and you may need to indicate text entries by beginning them with a double or single quote mark (" or ').] The equals sign (=) or other symbol used to indicate that a cell contains a formula must be the first character in the cell, with no spaces before it. The equation for the formula may be complex, but it must contain only symbol(s) and/or functions Excel understands for any operation(s) to be carried out, and it should not have any spaces in it. All references to particular cells must be to cells with known numerical contents. Parentheses should be carefully used to make clear to Excel the desired order of the mathematical operations you wish to carry out.

Inserting symbols that do not appear on your keyboard can be tricky. If you do not know or remember the keyboard shortcut (such as that for the degree symbol, °, as introduced previously) for a symbol you want to use, there is a more general, if slower, approach. In newer versions of Excel, use the **Insert**>**Symbol** menu option, selected by choosing **Symbol** from the **Insert** menu near the top of the screen; note, however, that there is nothing similar in older Windows versions. On older Mac versions, use the "Object Palette", accessible by clicking the second icon from the left on the top row of the "Formatting Palette" window, then click on the third icon from the left in the new row of buttons that appears at the top of that window. You will want to switch back to the "Formatting Palette" when you are done—it is the first icon in the top row of the window.

If you mis-type or mis-click (and this is inevitable!), you may find nothing happens when you press **enter**, or you may get an error message that may or may not be helpful. If you get into trouble, the delete (**Del**) or escape (**Esc**) keys may return the screen to its previous state (before the error occurred). If not (or if these buttons cause something else undesired to happen), using the **Undo** option (indicated by a toolbar icon with an arrow looping to the right and up and then back down and to the left, and/or found on the **Edit** menu—both in the upper left corner of the screen) will often put you back on track. A useful keyboard shortcut for **Undo** is **Ctrl-Z** (**command-Z** on a Mac)—which means you should press the **Z** key while holding down the **Ctrl** (or **command**) key on the keyboard. [*Note*: "Hovering" over an icon—that is, holding the mouse stationary over it for a few seconds—will bring up the meaning of that icon, as well as the keyboard shortcut for it (if it has one).] Do not be reluctant to seek help, either from your lab instructor or a fellow student more familiar with Excel than you are. Working with a spreadsheet can be frustrating at first, but learning to use one pays huge dividends.

Using Excel to Make Graphs

In some cases it is helpful to create graphs that show the relationships between experimentally measured variables—and a spreadsheet package can make this much easier to do than by hand.

Given two columns of data for two related variables, such as temperature and volume, Excel can be used to construct a graph (Excel calls it a "chart") that shows how the variables are related.

Before you proceed, click on **View** in the top menu line on the screen and select **Normal** from the options presented. Also click on a single *empty* cell somewhere in the spreadsheet, far from your data (such as E1), so that Excel does not assume anything about what you want a new chart to contain.

Next, choose **Chart** from the **Insert** menu near the top of the screen. If you are using a newer version of Excel, you will need to select the type of chart you want at this point—in science, the default selection you should make is **Scatter**, and specifically the flavor that displays only the data points. If you are using a version of Excel dating from before 2007 (your authors find these superior, frankly), choosing **Chart** from the **Insert** menu will bring up a screen showing several possible graph types, from which you will want to select **XY Scatter** and then click on **Next >**.

Now you need to specify the data you wish to use in the chart. In an older version of Excel, click on the **Series** tab; in a newer version, click on the chart and then choose **Select Data** from the **Chart Design** menu at the top of the screen. Click on **Add**, or on an older Mac version it may be a small + button. In the data entry regions that appear, labeled **Name**, **X Values**, and **Y Values**, delete the contents of any boxes that are not already empty by selecting (clicking and dragging over) the contents and pressing either the **backspace** or the **delete** (**Del**) key on the keyboard.

To specify the *y* values for the chart, put the cursor in the "Y Values" box and click and drag the mouse down over the cells containing the dependent variable data, in this case the volumes in column C (specifically, cells C3 through C12). When you release the mouse button, that data range will appear in the "Y Values" box (as "=Sheet1!$C$3:$C$12").

To specify the *x* values, put the cursor in the "X Values" box and click and drag the mouse down over the independent variable data, in this case the Kelvin temperatures in column B (specifically, cells B3 to B12). When you release the mouse button, that cell range should appear in the "X Values" box (as "=Sheet1!$B$3:$B$12"). Press **enter** and a graph showing the relationship between the variables will appear on the screen. The scale will probably be poorly chosen, but you will fix that soon.

If you are using a newer Windows version of Excel (from 2007 or later), click on **OK**, since you are done specifying the data you wish to use; then click on the first of the **Chart Layouts** that appear at the top of your screen, or in the newest versions click on **Add Chart Element** from the **Chart Design** menu. Now, no matter what version of Excel you are using, click on **Chart Title** (in the newest versions, specify **Above Chart** and then double-click the text that appears inside the new text box above the chart.) Type "One mole of an ideal gas at one atmosphere pressure" as the title of the graph (do not press the **enter** key yet). Now click once on the "X Axis title" box and type in the axis label for the independent variable, "Temperature {K}". Click once on the "Y Axis title" box and type the axis label for the dependent variable, "Volume {L}". (In the newest versions of Excel you will have to select each of these things in turn from the **Add Chart Element** pulldown, where they are under **Axis Titles** as **Primary Horizontal** and **Primary Vertical**, respectively, and then double-click the resulting text boxes to edit them.)

In newer versions of Excel, click on the **Chart Layout** tab at the top of the screen (in the newest versions, the **Add Chart Element** button), then click on **Gridlines**. With older versions of Excel, just click on the **Gridlines** tab of the "Chart Options" box you are currently in. Add or remove gridlines as seems reasonable.

If you are using an older version of Excel, you will now want to click on **Next >**, to get to **Step 4**, and decide where to place the graph, either on a separate page or as an object in the current page (usually, you will want the latter), before clicking on **Finish** to bring down the final graph, labeled and titled, but probably not scaled very well.

To fix the scale, _right_-click (press down on the side of the mouse _opposite_ that you normally click) on the numbers of the axis you wish to change and select **Format Axis**. Click on **Scale** or **Axis Options**, whichever is offered to you. Proceed by entering appropriate minimum and maximum axis limit values (250 to 500 for x, 20 to 40 for y, in this example). This is easier to do if you can see the chart and the data as you do so—you can move the data entry window out of the way, if necessary, by clicking and holding on the window name ("Format Axis") and dragging the window to a more convenient location. The graph will update as you make changes, which makes it easy to fine-tune your adjustments.

In a chart displaying a single series, such as this example, the legend serves no useful function and can be removed by clicking on it once and pressing **backspace** or **delete**. (The newest versions of Excel do not automatically put a legend in, as older ones did.) This frees up more room for the plot. Were a legend useful, it could be moved onto the plot region by clicking and dragging it there (in new versions of Excel, doing so after adding one from the **Add Chart Element** menu).

To change the physical dimensions and position of the graph you have created, click once on the edge of its frame, to select it. Then position the mouse over any of the handles that appear on the frame (a single black square in older versions of Excel, a collection of three tiny dots in middle-aged versions, or a hollow circle or square in the newest ones) and click and drag to change the dimension to which that handle corresponds. To move the entire graph, click on an empty region of the chart, just inside one of the upper handles, and click and drag. Especially with these operations, it is easy to have things go wrong; if they do, press the **Esc** key and then, if needed, use the **Undo** command.

If you wish to have Excel calculate the best line through your data, select your chart (click once on the edge of its frame) and then go to the **Trendline** menu (**Chart>Add Trendline. . .** with most older versions, **Chart Layout>Trendline** with some middle-aged ones, and under the **Add Chart Element** pulldown on the newest versions). You can specify that Excel should display the equation of this line (as well as customize several other aspects of it) from the **Options** tab at the top of the **Format Trendline** menu brought up by right-clicking on the trendline.

Appendix G2

An Introduction to Google Sheets

Google Sheets is spreadsheet software like Excel (see Appendix G1), but with several important differences. First, it is browser-based, meaning it can be run from almost any computing device with a web browser (and a large enough screen to be practical). Second, it uses cloud-based storage, which means the files it generates are available on any Internet-connected system (this is possible with the newest versions of Excel as well). Finally, but perhaps most important for education users, it is free. Google Sheets works much like Excel, but the details differ enough that we are including detailed instructions on how to use it to analyze and plot data, as we did for Excel in Appendix G1.

Start by going to Google Sheets in a web browser: sheets.google.com is the address. If you are not already logged into a Google account, you will need to do so—or create one—because Sheets stores your data in the cloud, and it needs to know who you are in order to store it as well as let you access it from other places. Once you are logged in, you will be able to create a new spreadsheet; in this case you want it to be a blank one, which you can open by clicking on the multicolored + sign in the lower right corner of the screen. When you choose that, an empty spreadsheet will open up, and you are ready to begin.

Basic Layout

You are now looking at a screen on which there is a grid containing many rectangular boxes, called <u>cells</u>. Each cell has an associated label, with a letter (above it) indicating its column and a number (to the left of it) indicating its row; the letter + number combination referring to a given cell is called a <u>cell reference</u>. If you click on cell B5, it will be highlighted. If you <u>double-click</u> in [move into and press the primary (usually left) mouse button twice, in quick succession] that cell, type a number or a word, and press the **enter** key on the keyboard, what you type will appear in the cell. You can change the width of the cell by clicking and holding on the grid line in question (called <u>clicking and dragging</u>) where it intersects with the rectangle containing a letter (called the <u>column heading</u>) up at the top. Double-clicking there will automatically adjust the width of the column to accommodate the widest entry in that column.

Setting Up Columns of Data

Start by entering the column heading "Temperature {°C}" in cell A1. (*Note*: To get the ° symbol on a PC, hold down the **Alt** key, type "248" *on the numeric keypad*, then release **Alt**; to get it on a Mac, hold down the **shift** and **option** keys and type an "8". With a Chromebook, hold down the **ctrl** and **shift** keys and type "U", then release all keys, type "b0", and press **enter**. There are other ways to get symbols, but this is the fast way!) Change the cell width, if necessary, to make the column heading fit. In cell A3, type in "20" and press the **enter** key. The highlighted cell will automatically become A4. Enter the formula "=A3+20" in that cell, press **enter**, and 40 should appear in A4. Click cell A4 *once* with the mouse, and then move the <u>cursor</u> (what the mouse moves around) to the lower right-hand corner of that cell. You will find that the cursor will change from a white arrow to a thin black plus symbol (if you see a hand icon, you are not in the right spot!). While the cursor is a black plus symbol, click and drag the cursor down to select all the cells into which you want to copy that formula; in this case, go down to cell A12. When you release the mouse button, the numbers 60, 80, 100, and so on, up to 200, will appear in column A. This method is very useful if you have a column of data related by a mathematical formula.

Now enter the heading "Temperature {K}" in cell B1, adjusting the width of that column as necessary to make it fit. In cell B3, enter the formula "=A3+273", press **enter**, and 293 will appear in cell B3. Using the same approach used for column A, copy this formula down into cells B4 through B12 to produce the Kelvin temperatures in column B for each of the Celsius temperatures in column A.

In cell C1, type the heading "Volume {L}". You will use the ideal gas law, $PV = nRT$, to find the volume (in liters) of $\underline{1}$ mole of gas at $\underline{1}$ atmosphere pressure and the Celsius temperatures in column A. By the ideal gas law, $V = nRT/P$; in this case, $n = \underline{1}$ mol, $P = \underline{1}$ atm, $R = 0.08206$ L·atm/(mole·K), and T is the Kelvin temperature. So in column C, V equals $\underline{1}*0.08206*T/\underline{1}$, or just $0.08206*T$. [*Note*: Some mathematical symbols used for mathematical operations in a spreadsheet do not correspond to those generally used elsewhere: + and − work for add and subtract, but * must be used for multiply, / for divide, and ^ for raising something to a power. We write 2^3 as 2^3 in a spreadsheet. A number such as 3.45×10^5 should be entered and will be displayed as 3.45E5 in a spreadsheet. Some mathematical operations are (only) available through <u>functions</u>: for example, "SQRT(2)" returns the square root of 2 in Google Sheets.]

To calculate the volumes in column C, first enter the gas constant into a cell and label it: type "$R =$" into cell A14, and "0.08206" in cell B14. You should specify the units for R, so in cell C14, enter "L atm/(mol K)". In cell C3, type the formula "=$B$14*B3" and press **enter**. (Here is a handy shortcut: rather than typing in B3 from the keyboard, you can click on that cell in the spreadsheet when you get to that point when typing in the formula. This can be very helpful when you are constructing your own formulas and see the cell you want to use: you can just click on it to get the right cell reference!) In cell C3 the number 24.0553 should appear (possibly with more or fewer digits shown, as by default Sheets knows nothing of significant figures and will just display as many digits as it has room to display). Highlight cell C3 by clicking on it once, position the cursor over the lower right-hand corner of the cell, and—when the black plus symbol appears— click and drag down to cell C12 and release. The volumes, in liters, of $\underline{1}$ mole of an ideal gas at the various temperatures should appear in column C.

You may wonder why you needed to use $B$14 instead of just B14 in the formula in cell C3. Try replacing $B$14 with B14 in the formula in cell C3, and see what happens! [A handy trick for doing this is to click on B14 in the formula and then press the **F4** key on the keyboard (you may also have to hold down **fn**) a few times.] At first, everything will look fine; but if you drag down to copy this cell, the program will increment not only B3 to B4, but also B14 to B15 in the copied formula, and zeroes will appear in column C because B15 and the cells below it are empty (and therefore treated as a value of zero in numerical formulas). In many formulas, you will find that you wish to have all or part of a cell reference remain unchanged when the formula is copied. You can accomplish this by preceding the letter, the number, or both in a cell reference with a dollar sign ($).

Some Comments on Google Sheets

Google Sheets assigns the contents of each cell to one of three categories: number, text, or formula. The number and text categories cause trouble when one is mistaken for the other. To avoid this, when entering numerical values you intend to use in calculations or a plot, be certain to enter *only* the number—do not, for example, add units to it in the same cell. (Specifying units is an excellent habit! But to do so in a spreadsheet, use a column heading or a next-door-neighbor cell—as demonstrated in the preceding steps.) Formulas most often cause trouble when Google Sheets does not realize you are entering one. You must get into the habit of beginning every formula you enter into Google Sheets with =, +, or −; otherwise Google Sheets will interpret your entry as text. [Some spreadsheets assume a formula, and you may need to indicate text entries by beginning them with a double or single quote mark (" or ').] The equals sign (=) or other symbol used to indicate that a cell contains a formula must be the first character in the cell, with no spaces before it. The equation for the formula may be complex, but it must contain only symbol(s) and/or functions that Sheets understands for any operation(s) to be carried out, and it should not have any spaces in it. All cell references in equations must be to cells with known numerical contents. Parentheses should be carefully used to make clear to Google Sheets the desired order of the mathematical operations you wish to carry out.

Inserting symbols that do not appear on your keyboard can be tricky. If you do not know or remember the keyboard shortcut (such as that for the degree symbol, °, as introduced previously) for a symbol you want to use, there is a more general, if much slower, approach. Open Google Docs, at docs.google.com, and use the **Insert**>**Special characters** menu option there, selected by choosing **Special characters** from the **Insert** menu near the top of the screen. Find the character you want and click on it, then close the character selection window. Select the character on the screen and copy it into your Google Sheet. At the time of this writing, this functionality is not available directly in Google Sheets itself—but it may eventually appear.

If you mis-type or mis-click (and this is inevitable!), you may find nothing happens when you press **enter**, or you may get an error message that may or may not be helpful. If you get into trouble, the delete (**Del**) or escape (**Esc**) keys may return the screen to its previous state (before the error occurred). If not (or if these buttons cause something else undesired to happen), using the **Undo** option (indicated by a toolbar icon with an arrow to the left, looping up and then back down, or found on the **Edit** menu—both in the upper left corner of the screen) will often put you back on track. A useful keyboard shortcut for **Undo** is **Ctrl-Z** (**command-Z** on a Mac)—which means you should press the **Z** key while holding down the **Ctrl** (or **command**) key on the keyboard. [*Note*: "Hovering" over an icon—that is, holding the mouse stationary over it for a few seconds—will bring up the meaning of that icon, as well as the keyboard shortcut for it (if it has one).] Do not be reluctant to seek help, either from your lab instructor or a fellow student more familiar with Google Sheets than you are. Working with a spreadsheet can be frustrating at first, but learning to use one pays huge dividends.

Using Google Sheets to Make Graphs

In some cases it is helpful to create graphs that show the relationships between experimentally measured variables—and a spreadsheet package can make this much easier to do than by hand.

Given two columns of data for two related variables, such as temperature and volume, Google Sheets can be used to construct a graph (Sheets calls it a "chart") that shows how the variables are related.

Start by selecting (clicking and dragging over) all the cells containing the data for the chart: in this case cells B3 to B12 and C3 to C12, the shorthand for which is B3:C12. Now choose **Chart** from the **Insert** menu near the top of the screen. Change the **Chart type** to "Scatter" by clicking on the default type, **Column chart**, in the pull-down menu and picking the **Scatter chart** option instead. (You may have to scroll down through the options, as **Scatter chart** is not near the top. Note that in science, you will almost always want your **Chart type** to be a "Scatter" chart.) The next steps will be a little easier if you move the chart out of the way, so click once on the chart, then click and hold while dragging it to the right, so that it is no longer over any of your data.

In this case Sheets has correctly guessed what data belongs on each axis, but it will not always get it right! Here is how you could change things in such a case: to specify the y values for a chart, click in the shaded box with rounded ends under **Series** and then on the small gray 2×2 grid that appears (labeled "Select a data range" if you hover over it). When the data entry box appears, click and drag the mouse down over the cells containing the dependent variable data, in this case the volumes in column C (specifically, cells C3 through C12). When you release the mouse button, that cell range should appear in the box (as "Sheet1!C3:C12"). Click **OK** to confirm this is what you want. To specify the x values, click in the shaded box with rounded ends under **X-axis** and then click on the gray 2×2 grid that appears. When the data entry box appears, click and drag the mouse down over the cells containing the dependent variable data, in this case the Kelvin temperatures in column B (specifically, cells B3 to B12), or manually enter "B3:B12" in the box: both accomplish the same result. Then click on **OK**.

Now that the chart contains the data of interest, it is time to fine-tune its appearance. In the "Chart editor" sidebar, click the **Customize** tab, directly under the sidebar name and to the right of the **Setup** tab. Click on the **Chart & axis titles** entry to open it up. Now, click in the box under "Title text" and type "One mole of an ideal gas at one atmosphere pressure" as the title of the graph. Next, click on the pull-down menu that says **Chart title** and change it to **Horizontal axis title**. Click under "Title text" again and this time type in the axis label for the independent variable, "Temperature {K}". Change **Horizontal axis title** to **Vertical axis title** and type in the axis label for the dependent variable, "Volume {L}". Finally, click on the **Gridlines and ticks** subheading (you may have to scroll down in the "Chart Editor" sidebar to see it) and add or remove gridlines as seems reasonable.

To fine-tune the scale of your plot, click on the **Horizontal axis** and/or **Vertical axis** subheading, as appropriate. Sheets usually does a pretty good job of automatic scaling, but try entering appropriate fixed maximum and minimum axis limit values (250 to 500 for x, 20 to 40 for y, in this example). The graph will update as you make changes, which makes it easy to fine-tune your adjustments.

In a chart displaying a single series, such as our example, a legend serves no useful function and the one automatically placed on the chart can be removed by clicking on the Legend subheading and setting the Position to "None".

To change the physical dimensions and position of the graph you have created, click once just inside its frame, to select it. Then position the mouse over any of the <u>handles</u> that appear on the frame (little filled-in squares at the corners and in the middle of each side) and click and drag to change the dimension to which that handle corresponds. To move the entire plot, click on an empty region of the chart, just inside one of the upper handles, and click and drag. Especially with these operations, it is easy to have things go wrong; if they do, press the **Esc** key and then, if needed, use the **Undo** command.

If you wish to have Google Sheets calculate the best line through your data, double-click on one of the data points of the series you want a trendline for (single-click if the "Chart editor" sidebar is already open), and Sheets will open the "Chart editor" sidebar and select the **Series** submenu there. Check the **Trendline** checkbox, and the best-fit line will appear. You can specify that Sheets should display the equation of this line (as well as customize several other aspects of it) from further down in the **Series** submenu.

In a chart displaying a single series, such as this example, a legend is not necessary. But if you add the equation of a trendline or make a chart that benefits from a legend, note that the legend can be moved onto the plot region by clicking and dragging it there, which allows the plot region to be kept as large as possible.

Appendix H

Statistical Treatment of Laboratory Data

Whenever there are multiple determinations of the same measurement (called <u>replicate measurements</u>, or just <u>replicates</u>), it is customary for scientists to use <u>statistics</u> (a specialized subfield of math) to arrive at a best estimate, as well as some measure of the variation of the replicates around that best estimate—which indicates something about the reliability of the experimental results. In chemistry, the kinds of experiments treated in this way generally involve analyzing a substance to determine the proportions of its chemical constituents, often by either titration or the weighing of precipitates (technical term: <u>gravimetry</u>). This branch of chemical science is referred to as <u>quantitative analysis</u>.

It is necessary to have some way to describe the best estimate of the true value that can be obtained from replicate data. You may have heard terms frequently used to express the distribution of test scores—the (arithmetic) mean, median, and mode, along with the range. Consider a set of laboratory data, such as four independent measurements of the percent iron by mass in the same ore sample:

Trial number	$\%_{mass}$ Fe
1	15.66
2	15.79
3	15.22
4	15.66

Inspecting this dataset reveals that the <u>median</u> value (the result with the number of replicates above it equal to the number of replicates below it) is 15.66%, the <u>mode</u> (the value having the largest number of replicates) is 15.66%, and the <u>mean</u> (the arithmetic mean, or "average") is 15.58%. The mean is calculated by adding up all of the percentage values (62.33%) and dividing by the number of replicates (4). The <u>range</u> (difference between the highest and lowest values) is 0.57%.

You may have used the terms "precision" and "accuracy" in a variety of settings. In science, <u>precision</u> is a measure of the <u>reproducibility</u> of a measurement—the agreement among several determinations of the same quantity. <u>Accuracy</u> describes how close a measurement is to the accepted (or true) value. It appears that this dataset has decent precision, but nothing can be said about its accuracy in the absence of additional information. In this course, your instructor will have access to accepted values for most of the quantities you measure. If the known value for this particular iron ore were 15.88%, the accuracy of this dataset would be poor—all of the measured values fall below the true value, suggesting a <u>systematic error</u>. Your instructor may use a combination of your accuracy and your precision in grading quantitative analysis experiments.

A common way to quantify the precision of a dataset is to evaluate its <u>standard deviation</u>, s, which is the average size of the deviations of the individual measurements from the mean value. Standard deviation is calculated using the following formula:

$$s = \sqrt{\frac{\Sigma (x_i - \bar{x})^2}{N - 1}} \qquad (1)$$

In Equation 1, s is the standard deviation, N is the number of replicates (trials), x_i is the value of an individual measurement, $\bar{x}$ is the mean value, and Σ is a mathematical symbol denoting a summation: the addition of the members of a series of terms. In this example there are four replicate measurements, so $N = 4$. The mean value $\bar{x}$ is 15.58%, and the index of the summation, i, is the trial number, such that $x_1 = 15.66\%$, $x_2 = 15.79\%$, and so on. The first term being squared for the summation will be $x_1 - \bar{x} = 15.66\% - 15.58\% = 0.08\%$, the second term will be $x_2 - \bar{x} = 15.79\% - 15.58\% = 0.21\%$, and so on.

The following table lists values for the intermediate steps in this calculation:

Trial	$x_i - \bar{x}$	$(x_i - \bar{x})^2$
1	0.08	0.0064
2	0.21	0.0441
3	−0.36	0.1296
4	0.08	0.0064
		$\Sigma = 0.1865$

Dividing 0.1865 by $(4 - 1 =)3$, and then taking the square root, yields 0.25 as the standard deviation, s, for this dataset. A small value of s relative to the size of $\bar{x}$ is an indication of good precision. While the standard deviation can always be calculated in this way, you will most likely find it more convenient to use a calculator (most of which are capable of evaluating statistics these days) or spreadsheet. With both Excel and Google Sheets, placing the cell range containing the replicate data inside the STDEV() function will return the standard deviation for that dataset, s, and the AVERAGE() function will return the mean, $\bar{x}$.

In this dataset the result from Trial 3, 15.22%, is further from $\bar{x}$ than any of the other replicates: it is an <u>outlier</u>. You might wonder whether you should exclude an outlier from your calculations. If you are the experimenter who gathered the data, you can exclude an outlier if a known mistake was made, such as spilling some of a solution, or if you know you definitely made a weighing error. However, if such a personal error is not known to have been made, an outlier should generally not be rejected just for being an outlier. There is a statistical test that serves as an unbiased guide as to whether to reject an outlier. This test, called the Q test, is based on the assumption that if it were sufficiently large the dataset would approach a <u>normal distribution</u>: the results expected if the deviations from the mean are purely random. [This is not always the case! For example, if transferring all of a solution to a flask is the main source of error in a procedure, the error associated with that procedure's dataset is likely to be somewhat systematic: you can only transfer too little of the solution, not too much (more than all of it).] If you believe a dataset would resemble a normal distribution if many more replicates were run, then a <u>rejection quotient</u>, Q_{rej}, can be calculated using Equation 2:

$$Q_{rej} = \frac{|x_q - x_n|}{r} \tag{2}$$

In Equation 2, x_q is the questionable value, x_n is the nearest replicate value to it, and r is the range of the dataset, including the questionable value. Note that the absolute value of the difference is used. The calculated rejection quotient, Q_{rej}, is then compared with the <u>critical quotient</u>, Q_c, listed in the following table:

N	2	3	4	5	6	7	8	9	10
Q_c	—	0.94	0.76	0.64	0.56	0.51	0.47	0.44	0.41

The value in the top row, N, is the number of replicates in the dataset, and Q_c represents a threshold at the 90% confidence level that the value x_q, the questionable value, can be discarded. (Specifically, that x_q is not just a random variation but the result of some unknown, but very real, error.) If $Q_{rej} > Q_c$, the questionable value can be discarded with a 90% degree of confidence provided that—if expanded to include many more replicates—the dataset would resemble a normal distribution. Note that a Q test can never eliminate a result from a dataset consisting of only two replicates, and it is very rare for it to do so with a dataset consisting of just three replicates. The Q test is most often useful when you have many replicates, with a clear outlier.

We have identified trial 3 in the iron percentages dataset, indicating 15.22% Fe, as the questionable one. The value nearest to 15.22% in the dataset is 15.66%. We calculate $Q_{rej} = |15.22 - 15.66|/0.57 = 0.44/0.57 = 0.77$, and find that the Q_c value for four trials ($N = 4$) is 0.76. Therefore $Q_{rej} > Q_c$ (just barely), and we can discard the 15.22% Fe value with a 90% degree of confidence that it is the result of a mistake, subject to the validity of the assumption that the variation pattern in this measurement follows a normal distribution.

With the 15.22% value excluded, we can calculate a new mean value, $\bar{x}$, of 15.70% Fe with a new standard deviation, s, of 0.075%. Note the significantly smaller standard deviation and the improved (though still far from stellar) accuracy. (This $\bar{x}$ is closer to the accepted value of 15.88%.)

This is a very brief introduction to the statistical analysis of experimental data. You may wish to do additional reading to gain further insight. Analytical chemistry textbooks usually have a section on statistics in the first few chapters, or in an appendix at the end. Other sources include any introduction to statistics, many of which can be found online. You may find that your physics and/or biology courses incorporate a treatment of statistics into their laboratory portions.